# Control Systems: Theory and Applications

# Control Systems: Theory and Applications

Edited by Chester Mann

www.clanryeinternational.com

Clanrye International,
750 Third Avenue, 9th Floor,
New York, NY 10017, USA



ISBN: 978-1-63240-626-2

**Cataloging-in-Publication Data**

Control systems : theory and applications / edited by Chester Mann.
p. cm.
Includes bibliographical references and index.
ISBN 978-1-63240-626-2
1. Automatic control. 2. Control theory. I. Mann, Chester.
TJ213 .C66 2017
629.8--dc23

For information on all Clanrye International publications
visit our website at www.clanryeinternational.com

Printed in the United States of America.

# Contents

# Preface

Control systems are devices that follow basic to complex protocol in order to control other devices or systems. This book on control systems deals with the complex set of advanced theories associated with already existing control systems and practices related to them. While understanding the long-term perspectives of the topics, the book makes an effort in highlighting their impacts as a modern tool for the growth of the discipline. For all readers who are interested in control systems, the case studies included in this book will serve as excellent guide to develop a comprehensive understanding. This book will be of great help to students and experts in the fields of systems theory, automation and actuators.

The researches compiled throughout the book are authentic and of high quality, combining several disciplines and from very diverse regions from around the world. Drawing on the contributions of many researchers from diverse countries, the book's objective is to provide the readers with the latest achievements in the area of research. This book will surely be a source of knowledge to all interested and researching the field.

In the end, I would like to express my deep sense of gratitude to all the authors for meeting the set deadlines in completing and submitting their research chapters. I would also like to thank the publisher for the support offered to us throughout the course of the book. Finally, I extend my sincere thanks to my family for being a constant source of inspiration and encouragement.

**Editor**

# 1

# A solar PV augmented hybrid scheme for enhanced wind power generation through improved control strategy for grid connected doubly fed induction generator

Adikanda Parida[1]* and Debashis Chatterjee[1]

*Corresponding author: Adikanda Parida, Department of Electrical Engineering, Jadavpur University, Kolkata, India
E-mail: adikanda_2003@yahoo.co.in

Reviewing editor: Siew Chong Tan, University of Hong Kong, Hong Kong

**Abstract:** In this paper, a wind power generation scheme using a grid connected doubly fed induction generator (DFIG) augmented with solar PV has been proposed. A reactive power-based rotor speed and position estimation technique with reduced machine parameter sensitivity is also proposed to improve the performance of the DFIG controller. The estimation algorithm is based on model reference adaptive system (MRAS), which uses the air gap reactive power as the adjustable variable. The overall generation reliability of the wind energy conversion system can be considerably improved as both solar and wind energy can supplement each other during lean periods of either of the sources. The rotor-side DC-link voltage and active power generation at the stator terminals of the DFIG are maintained constant with minimum storage battery capacity using single converter arrangement without grid-side converter (GSC). The proposed scheme has been simulated and experimentally validated with a practical 2.5 kW DFIG using dSPACE CP1104 module which produced satisfactory results.

**Subjects: Power Engineering; Renewable Energy; Systems & Controls**

**Keywords: wind–solar PV hybrid generation; model reference adaptive system; doubly fed induction generator; DC micro-grid**

## ABOUT THE AUTHORS

Adikanda Parida is working as an assistant professor in the Department of Electrical Engineering at North Eastern Regional Institute of Science and Technology, Nirjuli, Arunachal Pradesh, India. He is currently doing his research work in Department of Electrical Engineering, Jadavpur University, Kolkata, West Bengal, India. His areas of interest are Speed Sensorless Control of Doubly Fed Induction Generator, Wind-Solar PV hybrid Generation, Electrical Drives, and Energy Management.

Debashis Chatterjee is working as a professor in the Department of Electrical Engineering at Jadavpur University, Kolkata, West Bengal, India. His areas of interest are Speed Sensorless Control of Induction Motors, Speed sensorless control of Induction generators, Microgrid Control, Efficiency optimization of electric drive system, Electric vehicular drive, and Development of Solar/wind hybrid system and synchronization to grid

## PUBLIC INTEREST STATEMENT

The solar PV augmented wind energy conversion system presented in this paper addresses power generation issues during low-wind speed situations. The scheme can also have enormous potential to generate low-cost power in the remote region where the grid is absent and the average rotor speed remains in the sub-synchronous region with availability of solar power. The continuity of power generation both in the grid connected and standalone mode of operation is the major advantage of the proposed scheme. Moreover, the quality of power delivered to the grid or load can be considerably improved with the proposed scheme.

## 1. Introduction

The wind power generation is gaining increasing global importance due to environmental safety and growing energy crisis. Harnessing the generally fluctuating power from wind and maintaining continuous exchange of power with the utility grid is found to be a challenging task for the researchers in the recent past. The proposed solar PV augmented wind energy conversion system addresses the DFIG control issues under such situations in a cost-effective manner.

The wind turbine produces power in accordance with the available wind speeds with the help of existing technology (Yang & Yang, 2010; Zhang, Chen, & Hu, 2014). As the penetration of DFIG-based wind firms to the existing grids is significantly increasing day by day, the fluctuating power fed to the grid would adversely affect the power quality of the interconnected or weak or isolated grids as reported in Chen and Spooner (2001), Díaz-González, Sumper, Gomis-Bellmunt, and Bianchi (2013), Kanellos and Hatziargyriou (2012). To overcome such problems, comprehensive measures have been suggested in Lu, Chang, Lee, and Wang (2009), Zhou and Francois (2011) with the augmentation of power storage devices. The continuity of generation from DFIG-based wind energy conversion system which can be achieved through maximum power tracking control mechanism with augmentation of multiple renewable energy sources is presented in Mendis, Muttaqi, et al. (2014), Mendis, Muttaqi, Perera, and Kamalasadan (2015) for remote area power supply. A similar type of technique is proposed in Kou, Liang, Gao, and Gao (2015) for grid connected operation. However, increased complexity in terms of coordinated control and cost of energy is a subject of concern for practical implementation of such topologies. An appropriate scheme for rated power generation at generator bus is proposed in Vijayakumar, Tennakoon, Kumaresan, and Ammasai Gounden (2013), Daniel and AmmasaiGounden (2004) and Shukla and Tripathi (2015). However, the presented scheme does not clarify the operation over wide speed range or proper sizing methodology of the storage batteries under such situations.

The precise control over generated power requires enhanced performance of the rotor-side converter (RSC) controller, which can be achieved through accurate information of rotor position angle. The rotor position and speed estimation techniques reported in Cardenas, Pena, Proboste, and Asher (2005), Cardenas and Pena (2004), Cardenas, Pena, et al. (2008), Cardenas, Pena, Clare, Asher, and Proboste (2008) are based on model reference adaptive system. The scheme presented in Cardenas et al. (2005) is implemented using stator flux as adjustable variable, while Cardenas and Pena (2004) considers rotor flux for the same. The estimation technique adapted in Pena, Cardenas, Proboste, Asher, and Clare, (2008) is different from Cardenas et al. (2005) and Cardenas and Pena (2004) which considers rotor current as adjustable variable. The performance of the techniques adapted in Cardenas et al. (2005), Cardenas and Pena (2004), Cardenas et al. (2008) is analyzed in Cardenas et al. (2008). The reported techniques (Karthikeyan, Nagamani, & Ilango, 2012; Karthikeyan, Nagamani, Ray Chaudhury, & Ilango, 2012) are implemented with the use of differentiators based on open-loop methods. This can introduce severe errors for estimation of rotor speed and position due to noise in the input signals. Moreover, all these existing techniques have almost the similar snag of having strong dependency on various machine parameter variations.

The main advantageous features of the proposed scheme can be summarized as,

(1) The rotor position estimation technique for the proposed scheme is almost independent of machine parameter variations, hence accurate.
(2) Continuous and consistent active power generation over wider wind speed range in comparison to the existing schemes.
(3) The single converter topology for DFIG considerably reduces the operational and installation cost of the system.
(4) No circulation of generated power between stator and rotor circuits of DFIG. Thus, the proposed scheme facilitates a reduced associated continuous power loss in the system.

Proper experiments and simulations have been performed on a practical 2.5 kW DFIG to validate the proposed topology of wind energy conversion system (WECS).

## 2. Scheme for the proposed system

### 2.1. System description

The schematic of the proposed hybrid generation scheme is shown in Figure 1. Stator terminals are directly connected to the grid and rotor windings are connected to RSC through slip rings. The RSC controls the rotor power of DFIG independent of the utility grid, unlike the existing widely used schemes. The rotor power management is accomplished through the following four modes.

***Mode-I: Low wind speed and sufficient solar power***

In this sub-synchronous mode of operation, the rotor power is supplied directly from solar panel. If solar energy generated is more than rotor power requirement, the excess power will be delivered to the DC micro-grid as shown in Figure 1 after storing in the battery.

***Mode-II Both wind and solar power is low***

***In this sub-synchronous mode of operation, the rotor power needed is supplied from the battery***

***Mode-III: Large wind power and low solar power***

In this mode of operation, DFIG will deliver power through both stator and rotor. The rotor power can be directly stored in battery. If rotor power is very large, then DC micro-grid will be switched to evacuate the excess power available in the DC-link as shown in Figure 1.

***Mode-IV Both wind power and solar power is high***

In this case the DFIG will operate in the super-synchronous region and the power from rotor is added with the energy from solar PV and delivered to the DC micro-grid after charging the storage battery.

### 2.2. Rotor-side converter control

In the proposed scheme, RSC control signals are generated as shown in Figure 1 using measured values of stator voltage, stator current, rotor voltage, and rotor current as the input. Orienting stator flux $\psi_s$ along synchronously rotating d-axis, $\psi_s^e = \psi_{ds}^e$, thus $\psi_{qs}^e = 0$. Where, $\psi_{ds}^e, \psi_{qs}^e$ are the components of the stator flux in synchronous reference frame. The rated active and reactive power at the stator terminals of DFIG can be expressed as (Marques, Pires, & Sousa, 2011),

$$\left.\begin{aligned} P_s &= -V_{qs}^e \frac{L_m}{L_s} i_{qr}^e \\ Q_s &= \frac{(V_{qs}^e)^2}{\omega_e L_s} - \frac{L_m}{L_s} V_{qs}^e i_{dr}^e \end{aligned}\right\} \tag{1}$$

The d-axis rotor current reference $(i_{dr}^e)^*$ is generated by comparing estimated value of stator active power $\hat{P}_s$ with its reference value $P_s^*$. Similarly, comparing estimated and reference values of reactive power of the grid, the q-axis rotor current reference $\left(i_{qr}^e\right)^*$ is generated. The d-q axis rotor voltage references, $(V_{dr}^e)^*$, $\left(V_{qr}^e\right)^*$, can be expressed as,

$$(V_{dr}^e)^* = (V_{dr}^e)' - (\omega_{slip})\sigma L_r i_{qr}^e \tag{2}$$

$$\left(V_{qr}^e\right)^* = \left(V_{qr}^e\right)' + (\omega_{slip})\left(\left(L^2 m/L_s\right)i^e ms + \sigma L_r i_{dr}^e\right) \tag{3}$$

where, $(V_{dr}^e)'$ and $\left(V_{qr}^e\right)'$ are the control outputs of PI controllers PI_3 and PI_4 respectively. The PI controller outputs $(V_{dr}^e)'$ and $\left(V_{qr}^e\right)'$ are used to generate rotor d-q axes voltage references $(V_{dr}^e)^*$, $\left(V_{qr}^e\right)^*$ using (2) and (3).

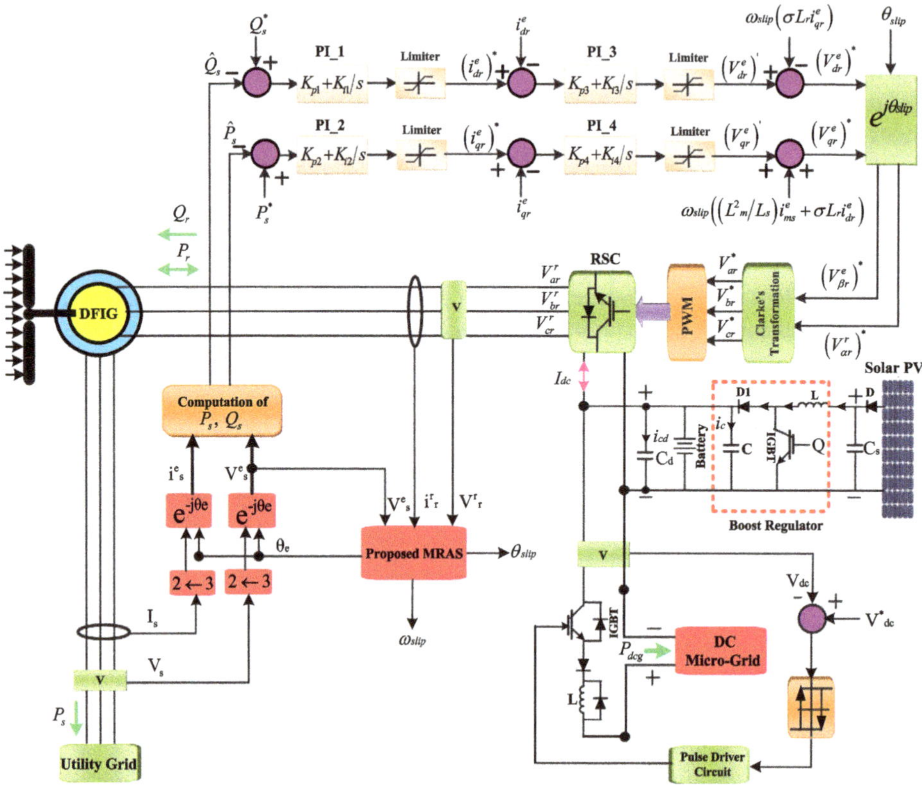


**Figure 1. The proposed hybrid scheme.**

### *2.3. Active power flow of the proposed wind–solar PV hybrid system*

The dynamics of the active power exchange of the proposed hybrid generation scheme is shown in Figure 2(a). In case of existing schemes reported in Karthikeyan, Nagamani, and Ilango (2012) and Karthikeyan, Nagamani, Ray Chaudhury, et al. (2012), during sub-synchronous operation, the requirement of DFIG rotor power $P_r$ is supplemented from the grid. Thus, the net power supplied to the grid by the generation system will be reduced considerably as shown in Figure 2(b). For the proposed scheme, the net power output from the generation system has been enhanced. This is due to the solar PV system supplements the rotor power pool in isolation with the grid. For the analysis, Pg signifies the air gap power, $P_{Grid}$ signifies the output power fed to the grid, and $P_m$ signifies the shaft mechanical power input. For the given machine specified in Appendix 1, Table 2, experiments were conducted with varying the rotor speed between 1,800 rpm and 1,050 rpm. Representing $V_{dc}$ and $i_{dc}$ as the measured DC-bus voltage and current, respectively, the rotor power $P_r$ is computed as $P_r = V_{dc} \times i_{dc}$.

The generated electrical power at stator terminals is measured as $P_s = V^s_{\alpha s} \times i^s_{\alpha s} + V^s_{\beta s} \times i^s_{\beta s}$. In this analysis, the losses incurred in the rotor-side converter, stator copper, and core losses are neglected for simplicity. The shaft mechanical power input can be calculated from the measured DC motor input power for known efficiency.

For each rotor slip, the corresponding values of $P_s$, $P_r$, and $P_m$ are computed and plotted as shown in Figure 2(b). From Figure 2(b), in the sub-synchronous speed region, it can be observed that the injected rotor power increases with increasing slip due to low shaft power. The net output power for the existing schemes is obtained by subtracting the rotor injected power $P_r$ from the power output $P_s$ at the stator side. It can be also observed that, for the proposed scheme, the net output power at the stator more than the existing conventional scheme due to the addition of solar PV source at the input of the rotor-side converter.

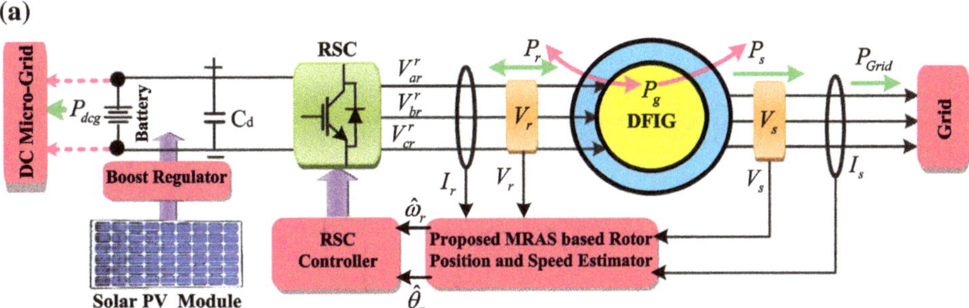


**Figure 2(a). The active power flow in proposed scheme.**

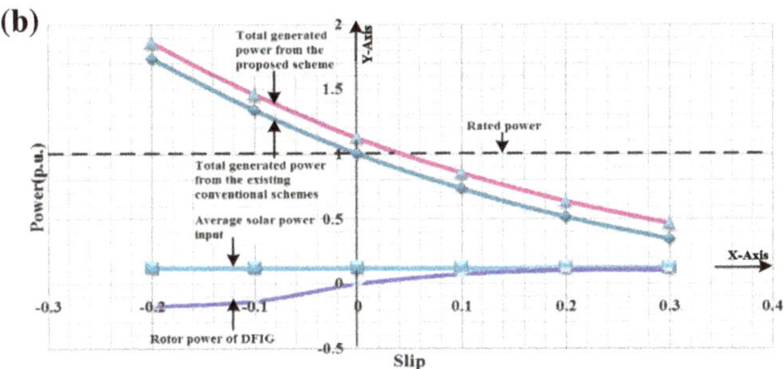


**Figure 2(b). Power distribution of the proposed and existing schemes.**

## 3. The proposed rotor speed and position estimator

### 3.1. Estimator modeling

In the proposed scheme, the rotor speed and position estimation is carried out with measured stator and rotor voltages and currents without using any differentiators and integration of low-frequency signals. The reference model variable is developed based on the measured values of rotor voltage and currents. The adjustable model variable requires the measured stator inputs and the rotor currents transformed to the synchronous reference frame as shown in Figure 3(b). With $v^s_{ds}$, $v^s_{qs}$ and $i^s_{ds}$, $i^s_{qs}$ being the measured values of stator voltages and currents in stationary reference frame, stator flux linkages $\psi^s_{ds}$, $\psi^s_{qs}$ can be expressed as (Leonhard, 2001),

$$\left.\begin{aligned}\psi^s_{ds} &= \int \left(v^s_{ds} - R_s i^s_{ds}\right)dt \\ \psi^s_{qs} &= \int \left(v^s_{qs} - R_s i^s_{qs}\right)dt\end{aligned}\right\} \tag{4}$$

The synchronous frequency of the stator flux is computed using the relation,

$$\omega_e = \left(\psi^s_{ds}\cdot\left(\dot{\psi}^s_{qs}\right) - \psi^s_{qs}\cdot\left(\psi^s_{ds}\right)\right) \Big/ \left(\psi^s_s\right)^2 \tag{5}$$

The stationary reference frame parameters can be converted into synchronous reference frame $(d^e - q^e)$ with $d^e$-axis aligned to stator flux axis. Therefore, the unit vectors can be computed as,

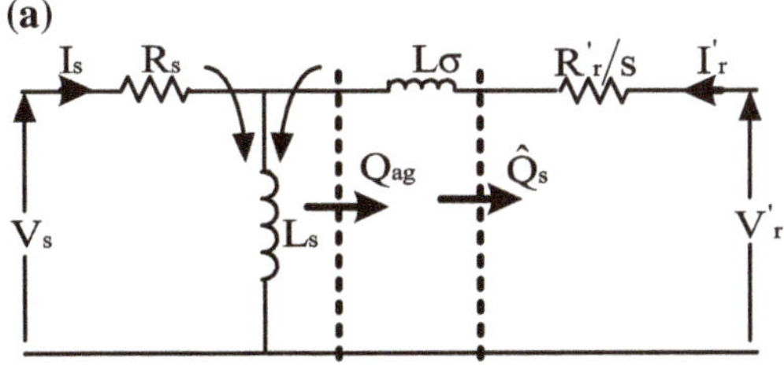


**Figure 3(a). Inverse-T model equivalent circuit of DFIG.**

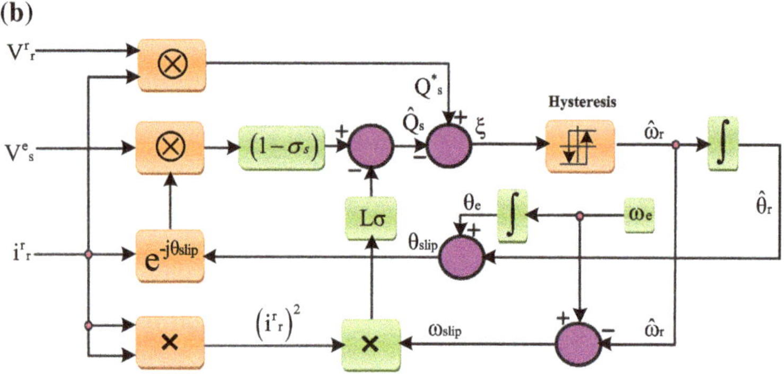


**Figure 3(b). The proposed rotor position and speed estimator.**

$\cos\theta_e = \psi^e_{ds}/\psi^e_{ds}\psi^e_s\psi^e_s$, $\sin\theta_e = \psi^e_{qs}\Big/\psi^e_{qs}\psi^e_s\psi^e_s$. With the help of unit vectors, the stator voltages can be transformed from stationary to synchronous reference frame as,

$$\begin{bmatrix} V^e_{ds} \\ V^e_{qe} \end{bmatrix} = \begin{bmatrix} \cos\theta_e & -\sin\theta_e \\ \sin\theta_e & \cos\theta_e \end{bmatrix} \begin{bmatrix} V^s_{ds} \\ V^s_{qs} \end{bmatrix} \tag{6}$$

The no load ($i^r_r \approx 0$) reactive power drawn by the DFIG at steady state neglecting stator resistive voltage drop can be obtained from Figure 3(a), as,

$$Q_m = (1/\omega_e L_s)\left[(v^e_s)^2\right] \tag{7}$$

Therefore, with, $(v^e_s)^2 = (v^e_{ds})^2 + \left(v^e_{qs}\right)^2$,

$$Q_m = (1/\omega_e L_s)\left[(v^e_{ds})^2 + \left(v^e_{qs}\right)^2\right] \tag{8}$$

Stator reactive power of DFIG at any load can be expressed as,

$$Q_s = \left[v^e_{qs} i^e_{ds} - v^e_{ds} i^e_{qs}\right] \tag{9}$$

Thus the air gap reactive power $Q_{ag}$ can computed from (8) and (9) as,

$$Q_{ag} = \left[v^e_{qs} i^e_{ds} - v^e_{ds} i^e_{qs}\right] - (1/\omega_e L_s)\left[(v^e_{ds})^2 + \left(v^e_{qs}\right)^2\right] \tag{10}$$

The d-q axis stator currents $i^e_{ds}, i^e_{qs}$ in synchronous reference frame can be expressed as,

$$\begin{bmatrix} i^e_{ds} \\ i^e_{qs} \end{bmatrix} = \frac{1}{L_s}\begin{bmatrix} \psi^e_{ds} \\ \psi^e_{qs} \end{bmatrix} - \frac{L_m}{L_s}\begin{bmatrix} i^e_{dr} \\ i^e_{qr} \end{bmatrix} \tag{11}$$

where, $i^e_{dr}$ and $i^e_{qr}$ of (11) can be expressed as,

$$\begin{bmatrix} i^e_{dr} \\ i^e_{qr} \end{bmatrix} = \begin{bmatrix} \cos\theta_{slip} & \sin\theta_{slip} \\ -\sin\theta_{slip} & \cos\theta_{slip} \end{bmatrix} \begin{bmatrix} i^r_{dr} \\ i^r_{qr} \end{bmatrix} \tag{12}$$

where, $i^r_{dr}, i^r_{qr}$ are the measured rotor currents in rotor reference frame.

Substituting $i^e_{ds}$ and $i^e_{qs}$ from (11) in (10) gives,

$$Q_{ag} = (L_m/L_s)\left[v^e_{ds} i^e_{qr} - v^e_{qs} i^e_{dr}\right] + \left[\left(v^e_{qs}\psi^e_{ds}/L_s\right) - \left(v^e_{ds}\psi^e_{qs}/L_s\right)\right] - (1/\omega_e L_s)\left[(v^e_{ds})^2 + \left(v^e_{qs}\right)^2\right] \tag{13}$$

Neglecting stator resistance drop, $\psi^e_{ds}, \psi^e_{qs}$ of (13) can be expressed as,

$$\begin{bmatrix} \psi^e_{ds} \\ \psi^e_{qs} \end{bmatrix} = (1/\omega_e)\begin{bmatrix} V^e_{qs} \\ -V^e_{ds} \end{bmatrix} \quad (14)$$

Considering (13) and (14),

$$Q_{ag} = (L_m/L_s)\left[v^e_{ds}i^e_{qr} - v^e_{qs}i^e_{dr}\right] = (1-\sigma_s)\left[v^e_{ds}i^e_{qr} - v^e_{qs}i^e_{dr}\right] \quad (15)$$

The inverse T-model equivalent circuit of DFIG is shown in Figure 3(a), where,

$$\left.\begin{matrix} L_\sigma \approx L_l r + Lls; i'_r = (L_m/L_s)i^r_r; V'_r = (L_s/L_m)V^r_r; \\ R'_r = (L_s/L_m)R_r; L\sigma = (1+\sigma_s)^2 L_r - L_s. \end{matrix}\right\} \quad (16)$$

The adjustable model reactive power $\hat{Q}_s$ as shown in Figure 4(a) can be expressed as,

$$\hat{Q}_s = (1-\sigma_s)\left[v^e_{ds}i^e_{qr} - v^e_{qs}i^e_{dr}\right] - (i^r_r)^2\omega_{slip}L\sigma \quad (17)$$

The reference model stator reactive power ($Q^*_s$) is calculated in the proposed algorithm with a concept that the rotor-side reactive power is same as the stator reactive power crossing the air gap. Considering inverse T-model equivalent circuit of DFIG at steady state, $Q^*_s$ can be obtained in the rotor reference frame as,

$$Q^*_s = \left[v^r_{qr}i^r_{dr} - v^r_{dr}i^r_{qr}\right] \quad (18)$$

Here in (18), $V^r_{dr}, V^r_{qr}$ are the measured rotor voltages in rotor reference frame. Expression (17) generates the adjustable model variable as a function of measured stator voltage, rotor current, and adjustable parameters of rotor position angle $\theta_r$. Slip speed $\omega_{slip} = \omega_e - \hat{\omega}_r$ and $\omega_e$ is computed using (5). An error signal $\xi$ is generated based on difference of $Q_s$ calculation from (17) and (18) which is given by,

$$\xi = Q^*_s - \hat{Q}_s \quad (19)$$

An adjustable mechanism designed by a hysteresis controller to drive the error $\xi$ computed using (19) to zero. The output of adjustable mechanism is the estimated rotor speed and is integrated for rotor position $\theta_r$ estimation.

### 3.2. Parameter Sensitivity

For most of the existing rotor speed and position estimation techniques, the adjustable variables are rotor current, stator flux, or rotor flux. These adjustable variables can be expressed as,

i. Rotor current, $\hat{i}_r = \frac{\psi_s - L_s i_s}{L_m} e^{-j\hat{\theta}_r}$
ii. Stator flux, $\hat{\psi}_s = L_s\left(i_s + \frac{L_m}{L_s}\right)i_r$ and
iii. Rotor flux $\hat{\psi}_r = L_r\left(i_r + \frac{L_m}{L\,r}\right)i_s$.

From the above equations, the adjustable variables are directly sensitive $L_m$ variations.

The direct sensitivity of the adjustable variables with $L_m$ will considerably affect the performance of the DFIG controller as the same usually varied during DFIG operation. In proposed control scheme, the adjustable variable $\hat{Q}_s$ seen from (17) is very weakly coupled to $L_m$. From (17), the term directly related with $L_m$ is $(1-\sigma_s)$. For any variation $\Delta L_m$, the new magnetizing inductance is given by $(L_m + \Delta L_m)$. Therefore, $(1-\sigma_s)$ can be modified to,

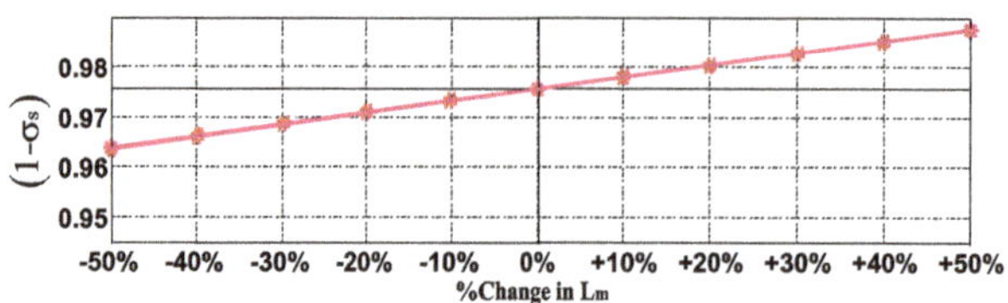


**Figure 4(a). Variation of $(1-\sigma_s)$ with change in $L_m$.**

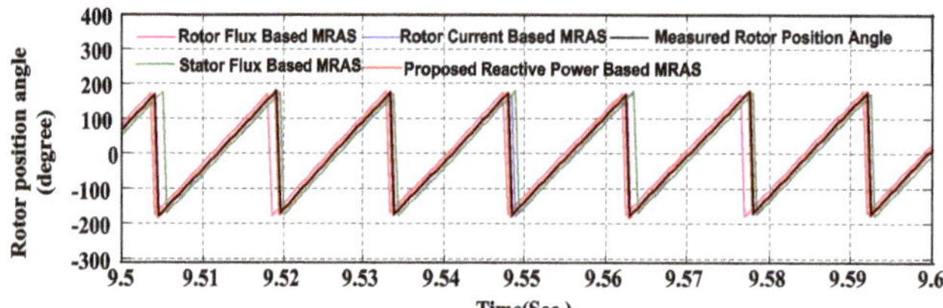


**Figure 4(b). Comparison of different rotor position estimation schemes with $L_m$ variation of −50%.**

$$\left(1-\left(\sigma_s+\Delta\sigma_s\right)\right)=\frac{\left(L_m+\Delta L_m\right)}{\left(L_m+\Delta L_m+Lls\right)}=\frac{\left(1+\frac{\Delta L_m}{L_m}\right)}{\left(1+\frac{\Delta L_m+Lls}{L_m}\right)} \tag{20}$$

The variation of the adjustable variable $\hat{Q}_s$ is plotted and shown in Figure 4(a) for $L_m$ deviation between −50% and +50%. It is observed from Figure 4(a) that the maximum variations of $(1-\sigma_s)$ are +1.23% and −1.2% for $L_m$ deviations of +50% and −50%. This error is acceptable. The rotor position estimation for $L_m$ variation of −50% and +50% is shown in Figures 4(b) and 4(c), respectively, using widely used techniques including the proposed one. The rotor position estimation error in radian for the same is shown in Figure 4(d). Observation of Figures 4(b)–4(d) shows the superiority of the proposed technique over the existing schemes.

### 3.3. Stability analysis

The block diagram for the stability analysis is shown in Figure 5(a). The distribution of stator and rotor space vectors are shown in Figure 5(b) when rotor is at sub-synchronous speed and in Figure 5(c) for super-synchronous speed with $\mu$ as the angle between $Vs$ and $Ir$ vectors (Wu, Lang, & Zargari, 2011). Assuming, $\sigma_s \approx 0$, (17) can be written as,

$$\hat{Q}_s=\left(1-\sigma_s\right)\left[v^e_{ds}i^e_{qr}-v^e_{qs}i^e_{dr}\right]-\left(i^r_r\right)^2\omega_{slip}L\sigma=\left(1-\sigma_s\right)\left[Vs\otimes Ir\right]-\left(i^r_r\right)^2\omega_{slip}L\sigma \tag{21}$$

where $\otimes$ is the symbol of cross product. From the vector diagram in Figures 5(b) and 6(c), (21) can be rewritten to obtain the estimated reactive power magnitude as,

$$\hat{Q}_s=K_{f1}\left[V_sI_r\sin(\gamma)\right]-K_{f2} \tag{22}$$

Assuming a disturbance of $\Delta\theta_r$ in estimation of rotor position $\theta_r$, the new space position angle between $Vs$ and $Ir$ will be $\gamma+\Delta\theta_r$. This will modify the estimated torque $\hat{Q}_s$ to $\hat{Q}_s+\Delta\hat{Q}_s$.

From (22),

$$\hat{Q}_s+\Delta\hat{Q}_s=K_{f1}\left[V_sI_r\sin(\gamma+\Delta\theta_r)\right]-K_{f2}=K_{f1}V_sI_r\left[\sin\gamma.\cos(\Delta\theta_r)+\cos\gamma.\sin(\Delta\theta_r)\right]-K_{f2} \tag{23}$$

For $\Delta\theta_r$ to be small, (23) can be written as, $\hat{Q}_s+\Delta\hat{Q}_s=K_{f1}V_sI_r\left[\sin\gamma+\Delta\theta_r.\cos\gamma\right]-K_{f2}$ (24)

The difference between (24) and (22) can be expressed as, $\Delta\hat{Q}_s=K_{f1}V_sI_r[\cos\gamma]\Delta\theta_r$ (25)

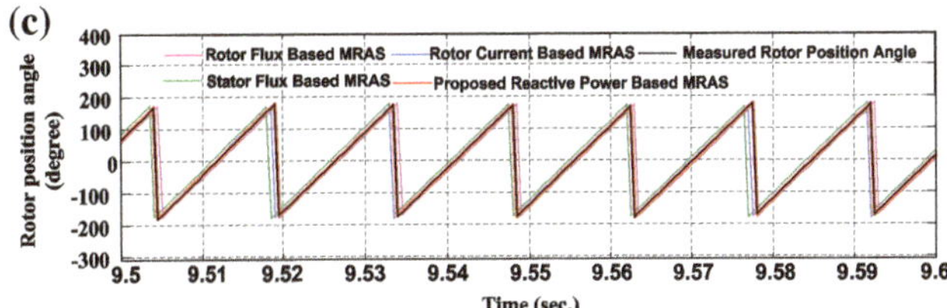


**Figure 4(c). Comparison of different rotor position estimation schemes with $L_m$ variation of +50%.**

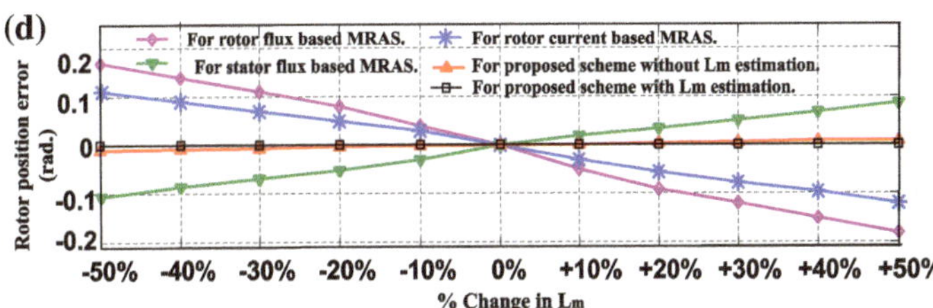


**Figure 4(d). Comparison of rotor position estimation errors associated with the proposed controller with the existing schemes.**

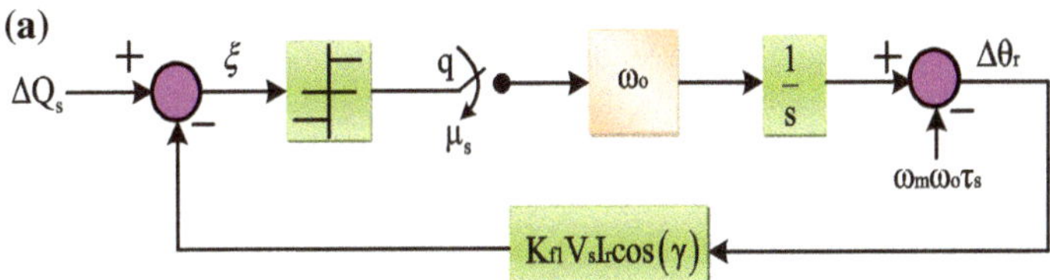


**Figure 5(a). Block diagram for the small perturbations.**

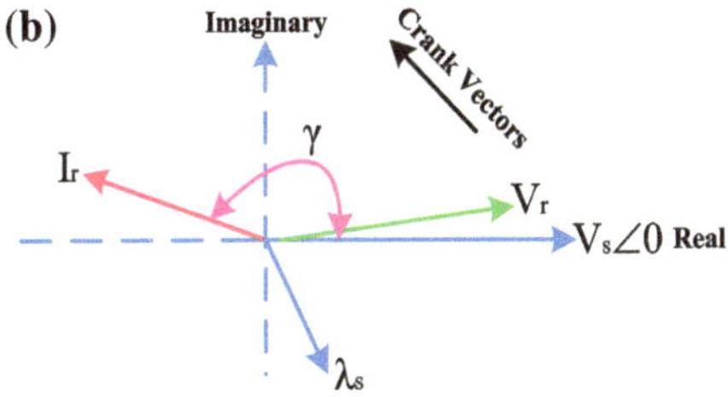


**Figure 5(b). Distribution of stator and rotor space.**

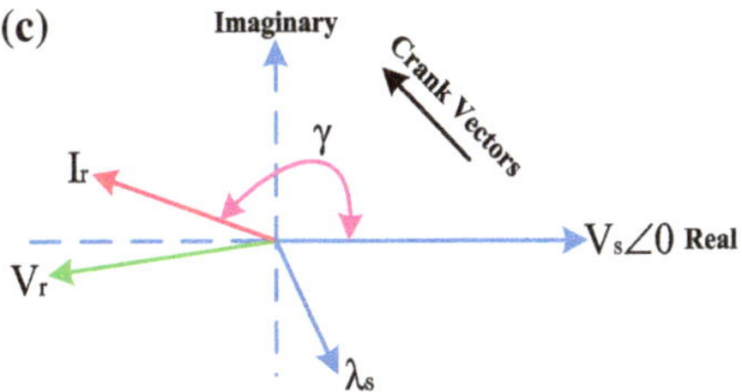


**Figure 5(c). Distribution of stator and rotor space vectors at sub-synchronous speed vectors at super-synchronous speed.**

Therefore from Figure 5(a), $\xi = K_{f1} V_s I_r [\cos \gamma] \Delta\theta_r$ (26)

$$\frac{d\Delta\theta_r}{dt} = \omega_o (q - \omega_m) \tag{27}$$

From (26),

$$\frac{d\xi}{dt} = K_{f1}V_sI_r[\cos\gamma]\frac{d\Delta\theta_r}{d}t \tag{28}$$

Substituting from (27) in (28),

$$\frac{d\xi}{dt} = K_{f1}V_sI_r[\cos\gamma]\omega_o(q - \omega_m) \tag{29}$$

When the hysteresis controller output slides between (0 to 2) per unit, condition for stability is analyzed as (Marques et al., 2011; Silva & Pinto, 2011),

For $\xi < 0; q = 0;$

$$\frac{d\xi}{d}t = K_{f1}V_sI_r[\cos\gamma]\omega_o(0 - \omega_m) \tag{30}$$

$\xi > 0; q = 2;$

$$\frac{d\xi}{d}t = K_{f1}V_sI_r[\cos\gamma]\omega_o(2 - \omega_m) \tag{31}$$

To satisfy condition for system stability, from (30), (31), $\cos\gamma < 0, V_s \neq 0, I_r \neq 0$. From Figures 5(b) and 5(c), $\frac{\pi}{2} < \gamma < \frac{3\pi}{2}$ for rotor speed from sub-synchronous to super-synchronous including synchronous speed at all loads. Therefore, irrespective of the load at the machine terminal, the estimation model is stable at all operating points.

## 4. Power generation economics

To justify the economical feasibility of the proposed scheme over the existing systems reported in Vijayakumar et al. (2013) and Cardenas et al. (2005), an economic analysis is performed for all the systems. The average availability of solar power is assumed to be between 9 am and 4 pm (http://www.nrel.gov/rredc). The capacity of solar panel is so considered that it can supply power to both DFIG rotor and to the battery during day time and low wind speed situations. Considering an average slip magnitude of 0.2 for the 2.5 kW DFIG in sub-synchronous and -0.15 in super-synchronous operating region, a 1 kW capacity of solar unit is taken so that both the charging of the battery and supplementing power to DFIG rotor can take place during sub-synchronous operation. The rated rotor voltage of 150 V for the DFIG will require a DC-bus and the battery voltage of 192 V. Considering charging–discharging cycle shown in Figure 6, a battery capacity of 50Ah is selected for the system. However, the capacity of both the solar unit and battery can be selected for different locations based on operation cycles. A comparison of the proposed and existing schemes has been made using HOMER Pro-3.2 (Hybrid Optimization Model for Electric Renewables) software. For the proposed system, an extra investment is required due to the PV modules along with the charging unit and the storage battery.

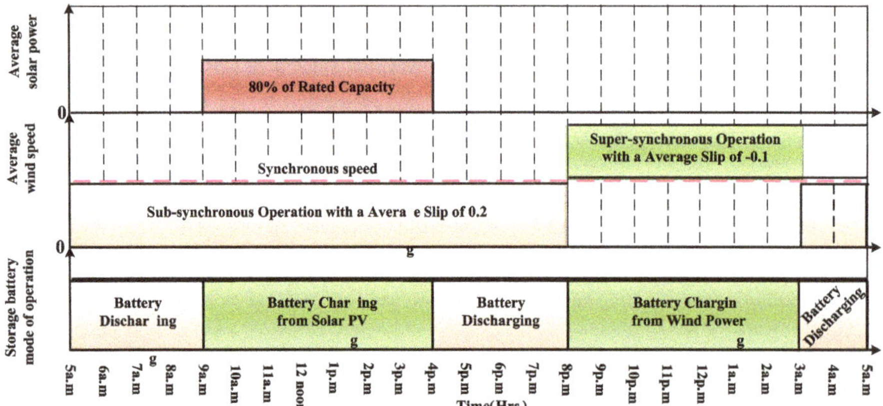


**Figure 6. Generation profile of the proposed wind–solar hybrid system.**

On the other hand, the conventional scheme with dual converter will require additional grid-side converter and transformer for grid inter connection. The other scheme with only one converter and battery backup will need a large storage battery capacity to cope up the situation of the system to be in the sub-synchronous region for the given cycle. Inclusion of both PV module and storage battery in the proposed scheme will result in additional generation of power compared to the existing schemes and results in lesser size of storage battery. On this basis, the three schemes are simulated and the results for the important parameters are tabulated in Table 1.

The extra investment for the proposed scheme compared to the existing schemes with back-to-back converter can be paid back within 3.5 years and generate profit beyond this period up to the first replacement schedule of storage battery. On the other hand, the larger battery required for the existing scheme with single converter and battery will result in larger unit cost of energy as shown in Table-1. Where, the life of the PV modules and battery is considered to be 20 years and 5 years, respectively, for the present calculations. Therefore, the proposed scheme is more reliable in terms of power generation and economically profitable than the existing schemes.

### 5. Simulation and Experimental Results

The proposed generation scheme is implemented with a 2.5 kW doubly fed induction generator with machine parameters as shown in Appendix 1, Table 2. The experimental arrangement is shown in Figure 7, where the measured values of stator voltage, stator currents, rotor voltage, and rotor currents are fed as inputs to the controller. The stator and rotor currents are sensed through Hall Effect sensors LTS25NP (LEM make), while the voltage sensing process has been accomplished through CV3-1000(LEM make) sensor. A 5Hp, 1,500 rpm, separately excited DC motor with necessary torque control arrangement is coupled to the DFIG can emulate the wind turbine.

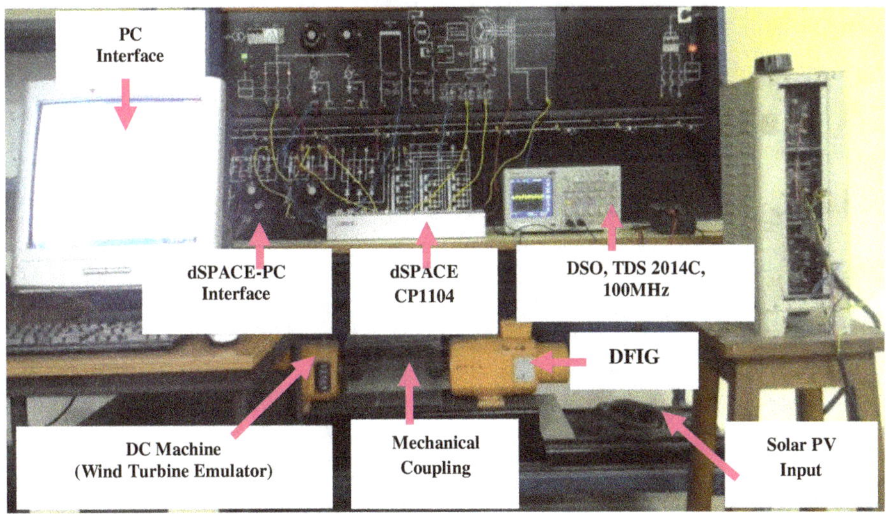


**Figure 7. Experimental setup for the proposed scheme.**

**Table 1. Comparative economic analysis of existing conventional system with the proposed system**

| Decisive factors type of the system | Proposed DFIG system with storage battery and PV modules | Existing DFIG system with single converter and storage battery (Vijayakumar et al., 2013) | Existing DFIG system with back to back converter (Cardenas et al., 2005) |
|---|---|---|---|
| Initial capital required ($) | 1500 | 1500 | 1800 |
| Cost of energy ($/kWh) | 0.008 | 0.009 | 0.01 |
| Net present cost ($) | 2047 | 2549 | 1532 |
| Operating cost ($/year) | 45 | 72 | 36 |

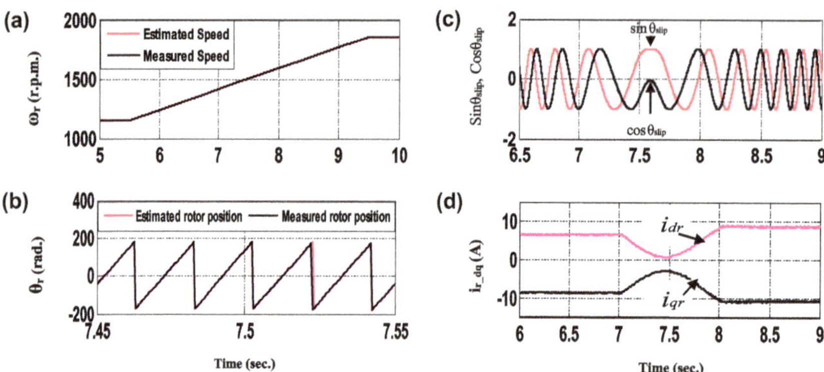


**Figure 8. Simulation results of rotor speed transient: (a) transition of rotor speed (rpm) through synchronous speed; (b) rotor position (rad.) estimation; (c) response of slip rotor position near synchronous speed; (d) response of rotor currents (A) near synchronous speed.**

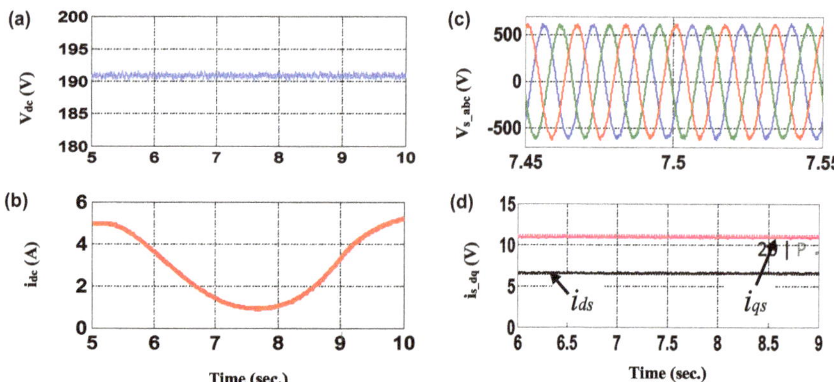


**Figure 9. (a) Response of dc-link voltage (V) during speed transition; (b) response of dc-link current (A) during speed transition; (c) response of stator voltage; (d) response of stator current.**

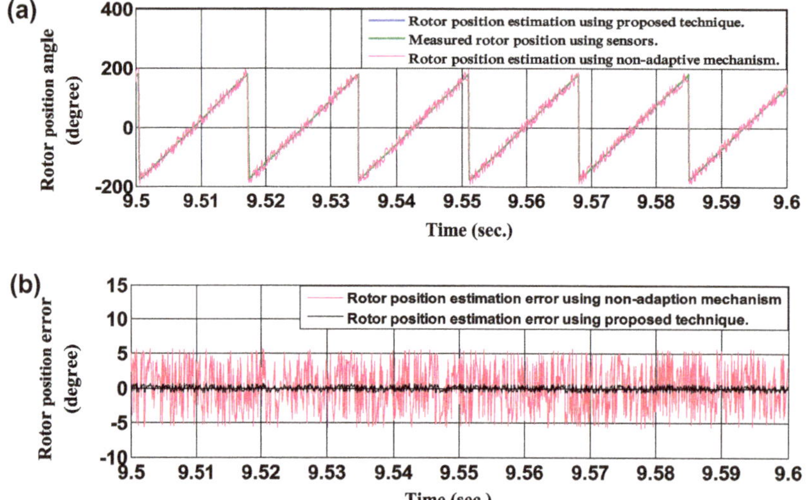


**Figure 10. (a) Comparison of the proposed rotor position estimation technique with a non-adaption type; (b) rotor position estimation error corresponding to (a).**

A closed-loop torque controller is used for the separately excited dc machine incorporating the wind turbine model. The shaft encoder gives the information regarding the speed and position of the shaft for the control of the prime mover. The proposed MRAS computes rotor speed and position and is compared with measured value for validation. The rotor speed and position information from

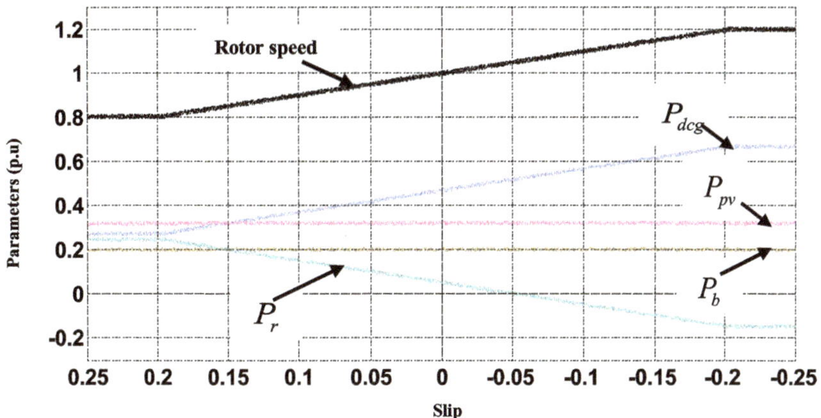


**Figure 11. Power sharing between soar PV augmented DC-link and the DFIG of the proposed scheme.**

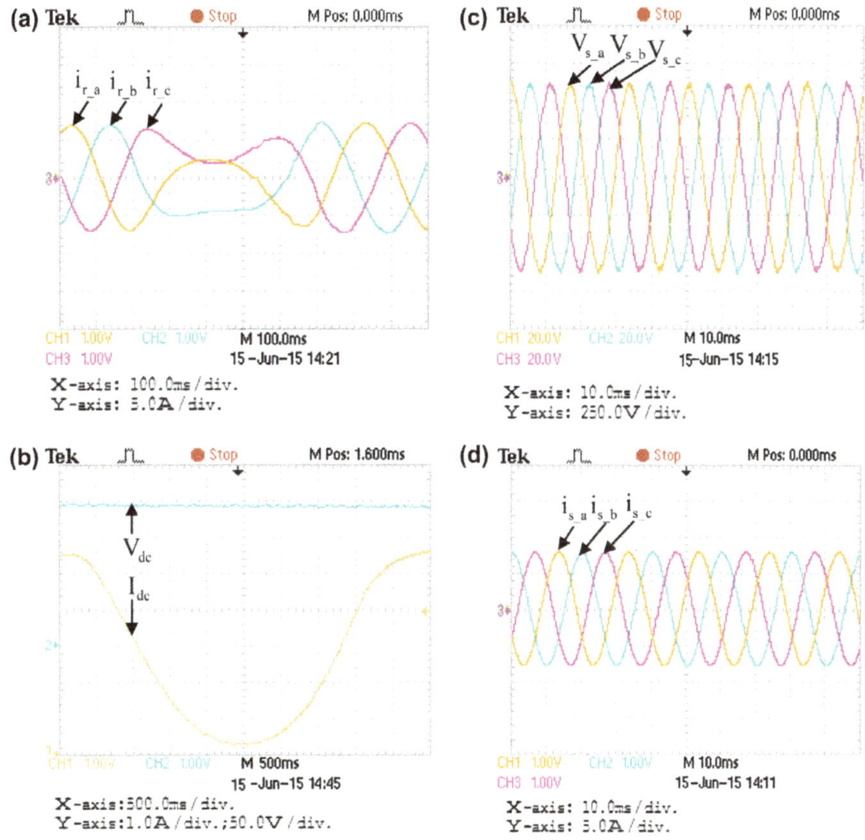


**Figure 12. Experimental results: (a) response of rotor currents (A) near synchronous speed; (b) response of dc-link voltage; (V) and dc-link current (A) during speed transition; (c) response of stator voltages (V); (d) response of stator currents (A).**

estimator are fed to the RSC controller. A dSPACE CP1104 module with a PC interface is used for implementation of the experimental system.

Initially, the machine was running at 1,150 rpm with the help of the DC motor drive system. Then the speed was slowly increased to 1,860 rpm in 4.5 s. The estimated speed through the proposed controller and the corresponding measured speed is shown in Figure 8(a), while the measured and estimated rotor position at near synchronous speed are shown in Figure 8(b). It can be observed from Figure 8(a) and Figure 8(b) that the proposed controller accurately estimates both the rotor position and speed during the speed transient. Figure 8(c) shows the computation of sin and cosine terms of slip angle which is used for demodulating the rotor voltage and current signals. The d-q rotor current $ir\_dq$ in synchronous reference frame is shown in Figure 8(d) corresponding to the speed variation as shown in Figure 8(a).

The response of the DC-link voltage and current can be observed from Figure 9(a) and Figure 9(b), respectively, during speed transition corresponding to Figure 8 (a). The stator voltage and current are shown in Figure 9(c) and Figure 9(d), respectively, which remain unbiased during the transition of rotor speed. Then the proposed scheme of position estimation was compared with the conventional rotor position estimation schemes realized through open-loop methods. The results are shown in Figure 10(a) and Figure 10(b) in which it can be observed that the proposed technique is superior to the existing technique. The proposed scheme have integraton of three type of surces e.g. battery, PV, and wind was tested for power sharing to the DC-bus. The wind speed varied from 0.8 to 1.2 p.u when the battery was fully charged and the solar insolation was at constant value. The power was measured at various points e.g. battery, PV module, DFIG rotor and DC micro-grid, and variations of each of them are recorded during thie speed transition. The results are shown in Figure 11. The battery, PV, DFIG rotor, and DC micro-grid powers are denoted by, $P_{pv}$, $P_b$, $P_{dcg}$, and $P_r$, respectively, in Figure 11. The DC micro-grid power can be given by, $P_{dcg}=P_{pv} + P_b - P_r$ as per the chosen convention of rotor power. It is observed from Figure 11 that when the DFIG is in sub-synchronous region and speed is increasing, the rotor power absorbed by the machine is reducing, and the DC micro-grid power is increasing. At the rotor speed of 1.0 p.u, the power delivered to rotor is low and the DC micro-grid power is almost equal to the sum of the battery power and the PV power which can be observed from Figure 11. After the speed crosses 1.0 p.u value i.e. in the super-synchronous region, the rotor delivers increasing power and as a result the DC micro-grid power is further increased.

Then, similar experiments were performed to validate the simulation results. The experimental results are shown in Figure 12 for the speed transition between 1,150 rpm and 1,860 rpm with the help of the prime mover. The rotor currents in rotor reference frame as shown in Figure 12(a), the rotor frequency is almost zero when the rotor speed is near syncronous speed region, thus the rotor current exhibits dc behavior. The DC-link voltage and currents during the speed transition is shown in Figure 12(b). The DC-link voltage remains constant while the DC-link current reduces near synchronous speed and again rises beyond synchronous speed. This is because, the rotor current at synchronous speed is dc and attains the minimum value. The stator voltage and current wave forms are given in Figure 12(c) and (d), respectively.

## 6. Conclusion

The proposed wind-solar hybrid generation scheme is successfully implemented for wide wind speed range. The continuity in the active power at stator terminals of DFIG can be maintained at minimum cost of energy. The single converter topology for the DFIG greatly reduces the operational and installation cost of the system. The overall generation reliability has been considerably increased through proper augmentation of solar PV system which supplements WECS during lean periods of available wind power. The performance of the system is considerably improved through accurate estimation of rotor speed and position of DFIG. Moreover, the proposed rotor speed and position estimation technique has negligible sensitivity to the machine parameter variations, which makes it more accurate. The proposed method is simple and can be implemented for grid-connected or grid-isolated modes through any already available drive compatible processors.

### List of symbols

$\psi_{ds}, \psi_{qs}$ Components of stator flux along d and q axes, respectively

$v_{ds}, v_{qs}$ Components of stator voltages along d and q axes, respectively

$v_{dr}, v_{qr}$ Components of rotor voltages along d and q axes, respectively

| | |
|---|---|
| $i_{ds}, i_{qs}$ | Components of stator currents along d and q axes, respectively |
| $i_{dr}, i_{qr}$ | Components of rotor currents along d and q axes, respectively |
| $L_m, L_s$ | Magnetizing, Stator self inductances of DFIG, respectively |
| $L_{ls}, L_{lr}$ | Stator and rotor leakage inductances of DFIG, respectively |
| $\gamma$ | Space angle between stator voltage and rotor current |
| $R_s$ | Resistance of stator |
| $P_s, Q_s$ | Active and reactive powers available at stator terminals of DFIG |
| $\omega_e, \omega_r$ | Angular velocity of stator magnetizing flux, Rotational speed of rotor |
| $\theta_r$ | Rotor position angle |
| $\sigma_s$ | Stator leakage factor |
| $K_p, T_i$ | PI controller constants |
| $\mu_s$ | Time constant of the delay introduced by sampling |
| $\omega_0$ | Electrical frequency of the machine |
| $\omega_{slip}$ | Slip speed of the machine |
| $\sigma$ | Leakage factor $= \left(1 - \frac{L_m^2}{L_s L_r}\right)$ |
| $ims$ | Magnetizing current of the machine |

**Funding**
The authors received no direct funding for this research.

**Author details**
Adikanda Parida[1]
E-mail: adikanda_2003@yahoo.co.in
Debashis Chatterjee[1]
E-mail: debashisju@yahoo.com
[1] Department of Electrical Engineering, Jadavpur University, Kolkata, India.

**References**

Cardenas, R., & Pena, R. (2004). Sensorless vector control of induction machines for variable-speed wind energy applications. *IEEE Transactions on Energy Conversion, 19*, 196–205. http://dx.doi.org/10.1109/TEC.2003.821863

Cardenas, R., Pena, R., Proboste, J., & Asher, G. (2005). MRAS observer for sensorless control of standalone doubly fed induction generators. *IEEE Transactions on Energy Conversion, 20*, 710–718. http://dx.doi.org/10.1109/TEC.2005.847965

Cardenas, R., Pena, R., Clare, J., Asher, G., & Proboste, J. (2008). MRAS observers for sensorless control of doubly-fed induction generators. *IEEE Transactions on Power Electronics, 23*, 1075–1084. http://dx.doi.org/10.1109/TPEL.2008.921189

Chen, Z., & Spooner, E. (2001). Grid power quality with variable speed wind turbines. *IEEE Transactions on Energy Conversion, 16*, 148–154. http://dx.doi.org/10.1109/60.921466

Daniel, S., & AmmasaiGounden, N. (2004). A novel hybrid isolated generating system based on PV fed inverter-assisted wind-driven induction generators. *IEEE Transactions on Energy Conversion, 19*, 416–422. http://dx.doi.org/10.1109/TEC.2004.827031

Díaz-González, F., Sumper, A., Gomis-Bellmunt, O., & Bianchi, F. D. (2013). Energy management of flywheel-based energy storage device for wind power smoothing. *Applied Energy, 110*, 207–219. http://dx.doi.org/10.1016/j.apenergy.2013.04.029

Kanellos, F. D., & Hatziargyriou, N. D. (2002). The effect of variable-speed wind turbines on the operation of weak distribution networks. *IEEE Transactions on Energy Conversion, 17*, 543–548. http://dx.doi.org/10.1109/TEC.2002.805224

Karthikeyan, A., Nagamani, C., Ray Chaudhury, A. B., & Ilango, G. S. (2012). Implicit position and speed estimation algorithm without the flux computation for the rotor side control of doubly fed induction motor drive. *IET Electric Power Applications, 6*, 243–252. http://dx.doi.org/10.1049/iet-epa.2010.0286

Karthikeyan, A., Nagamani, C., & Ilango, G. S. (2012). A versatile rotor position computation algorithm for the power control of a grid-connected doubly fed induction generator. *IEEE Transactions on Energy Conversion, 27*, 697–706. http://dx.doi.org/10.1109/TEC.2012.2199118

Kou, P., Liang, D., Gao, F., & Gao, L. (2015). Coordinated predictive control of DFIG-based wind-battery hybrid systems: Using non-Gaussian wind power predictive distributions. *IEEE Transactions on Energy Conversion, 30*, 681–695. http://dx.doi.org/10.1109/TEC.2015.2390912

Leonhard, W. (2001). *Control of electrical drives* (3rd ed., pp. 163–182). New York, NY: Springer.

Lu, M.-S., Chang, C.-L., Lee, W.-J., & Wang, L. (2009). Combining the wind power generation system with energy storage equipment. *IEEE Transactions on Industry Applications, 45*, 2109–2115.

Marques, G. D., Pires, V., & Sousa, S. (2011). A DFIG sensorless rotor-position detector based on a hysteresis controller. *IEEE Transactions on Energy Conversion, 26*, 9–17. http://dx.doi.org/10.1109/TEC.2010.2070507

Mendis, N., Muttaqi, K. M., et al. (2014). Management of low- and high-frequency power components in demand-generation fluctuations of a DFIG-based wind-dominated RAPS system using hybrid energy storage. *IEEE Transactions on Industry Applications, 50*, 2258–2268. http://dx.doi.org/10.1109/TIA.2013.2289973

Mendis, N., Muttaqi, K. M., Perera, S., & Kamalasadan, S. (2015). An effective power management strategy for a wind-diesel-hydrogen-based remote area power supply to meet fluctuating demands under generation uncertainty. *IEEE Transactions on Industry Applications, 51*, 1228–1238. http://dx.doi.org/10.1109/TIA.2014.2356013

Pena, R., Cardenas, R., Proboste, J., Asher, G., & Clare, J. (2008). Sensorless control of doubly-fed induction generators using a rotor-current-based MRAS observer. *IEEE Transactions on Industrial Electronics, 55*, 330–339.

Shukla, R. D., & Tripathi, R. K. (2015). Isolated wind power supply system using double-fed induction generator for remote areas. *Energy Conversion and Management, 96*, 473–489. http://dx.doi.org/10.1016/j.enconman.2015.02.084

Silva, J. F., & Pinto, S. F. (2011). *Power electronics hand book, circuits, devices, and applications* (3rd ed., p. 1060). Oxford: Elsevier.

Vijayakumar, K., Tennakoon, S. B., Kumaresan, N., & Ammasai Gounden, N. G. (2013). Real and reactive power control of hybrid excited wind-driven grid-connected doubly fed induction generators. *IET Power Electronics, 6*, 1197–1208. http://dx.doi.org/10.1049/iet-pel.2012.0709

Wu, B., Lang, Y., Zargari, N., et al. (2011). *Power conversion and control of wind energy systems* (pp. 237–274). Wiley. http://dx.doi.org/10.1002/9781118029008

Yang, L., Yang, G. Y., Xu, Z., Dong, Z. Y., Wong, K. P., & Ma, X. (2010). Optimal controller design of a doubly-fed induction generator wind turbine system for small signal stability enhancement. *IET Generation, Transmission & Distribution, 4*, 579–597.

Zhang, Y., Chen, Z. H. W., & Hu, W. (2014). Flicker mitigation by individual pitch control of variable speed wind turbines with DFIG. *IEEE Transactions on Energy Conversion, 29*, 20–28. http://dx.doi.org/10.1109/TEC.2013.2294992

Zhou, T., & Francois, B. (2011). Energy management and power control of a hybrid active wind generator for distributed power generation and grid integration. *IEEE Transactions on Industrial Electronics, 58*, 95–104. http://dx.doi.org/10.1109/TIE.2010.2046580

## Appendix 1.

**Table 2. Design parameters**

| Element | Parameter | Element | Parameter | Element | Parameter |
|---|---|---|---|---|---|
| DFIG | | Magnetizing inductance | 0.44H | Ratings | 1000 PW |
| Stator | Y-connected, 415 V (L-L) | RSC | | Output voltage | 12 V |
| Rotor | Y-connected, 150 V (L-L) | Type | 3 phase, Y-connected | Output current | 7.5A |
| Rated power | 2.5 kW | Rated power | 1 kW | Operating temperature | 25–35°C |
| No. of poles | 4 | DC-link voltage, $V_{dc}$ | 200 V | STORAGE BATTERY | Rechargeable |
| Rated speed | 1440 rpm | DC-link current, $I_{dc}$ | 7.5A | Type | Pb-acid |
| Stator resistance | 0.4ohm | AC voltage, $V_{ac}$ | 150 V(L-L) | Total capacity | 50Ah |
| Rotor resistance | 0.45ohm | SOLAR PV MODULE | | Output voltage/unit | 12 V |
| Stator self inductance | 0.015H | Type Polycrystalline | | | |
| Rotor self inductance | 0.015H | | | | |

# 2

# Event-triggered decentralized robust model predictive control for constrained large-scale interconnected systems

Ling Lu[1], Yuanyuan Zou[1*] and Yugang Niu[1]

*Corresponding author: Yuanyuan Zou, Key Laboratory of Advanced Control and Optimization for Chemical Process, East China University of Science & Technology, Ministry of Education, Shanghai 200237, China
E-mail: yyzou@ecust.edu.cn

Reviewing editor: James Lam, University of Hong Kong, Hong Kong

**Abstract:** This paper considers the problem of event-triggered decentralized model predictive control (MPC) for constrained large-scale linear systems subject to additive bounded disturbances. The constraint tightening method is utilized to formulate the MPC optimization problem. The local predictive control law for each subsystem is determined aperiodically by relevant triggering rule which allows a considerable reduction of the computational load. And then, the robust feasibility and closed-loop stability are proved and it is shown that every subsystem state will be driven into a robust invariant set. Finally, the effectiveness of the proposed approach is illustrated via numerical simulations.

**Subjects: Automation Control; Control Engineering; Dynamical Control Systems; Intelligent Systems**

**Keywords: constraint tightening method; decentralized control; event-triggered control; input-to-state stability; robust model predictive control**

## 1. Introduction

A class of complex large-scale systems composed of several interconnected subsystems has been receiving an increasing attention due to its various practical applications, e.g. power systems, chemical processes, and transportation systems (Hua, Leng, & Guan, 2012; Yan, Edwards, Spurgeon, &

## ABOUT THE AUTHORS

Ling Lu is currently pursuing her masters degree in East China University of Science and Technology. Her research interests include large scale system, event-triggered control, model predictive control, and their applications.

Yuanyuan Zou is an Associate Professor in East China University of Science and Technology. Her research interests include predictive control, network-based control systems, and distributed control systems.

Yugang Niu is a Professor and Vice-Dean in the East China University of Science and Technology. His research interests include stochastic systems, sliding mode control, wireless sensor network, congestion control, and smart grid.

## PUBLIC INTEREST STATEMENT

Model predictive control (MPC) is a popular and effective control method to handle the uncertainties and hard constraints on states and controls in the process industry. To deal with the computational complexity in complex large-scale systems, decentralized MPC strategy has been developed. Not only does it maintain the superior properties of MPC method, but it also provides some advantages such as easier maintenance, greater reliability, and less computational effort.

However, the time-triggered control scheme in traditional decentralized MPC algorithms will consume redundant computation resources. To overcome this problem, this paper considers the problem of event-triggered decentralized model predictive control for constrained large-scale linear systems subject to additive bounded disturbances. The proposed strategy can not only reduce the on-line computation load, but also achieve the alleviation of computational complexity.

Bleijs, 2004; Zhang & Liu, 2013; Zhang, Zhang, & Wang, 2014). In the control of large-scale systems, decentralized control structure is often the most appropriate control method for handling the computational complexity. Also, it has the advantages such as easier maintenance, greater reliability, and less computational effort (see e.g. Keviczky, Borrelli, & Balas, 2006; Riverso, Farina, & Trecate, 2013; Yan, Lam, Li, & Chen, 2000, and the references therein).

On the other hand, as a popular control technique, model predictive control (MPC) strategy can effectively handle the uncertainties and hard constraints on states and controls in the process industry. In recent years, many MPC synthesis algorithms that ensure closed-loop stability and robust convergence have been proposed (see e.g. Alessio, Barcelli, & Bemporad, 2011; Magni & Scattolini, 2006; Mayne, Rawlings, Rao, & Scokaert, 2000; Zou & Niu, 2013). Especially, the study of decentralized MPC algorithm for large-scale systems has attracted much attention (Mayne, 2014; (Raimondo, Magni, & Scattolini, 2007; Tran & Ha, 2014). Among them, decentralized MPC design was introduced in Tran and Ha (2014) for networks of linear systems with bounded coupling delay. The stability condition was derived for the constrained optimization problem and the issues of input and state constraints had been addressed by adopting decentralized MPC method. In Magni and Scattolini (2006), a stabilizing decentralized MPC algorithm was presented for nonlinear, discrete time systems under the assumption that no information can be exchanged between local control laws. The closed-loop stability was achieved based on the inclusion of a contractive constraint in the optimization problem. Alessio et al. (2011) proposed a decentralized MPC algorithm for constrained large-scale linear system and analyzed the asymptotic stability of closed-loop system. In particular, the decentralized MPC strategy for large-scale nonlinear system with bounded disturbances was considered in Raimondo et al. (2007), where each subsystem was locally controlled with a MPC algorithm ensuring the robust stability. However, it should be pointed out that the main mechanism in the aforementioned decentralized MPC works was based on time-triggered control scheme. That is, at each sampling instant, a finite horizon local optimization problem was solved on-line to determine the local optimal control sequence, in which only the first control signal would be applied to the subsystem. Apparently, this will consume redundant computational and communication resources, and even affect its applications for a case with limited resources and insufficient communication bandwidth. This motivates the research on event-triggered decentralized MPC algorithms.

The key feature of event-triggered control schemes is that the decision for the execution of control laws is not made periodically, but depending on the detailed system behaviors, such as the system state or the performance index (Dimarogonas, Frazzoli, & Johansson, 2012). At present, many developments have been reported on the event-triggered schemes (Dong, Wang, Alsaadi, & Ahmad, 2015; Dong, Wang, Ding, & Gao, 2015; Liu & Hao, 2013). In Liu and Hao (2013), a decentralized event-triggered scheme is proposed for networked control systems in order to reduce network traffic and computation resource. In Dong, Wang, Alsaadi, et al. (2015), an event-triggered robust distributed state estimation problem for sensor networks was studied, and in Dong, Wang, Ding, et al. (2015), an event-triggered H-infinity filter algorithm was presented to alleviate the unnecessary waste of communication resources. For event-triggered MPC, some related works can be found in Eqtamin, Dimarogonas, and Kyriakopoulos (2010), Lehmann, Henriksson, and Hohansson (2013), Eqtami, Dimarogonas, and Kyriakopoulos (2011a), Li and Shi (2014). In Eqtamin et al. (2010), an event-triggered MPC algorithm for discrete-time systems was presented, where the optimization problem was solved only when the triggering condition was violated. Eqtami et al. (2011a) considered the event-triggered robust MPC for both continuous and discrete-time uncertain nonlinear systems with additive disturbances, and derived the triggering rule according to the input-to-state stability (ISS) property. More recently, a class of interconnected large-scale system with bounded disturbances was considered in Eqtami, Dimarogonas, and Kyriakopoulos (2011b), whose key idea was that each subsystem was controlled by a local event-triggered robust model predictive controller. However, it is worthy to note that although the method in Eqtami et al. (2011b) can achieve the reduction on the number of the optimal control updating, there still exists high computational complexity in the optimization problem due to the uncertainties.

In this paper, we investigate the event-triggered decentralized predictive control problem based on constraint tightening approach to reduce both the times of solving optimization problem and computational complexity. By constructing a candidate control sequence and ISS stability, the event-triggered conditions are derived to determine whether the local predictive control optimization problem is solved. Moreover, the robust feasibility and closed-loop stability are proved to show the convergence of subsystem states.

The remainder of the paper is organized as follows. In Section 2, the problem statement for the large-scale system is presented. In Section 3, the main results, including the event-triggered decentralized model predictive controller and the proof of robust feasibility and robust stability are presented. Section 4 provides a numerical example to show the efficiency of the proposed algorithm.

**Notations**: $\mathbb{R}^n$ denotes the real $n$ dimensional Euclidean space, $\mathbb{R}^+$ denotes the positive real number. Given two vectors $x,\ y \in \mathbb{R}^n$, $x \geq y \Leftrightarrow x_i \geq y_i$, $i = 1, 2, \ldots, n$. For any vector $x \in \mathbb{R}^n$ and matrix $Q, ||x||_Q^2 = x^T Qx$. $\lambda_{max}(\cdot)$ represents the maximum eigenvalue of a real matrix. Given any two sets $A$, $B$ of $\mathbb{R}^n$, the operator "$\sim$" denotes the Pontryagin set difference, i.e. $A \sim B = \{a | a + b \in A, \forall b \in B\}$, while the operator $\oplus$ denotes the Minkowski set addition, i.e. $A \oplus B = \{a + b | a \in A, b \in B\}$.

## 2. Preliminary

### *2.1. System description*

Consider the linear discrete-time interconnected system composed of $M$ local subsystems

$$\begin{cases} x_i(k+1) = A_i x_i(k) + B_i u_i(k) + G_i y(k) + v_i(k), \\ y_i(k) = C_i x_i(k), \quad i \in \{1, 2, \ldots, M\}, \end{cases} \tag{1}$$

where $x_i(k) \in \mathbb{R}^{n_i}$ is the state of the $i$th subsystem, $u_i(k) \in \mathbb{R}^{m_i}$ is the control variable, $y_i(k) \in \mathbb{R}^{s_i}$ is the output, $v_i(k) \in \mathbb{R}^{n_i}$ is the additive bounded disturbance, and $G_i y(k)$ denotes the mutual influence of $M$ subsystems, where $y(k) \triangleq [y_1^T(k), y_2^T(k), \ldots, y_M^T(k)]^T \in \mathbb{R}^s$ with $s = \sum_{i=1}^M s_i$ is the overall output.

The output, input, and disturbance of the $i$th subsystem are assumed to satisfy the following constraints

$$u_i(k) \in U_i = \{u_i^{min} \leq u_i(k) \leq u_i^{max}\}, \tag{2}$$

$$y_i(k) \in Y_i = \{y_i^{min} \leq y_i(k) \leq y_i^{max}\}, \tag{3}$$

$$v_i(k) \in V_i = \{v_i^{min} \leq v_i(k) \leq v_i^{max}\}. \tag{4}$$

By letting $\tilde{y}_i(k) \triangleq G_i y(k)$, we obtain

$$\tilde{y}_i(k) \in \tilde{Y}_i = \{\tilde{y}_i^{min} \leq \tilde{y}_i(k) \leq \tilde{y}_i^{max}\}, \tag{5}$$

where $\tilde{y}_i^{min} \triangleq [(G_i \tilde{y}_1^{min})^T, (G_i \tilde{y}_2^{min})^T, \ldots, (G_i \tilde{y}_M^{min})^T]^T$, $\tilde{y}_i^{max} \triangleq [(G_i \tilde{y}_1^{max})^T, (G_i \tilde{y}_2^{max})^T, \ldots, (G_i \tilde{y}_M^{max})^T]^T$.

Define the following augmented vectors

$$\begin{cases} x(k) \triangleq [x_1^T(k), x_2^T(k), \ldots, x_M^T(k)]^T, \\ y(k) \triangleq [y_1^T(k), y_2^T(k), \ldots, y_M^T(k)]^T, \\ v(k) \triangleq [v_1^T(k), v_2^T(k) \ldots, v_M^T(k)]^T. \end{cases} \tag{6}$$

The whole system can be written as

$$\begin{cases} x(k+1) = Ax(k) + Bu(k) + Gy(k) + v(k), \\ y(k) = Cx(k), \end{cases} \tag{7}$$

where
$A = \text{diag}\{A_1, A_2, \dots, A_M\}$, $B = \text{diag}\{B_1, B_2, \dots, B_M\}$, $G = \text{diag}\{G_1, G_2, \dots, G_M\}$,
$C = \text{diag}\{C_1, C_2, \dots, C_M\}$.

### *2.2. Decentralized MPC formulation*

In the sequel, we present the decentralized MPC scheme based on the constraint tightened approach, in which each local MPC optimization control problem (OCP) is formulated based on the nominal subsystem corresponding to (1). Moreover, we take the sum of the interaction term $\tilde{y}_i(k)$ and the additive disturbance $v_i(k)$ as the perturbation $w_i(k) = \tilde{y}_i(k) + v_i(k) \in W_i, \quad \forall k$ with $W_i = \tilde{Y}_i \oplus V_i$.

Thus, we obtain the following nominal subsystem

$$\begin{cases} \bar{x}_i(k+1) = A_i\bar{x}_i(k) + B_iu_i(k), \\ \bar{y}_i(k) = C_i\bar{x}_i(k). \end{cases} \tag{8}$$

The following finite horizon optimization problem for the uncertain subsystem (1) is considered

$$J_i^*(k) = \min_{U_i(k)} \left\{ \sum_{j=0}^{N_i} (||y_i(k+j|k)||_{Q_i}^2 + ||u_i(k+j|k)||_{R_i}^2) + ||y_i(k+N_i+1|k)||_{P_i}^2 \right\}, \tag{9}$$

subject to

$$x_i(k+j+1|k) = A_ix_i(k+j|k) + B_iu_i(k+j|k), \tag{10}$$

$$y_i(k+i|k) = C_ix_i(k+j|k), \tag{11}$$

$$x_i(k|k) = x_i(k), \tag{12}$$

$$y_i(k+j|k) \in Y_i(j), \tag{13}$$

$$u_i(k+j|k) \in U_i(j), \tag{14}$$

$$x_i(k+N_i+1|k) \in X_{iF}, \tag{15}$$

where $N_i$ is the prediction horizon, $Q_i = q_i \cdot I_{s_i \times s_i}$, $q_i \in \mathbb{R}^+$, $R_i = r_i \cdot I_{m_i \times m_i}$, $r_i \in \mathbb{R}^+$.

The constraint sets $U_i(j)$ in (14) are defined by a tightening recursion

$$U_i(0) = U_i, \tag{16}$$

$$U_i(j+1) = U_i(j) \sim K_i(j)L_i(j)W_i, \quad \forall j \in \{0, 1, \dots, N_i - 1\}. \tag{17}$$

Similarly, the constraint sets $Y_i(j)$ in (13) are

$$Y_i(0) = Y_i, \tag{18}$$

$$Y_i(j+1) = Y_i(j) \sim C_iL_i(j)W_i, \quad \forall j \in \{0, 1, \dots, N_i - 1\}. \tag{19}$$

The matrices $K_i(j)$ and $L_i(j)$ denote the associated state transmission matrices under the following candidate policy

$$L_i(0) = I, \tag{20}$$

$$L_i(j+1) = (A_i + B_iK_i(j))L_i(j), \quad \forall j \in \{0, 1, \ldots, N_i - 1\}. \tag{21}$$

The terminal constraint set $X_{iF}$ in (15) is defined by

$$X_{iF} = \Omega_i \sim L_i(N_i)W_i, \tag{22}$$

where $\Omega_i$ is a robust control invariant admissible set under disturbances $L_i(N_i)W$, i.e. there exists a control law $H_ix_i$ satisfying

$$\forall x_i \in \Omega_i, A_ix_i + B_iH_ix_i + L_i(N_i)(G_iy(k) + v_i(k)) \in \Omega_i, \forall \tilde{y}(k) \in \tilde{Y}_i, v_i(k) \in V_i, \tag{23}$$

$$\forall x_i \in \Omega_i, C_ix_i \in Y_i(N_i), H_ix_i \in U_i(N_i). \tag{24}$$

*Remark 1* In order to simplify the computation of terminal constraint set $X_{iF}$, the nilpotent LQR policy (Richards & How, 2006) is adopted such that $L_i(N_i) = 0$. Hence, (23–24) can be rewritten as

$$\forall x_i \in X_{iF}, A_ix_i + B_iH_ix_i \in X_{iF}, \forall \tilde{y}_i \in \tilde{Y}_i, v_i \in V_i, \tag{25}$$

$$\forall x_i \in X_{iF}, C_ix_i \in Y_i(N_i), H_ix_i \in U_i(N_i). \tag{26}$$

Since the condition (25) and (26) do not involve the disturbance, it is much simpler for identifying a suitable set $X_{iF}$.

*Remark 2* In this work, a constraint tightened strategy is applied to each uncertain subsystem. Since only a nominal prediction model is used in the OCP and the effect of disturbances is considered by resorting to suitable restrictions of the constraints, the resultant computational complexity is avoided effectively.

In this work, our objective is to propose an event-triggered decentralized MPC algorithm based on the constraint tightened strategy such that system resources can be saved and each local OCP computational complexity caused by uncertainties can be reduced.

## 3. Event-triggered decentralized robust model predictive controller

In the traditional decentralized MPC strategy, the local optimal control law is usually applied to each subsystem at each sampling instant by solving on-line the local OCP. In this work, we propose an event-triggered decentralized MPC strategy, which determines the updating of control inputs according to a certain triggering condition. In other words, the optimal control law is applied to each subsystem only at its triggered time instant $k_i^t$. During the interval step $k_i^t + m_i$, $m_i \in \{1, 2, \ldots, N_i\}$ of any two successive triggering events $k_i^t$ and $k_i^{t+1}$, a candidate control sequence $\bar{U}_i(k_i^t + m_i)$ based on the optimal control sequence $U_i^*(k_i^t)$ at time $k_i^t$ is applied to the $i$th subsystem. Note that $k_i^t$ is the prior triggering step. Hence, it is important to provide an appropriate control sequence $\bar{U}_i(k_i^t + m_i)$ which satisfies specific constraints at time $k_i^t + m_i$. Based on the analysis of feasibility and robust stability, we further obtain the triggering condition for each subsystem.

### 3.1. Robust feasibility

In this case, the robust feasibility of the local constrained OCP is analyzed in the following theorem.

THEOREM 1 *Suppose that the local OCP has the optimal control sequence $U_i^*(k_i^t)$ at the triggered time $k_i^t$. The local OCP with the candidate control sequence $\bar{U}_i(k_i^t + m_i) = \{\bar{u}_i(k_i^t + m_i|k_i^t + m_i), \ldots, \bar{u}_i(k_i^t + m_i + N_i|k_i^t + m_i)\}$ is feasible at time $k_i^t + m_i$, $m_i \in \{1, 2, \ldots, N_i\}$, where*

$$\begin{cases} \bar{u}_i(k_i^t+m_i+j|k_i^t+m_i) = \bar{u}_i(k_i^t+m_i+j|k_i^t+m_i-1) + K_i(j)L_i(j)\tilde{y}_i(k_i^t+m_i-1) \\ \qquad + K_i(j)L_i(j)v_i(k_i^t+m_i-1), \qquad \forall j \in \{0,1,\ldots,N_i-1\}, \\ \bar{u}_i(k_i^t+m_i+j|k_i^t+m_i) = H_i\hat{x}_i(k_i^t+m_i+j|k_i^t+m_i), \qquad j = N_i. \end{cases} \tag{27}$$

*Proof* Assume that the local OCP is successfully solved on time $k_i^t$, the optimal control sequence $U_i^*(k_i^t)$ satisfying (14) is obtained with the corresponding optimal outputs $y_i^*(k_i^t+1+j|k_i^t)$, $j \in \{0,1,\ldots,N_i-1$ satisfying (13). The feasibility of the local OCP at time $k_i^t+m_i$, $m_i \in \{1,2,\ldots,N_i\}$ is ensured if the constraints (10–15) are satisfied.

Firstly, the feasibility of the local OCP at time $k_i^t+1$ is proved. The following candidate control sequence $\bar{U}_i(k_i^t+1) = \{\bar{u}_i(k_i^t+1|k_i^t+1),\ldots,\bar{u}_i(k_i^t+1+N_i|k_i^t+1)\}$ is constructed,

$$\begin{cases} \bar{u}_i(k_i^t+1+j|k_i^t+1) = u_i^*(k_i^t+1+j|k_i^t) + K_i(j)L_i(j)w_i(k_i^t), \ \forall j = 0,1,\ldots,N_i-1, \\ \bar{u}_i(k_i^t+1+j|k_i^t+1) = H_i\hat{x}_i(k_i^t+1+j|k_i^t+1), j = N_i, \end{cases} \tag{28}$$

where $w_i(k_i^t) = G_iy_i(k_i^t) + v_i(k_i^t)$.

With the candidate control sequence (28), we have

$$\hat{x}_i(k_i^t+1+j|k_i^t+1) = x_i^*(k_i^t+1+j|k_i^t) + L_i(j)w_i(k_i^t), \quad \forall j \in \{0,1,\ldots,N_i\}, \tag{29}$$

$$\begin{cases} \hat{y}_i(k_i^t+1+j|k_i^t+1) = C_i\hat{x}_i(k_i^t+1+j|k_i^t+1) = y_i^*(k_i^t+1+j|k_i^t) + C_iL_i(j)w_i(k_i^t), \quad \forall j \in \{0,1,\ldots,N_i-1\}, \\ \hat{y}_i(k_i^t+1+j|k_i^t+1) = C_i\hat{x}_i(k_i^t+1+j|k_i^t+1), \ j = N_i. \end{cases} \tag{30}$$

The feasibility at $k_i^t$ implies $x_i^*(k_i^t+N_i+1|k_i^t) \in X_{iF}$. From (22), we have

$$\hat{x}_i(k_i^t+1+N_i|k_i^t+1) = x_i^*(k_i^t+1+N_i|k_i^t) + L_i(N_i)w_i(k_i^t) \in \Omega_i. \tag{31}$$

In the sequel, we prove the candidate control sequence $\bar{U}_i(k_i^t+1)$ can satisfy constraints (13–15).

(i) **Constraint (13)**: $\hat{y}_i(k_i^t+1+j|k_i^t+1) \in Y_i(j)$ for $j \in \{0, 1, \ldots, N_i\}$.

The optimal outputs $y_i^*(k_i^t+1+j|k_i^t)$ satisfies (13) at $k_i^t$, so we obtain

$$y_i^*(k_i^t+1+j|k_i^t) \in Y_i(j+1), \ \forall j \in \{0,1,\ldots,N_i-1\}. \tag{32}$$

According to (18–19), we have

$$\hat{y}_i(k_i^t+1+j|k_i^t+1) \in Y_i(j), \ \forall j \in \{0,1,\ldots,N_i-1\}. \tag{33}$$

Since $\hat{x}_i(k_i^t+1+N_i|k_i^t+1) \in \Omega_i$, from (24) it yields

$$\hat{y}_i(k_i^t+N_i+1|k_i^t+1) = C_i\hat{x}_i(k_i^t+N_i+1|k_i^t+1) \in Y_i(N_i). \tag{34}$$

Combining (33) and (34), it can be further written as

$$\hat{y}_i(k_i^t+1+j|k_i^t+1) \in Y_i(j), \forall j \in \{0,1,\ldots,N_i\}. \tag{35}$$

(ii) **Constraint (14)**: $\bar{u}_i(k_i^t+1+j|k_i^t+1) \in U_i(j)$ for $j \in \{0, 1, \ldots, N_i\}$.

Considering that $U_i^*(k_i^t)$ is the optimal control sequence at $k_i^t$, it holds that

$$u_i^*(k_i^t+j|k_i^t) \in U_i(j), \ \forall j \in \{0,1,\ldots,N_i\}. \tag{36}$$

According to (16–17), we have

$$\bar{u}_i(k_i^t + 1 + j|k_i^t + 1) \in U_i(j), \forall j \in \{0, 1, \dots, N_i - 1\}. \tag{37}$$

Since $\hat{x}_i(k_i^t + 1 + N_i|k_i^t + 1) \in \Omega_i$, we obtain

$$\bar{u}_i(k_i^t + 1 + N_i|k_i^t + 1) = H_i\hat{x}_i(k_i^t + 1 + N_i|k_i^t + 1) \in U_i(N_i). \tag{38}$$

From (37) and (38), it can be obtained that

$$\bar{u}_i(k_i^t + 1 + j|k_i^t + 1) \in U_i(j), \forall j \in \{0, 1, \dots, N_i\}. \tag{39}$$

(iii) **Constraint (15)**: $\hat{x}_i(k_i^t + N_i + 2|k_i^t + 1) \in X_{iF}$.

From (31), we have $\hat{x}_i(k_i^t + 1 + N_i|k_i^t + 1) \in \Omega_i$, according to (23), the subsequent state must satisfy

$$A_i\hat{x}_i(k_i^t + N_i + 1|k_i^t + 1) + B_iH_i\hat{x}_i(k_i^t + N_i + 1|k_i^t + 1) + L_i(N_i)w_i \in \Omega_i.$$

By the definition of $X_{iF}$ in (22), we have

$$\hat{x}_i(k_i^t + N_i + 2|k_i^t + 1) = (A_i + B_iH_i)\hat{x}_i(k_i^t + N_i + 1|k_i^t + 1) \in X_{iF}, \tag{40}$$

which implies that the terminal constraint (15) at $k_i^t + 1$ is satisfied.

From the above it shows that the candidate control law $\bar{U}_i(k_i^t + 1)$ at time instant $k_i^t + 1$ can satisfy the constraints (10–15) and the local OCP is feasible at step $k_i^t + 1$. By means of similar arguments, the feasibility of the local OCP at subsequent time $k_i^t + m_i$, $m_i \in \{2, 3, \dots, N_i\}$ can be recursively proved. This completes the proof. □

### *3.2. ISS and triggering condition*

We choose the cost function in (9) as a candidate Lyapunov function for the $i$th subsystem, and define the difference of the feasible cost function as

$$\Delta J_{m_i}^i = \bar{J}_i(k_i^t + m_i) - \bar{J}_i(k_i^t + m_i - 1). \tag{41}$$

Note that $\bar{J}_i(k_i^t) = J_i^*(k_i^t)$.

Before we discuss the ISS stability and the triggering condition, the following results are presented.

THEOREM 2 *Consider the subsystem (1) subject to (3–5) and the control law (27), and suppose the matrix $P_i$ in (9) satisfies $C_i^TP_iC_i \geq C_i^TQ_iC_i + H_i^TR_iH_i + (A_i + B_iH_i)^TC_i^TP_iC_i(A_i + B_iH_i)$. The difference of the feasible cost functions between the time $k_i^t + m_i$ and $k_i^t + m_i - 1$ is bounded by*

$$\Delta J_{m_i}^i \leq \sum_{j=1}^{M} (\hat{\beta}_1^i ||x_j(k_i^t + m_i - 1)||) + \beta_2^i ||v_i(k_i^t + m_i - 1)|| - \alpha_i ||x_i(k_i^t + m_i - 1)||^2, \tag{42}$$

where

$$\alpha_i \triangleq q_i \cdot ||C_i||^2, \tag{43}$$

$$\hat{\beta}_1^i \triangleq \beta_1^i \cdot ||C_j||, \tag{44}$$

$$\beta_1^i \triangleq S_1^i\left(\sum_{j=0}^{N_i-1} ||L_i(j)||\right) + S_2^i\left(\sum_{j=0}^{N_i-1} ||L_i(j)||^2\right) + F_1^i \sum_{j=0}^{N_i-1} (||K_i(j)|| \cdot ||L_i(j)||) + F_2^i \sum_{j=0}^{N_i-1} (||K_i(j)||^2 \cdot ||L_i(j)||^2) + M_2^i \tag{45}$$

$$\beta_2^i \triangleq S_3^i\left(\sum_{j=0}^{N_i-1}||L_i(j)||\right)+S_4^i\left(\sum_{j=0}^{N_i-1}||L_i(j)||^2\right)+F_3^i\sum_{j=0}^{N_i-1}(||K_i(j)||\cdot||L_i(j)||)+F_4^i\sum_{j=0}^{N_i-1}(||K_i(j)||^2\cdot||L_i(j)||^2)+M_1^i, \quad (46)$$

with $\psi_i \triangleq \max\{||x_i||:x_i \in X_{iF}\}$, $\gamma_y^i \triangleq \max\{||y_i||:y_i \in Y_i(j),\ j \in \{0,1,\dots,N_i\}\}$, $\gamma_u^i \triangleq \max\{||u_i||:u_i \in U_i(j)$, $j \in \{0,1,\dots,N_i\}\}$, $\gamma_Y \triangleq \max\{||y||:y \in Y\}$, $\gamma_v^i \triangleq \max\{||v_i||:v_i \in V_i\}$, $S_1^i \triangleq 2q_i\cdot\gamma_y^i\cdot||C_i||\cdot||G_i||$, $S_2^i \triangleq q_i\cdot||C_i||\cdot||G_i||\cdot(||C_i||\cdot||G_i||\cdot\gamma_Y+||C_i||\cdot\gamma_v^i)$, $S_3^i \triangleq 2q_i\cdot\gamma_y^i\cdot||C_i||$, $S_4^i \triangleq q_i\cdot||C_i||^2(||G_i||\cdot\gamma_Y+\gamma_v^i)$, $F_1^i \triangleq 2r_i\cdot\gamma_u^i\cdot||G_i||$, $F_2^i \triangleq r_i\cdot||G_i||\cdot(||G_i||\cdot\gamma_Y+\gamma_v^i)$, $F_3^i \triangleq 2r_i\cdot\gamma_u^i$, $F_4^i \triangleq r_i\cdot(||G_i||\cdot\gamma_Y+\gamma_v^i)$, $M_1^i \triangleq \lambda_{\max}(C_i^TP_iC_i)\cdot\{||L_i(N_i)||^2\cdot||G_i||\cdot\gamma_Y+||L_i(N_i)||^2\cdot\gamma_v^i+2\psi_i\cdot||L_i(N_i)||\}$, $M_2^i \triangleq M_1^i\cdot||G_i||$.

*Proof* For $m_i = 1$, we have

$$\begin{aligned}
\Delta J_1^i &= \bar{J}_i(k_i^t+1)-J_i^*(k_i^t)\\
&=\sum_{j=0}^{N_i}\{||\bar{y}_i(k_i^t+1+j|k_i^t+1)||_{Q_i}^2+||\bar{u}_i(k_i^t+1+j|k_i^t+1)||_{R_i}^2\}+||\bar{y}_i(k_i^t+N_i+2|k_i^t+1)||_{P_i}^2\\
&-\sum_{j=0}^{N_i}\{||y_i^*(k_i^t+j|k_i^t)||_{Q_i}^2+||u_i^*(k_i^t+j|k_i^t)||_{R_i}^2\}-||y_i^*(k_i^t+N_i+2|k_i^t+1)||_{P_i}^2\\
&\le\{S_1^i\left(\sum_{j=0}^{N_i-1}||L_i(j)||\right)+S_2^i\left(\sum_{j=0}^{N_i-1}||L_i(j)||^2\right)\}||y(k_i^t)||+\{S_3^i\left(\sum_{j=0}^{N_i-1}||L_i(j)||\right)+S_4^i\left(\sum_{j=0}^{N_i-1}||L_i(j)||^2\right)\}\cdot||v_i(k_i^t)||\\
&+\left\{F_1^i\sum_{j=0}^{N_i-1}(||K_i(j)||\cdot||L_i(j)||)+F_2^i\sum_{j=0}^{N_i-1}(||K_i(j)||^2||L_i(j)||^2)\right\}||y(k_i^t)||+\left\{F_3^i\sum_{j=0}^{N_i-1}(||K_i(j)||\cdot||L_i(j)||)\right. \quad (47)\\
&\left.+F_4^i\sum_{j=0}^{N_i-1}(||K_i(j)||^2||L_i(j)||^2)\right\}||v_i(k_i^t)||+M_2^i||y(k_i^t)||+M_1^i||v_i(k_i^t)||-q_i||y(k_i^t)||^2\le\beta_1^i\sum_{j=0}^{M}(||C_j||\cdot||x_i(k_i^t)||)\\
&+\beta_2^i||v_i(k_i^t)||-\alpha_i||x_i(k_i^t)||^2=\sum_{j=0}^{M}(\hat{\beta}_1^i||x_i(k_i^t)||)+\beta_2^i||v_i(k_i^t)||-\alpha_i||x_i(k_i^t)||^2,
\end{aligned}$$

For $m_i = 2$, the difference (41) is

$$\begin{aligned}
\Delta J_2^i &= \bar{J}_i(y_i(k_i^t+2))-\bar{J}_i(y_i(k_i^t+1))\frac{n!}{r!(n-r)!}=\sum_{j=0}^{N_i}\{||\bar{y}_i(k_i^t+2+j|k_i^t+2)||_{Q_i}^2+||\bar{u}_i(k_i^t+2+j|k_i^t+2)||_{R_i}^2\} \quad (48)\\
&+||\bar{y}_i(k_i^t+N_i+3|k_i^t+2)||_{P_i}^2-\sum_{j=0}^{N_i}\{||\bar{y}_i(k_i^t+1+j|k_i^t+1)||_{Q_i}^2+||\bar{u}_i(k_i^t+1+j|k_i^t+1)||_{R_i}^2\}-||\bar{y}_i(k_i^t+N_i\\
&+2|k_i^t+1)||_{P_i}^2\le\left\{S_1^i\left(\sum_{j=0}^{N_i-1}||L_i(j)||\right)+S_2^i\left(\sum_{j=0}^{N_i-1}||L_i(j)||^2\right)\right\}||y(k_i^t+1)||+\left\{S_3^i\left(\sum_{j=0}^{N_i-1}||L_i(j)||)\right.\right.\\
&\left.\left.+S_4^i\left(\sum_{j=0}^{N_i-1}||L_i(j)||^2\right)\right\}||v_i(k_i^t+1)||+\{F_1^i\sum_{j=0}^{N_i-1}(||K_i(j)||\cdot||L_i(j)||)+F_2^i\sum_{j=0}^{N_i-1}(||K_i(j)||^2\cdot||L_i(j)||^2)\}\cdot||y(k_i^t+1)|\\
&+\left\{F_3^i\sum_{j=0}^{N_i-1}(||K_i(j)||\cdot||L_i(j)||)+F_4^i\sum_{j=0}^{N_i-1}(||K_i(j)||^2\cdot||L_i(j)||^2)\right\}\cdot||v_i(k_i^t+1)|+M_2^i||y(k_i^t+1)||\\
&+M_1^i||v_i(k_i^t+1)||-q_i||y(k_i^t+1)||^2\le\beta_1^i\sum_{j=0}^{M}(||C_j||\cdot||x_i(k_i^t+1)||)+\beta_2^i||v_i(k_i^t+1)||-\alpha_i||x_i(k_i^t+1)||^2\\
&=\sum_{j=0}^{M}(\hat{\beta}_1^i||x_i(k_i^t+1)||)+\beta_2^i||v_i(k_i^t+1)||-\alpha_i||x_i(k_i^t+1)||^2.
\end{aligned}$$

By adopting similar procedures as in (47) and (48), it is easily shown that $\Delta J_{m_i}^i$, $m_i \in \{1,2,\dots,N_i\}$ yields (42). That leads to the conclusion of Theorem 2 .□

In order to ensure the ISS property of the $i$th subsystem, the relevant Lyapunov function $J_i(k)$ must be decreasing for consecutive steps. Now, we are ready to provide the local triggering condition and prove the ISS property for each subsystem.Theorem 3Consider the subsystem (1) and the event-triggered decentralized robust MPC strategy. The local optimal control law is applied only when the following triggering condition is violated

$$\sum_{j=1}^{M}(\beta_1^{ji}||x_j(k_i^t+m_i-1)||)+\beta_2^i||v_i(k_i^t+m_i-1)||\leq\sigma_i\cdot\alpha_i||x_i(k_i^t+m_i-1)||^2, 0<\sigma_i<1, m_i\in\{1,2,\ldots,N_i\}. \tag{49}$$

Otherwise, the control sequence given by (27) is applied to the subsystem. Using this event-triggered control scheme, the subsystem is ISS and the subsystem state will be driven to a robust invariant set.

*Proof* Theorem 1 provides a candidate control sequence for the $i$th subsystem at $k_i^t+m_i$, $m_i\in\{1,2,\ldots,N_i\}$ based on the optimal solution computed at triggered step $k_i^t$. Theorem 2 presents the bounded difference of the Lyapunov function $J_i(y_i)$ between steps $k_i^t+m_i$ and $k_i^t+m_i-1$.

If the disturbance term satisfies (49), we have

$$\sum_{j=1}^{M}(\beta_1^{ji}||x_j(k_i^t+m_i-1)||)+\beta_2^i||v_i(k_i^t+m_i-1)||\leq\sigma_i\cdot\alpha_i||x_i(k_i^t+m_i-1)||^2, 0<\sigma_i<1,$$

Substituting (49) into (42), we obtain

$$\Delta J_{m_i}^i=\bar{J}_i(y_i(k_i^t+m_i))-\bar{J}_i(y_i(k_i^t+m_i-1))\leq(\sigma_i-1)\cdot\alpha_i||x_i(k_i^t+m_i-1)||^2<0.$$

Therefore, the Lyapunov function is strictly decreasing. It also implies theith subsystem is ISS. Moreover, the subsystem state $x_i(k)$ must enter its robust invariant admissible set $\Omega_i$ in finite time and remain there in the subsequent time. This completes the proof. □

*Remark 3* In this work, the interconnections between subsystems are treated as perturbation. Therefore, the overall system (7) is ISS since the ISS property of each subsystem (1) is derived under the event-triggered decentralized robust MPC strategy.

## 4. Simulation results

Consider the overall system that consists of three linear discrete-time subsystems in the form of (1–4) with the following parameters:

$$A_1=A_2=A_3=\begin{bmatrix}1&1\\0&1\end{bmatrix}, B_1=B_2=B_3=\begin{bmatrix}0.5&1\end{bmatrix}^T,$$

$$C_1=\begin{bmatrix}0.5&0.6\end{bmatrix}, C_2=\begin{bmatrix}0.4&0.5\end{bmatrix}, C_3=\begin{bmatrix}0.6&0.6\end{bmatrix},$$

$$G_1=\begin{bmatrix}0.0006&0.0005&0.0001\\0&0.0005&0.0002\end{bmatrix}, G_2=\begin{bmatrix}0.0005&0&0.0004\\0&0.0005&0.0006\end{bmatrix},$$

$$G_3=\begin{bmatrix}0.0005&0&0.0002\\0.0001&0.0005&0\end{bmatrix}.$$

subject to the following constraints

$$|u_1|=|u_2|=|u_3|\leq 1.0, |y_1|=|y_2|=|y_3|\leq 10,$$

$$|v_{11}|=|0.1\sin(x_{11})|\leq 0.1, |v_{12}|=|0.1\sin(x_{12})|\leq 0.1,$$

$$|v_{21}|=|0.1\sin(x_{21})|\leq 0.1, |v_{22}|=|0.1\sin(x_{22})|\leq 0.1,$$

$|v_{31}| = |0.1\sin(x_{31})| \le 0.1, |v_{32}| = |0.1\sin(x_{32})| \le 0.1.$

The cost functions are defined by (10) with $Q_1 = I_{2\times 2}$, $R_1 = 0.01$, $Q_2 = 0.8I_{2\times 2}$, $R_2 = 0.03$, $Q_3 = 0.8I_{2\times 2}$, $R_3 = 0.02$. The prediction horizons of each subsystem are chosen as $N_1 = N_2 = N_3 = 8$, then the candidate matrices can be derived off-line using the LQR nilpotent policy. The constraints (13–14) of each subsystem can be calculated off-line using the tightening recursion (16–19). With these results, it is ready to execute the on-line optimization. The initial states of each subsystem are $x_{10} = [5.2, -5.4]^T$, $x_{20} = [-5.2, 5.0]^T$, $x_{30} = [5.0, -5.4]^T$, and the simulation step is $T = 100$.

The tightened constraints $U_i(j)$, $Y_i(j)$ of each local OCP are as follows:

Subsystem 1:

$$\begin{cases} U_1(0) = [-1.0, 1.0], & Y_1(0) = [-10, 10], \\ U_1(1) = [-0.7841, 0.7841], & Y_1(1) = [-9.8798, 9.8798], \\ U_1(2) = [-0.7131, 0.7131], & Y_1(2) = [-9.8700, 9.8700], \\ U_1(3) = [-0.6878, 0.6878], & Y_1(3) = [-9.8661, 9.8661], \\ U_1(4) = [-0.6795, 0.6795], & Y_1(4) = [-9.8647, 9.8647], \\ U_1(5) = [-0.6770, 0.6770], & Y_1(5) = [-9.8641, 9.8641], \\ U_1(j) = [-0.6752, 0.6752], & Y_1(j) = [-9.8635, 9.8635], \\ & j \ge 6. \end{cases}$$

Subsystem 2:

$$\begin{cases} U_2(0) = [-1.0, 1.0], & Y_2(0) = [-10, 10], \\ U_2(1) = [-0.7845, 0.7845], & Y_2(1) = [-9.9009, 9.9009], \\ U_2(2) = [-0.7203, 0.7203], & Y_2(2) = [-9.8935, 9.8935], \\ U_2(3) = [-0.6933, 0.6933], & Y_2(3) = [-9.8892, 9.8892], \\ U_2(4) = [-0.6844, 0.6844], & Y_2(4) = [-9.8877, 9.8877], \\ U_2(5) = [-0.6817, 0.6817], & Y_2(5) = [-9.8871, 9.8871], \\ U_2(j) = [-0.6799, 0.6799], & Y_2(j) = [-9.8866, 9.8866], \\ & j \ge 6. \end{cases}$$

Subsystem 3:

$$\begin{cases} U_3(0) = [-1.0, 1.0], & Y_3(0) = [-10, 10], \\ U_3(1) = [-0.7907, 0.7907], & Y_3(1) = [-9.8722, 9.8722], \\ U_3(2) = [-0.7256, 0.7256], & Y_3(2) = [-9.8682, 9.8692], \\ U_3(3) = [-0.7000, 0.7000], & Y_3(3) = [-9.8689, 9.8689], \\ U_3(4) = [-0.6917, 0.6917], & Y_3(4) = [-9.8687, 9.8687], \\ U_3(5) = [-0.68927, 0.6892], & Y_3(5) = [-9.8685, 9.8685], \\ U_3(j) = [-0.6874, 0.6874], & Y_3(j) = [-9.8679, 9.8679], \\ & j \ge 6. \end{cases}$$

The trajectories of states and outputs are given in Figures 1 and 2, which show the convergence of the subsystems under the proposed event-triggered decentralized MPC framework. To verify the reduction on the number of updating control laws, the triggering instants of each subsystem are plotted in Figure 3.

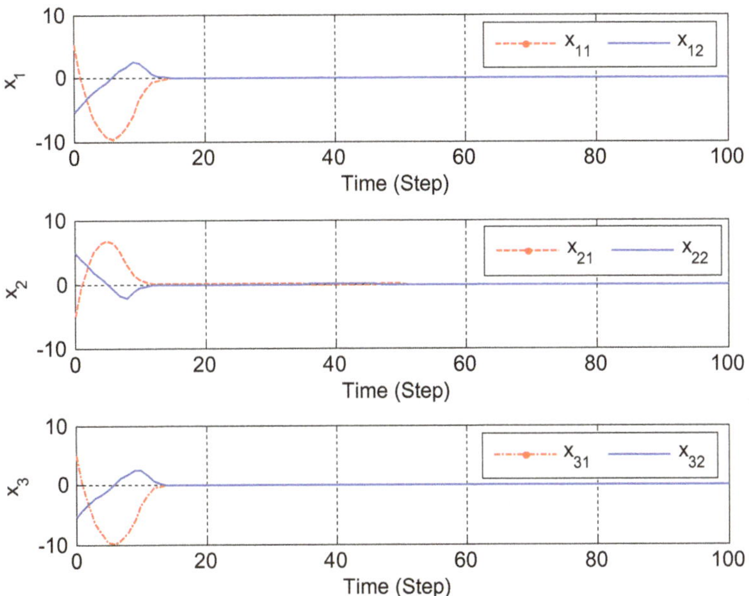


**Figure 1. State trajectories.**

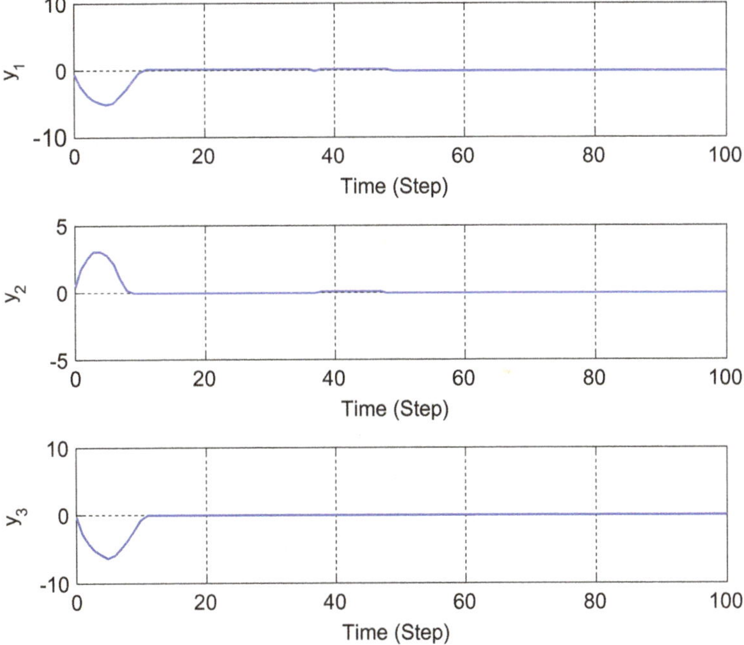


**Figure 2. Output trajectories.**

## 5. Conclusions

In this work, we have provided an event-triggered decentralized robust model predictive controller for a class of constrained linear discrete-time system with additive bounded disturbances. The proposed strategy can not only reduce the on-line computation load, but also achieve the alleviation of computational complexity. It should be pointed out that the systems under consideration in this work are assumed to have full knowledge of states. Actually, it is often difficult to measure the system state in practical application; the event-triggered output feedback MPC strategy will be further considered in future research.

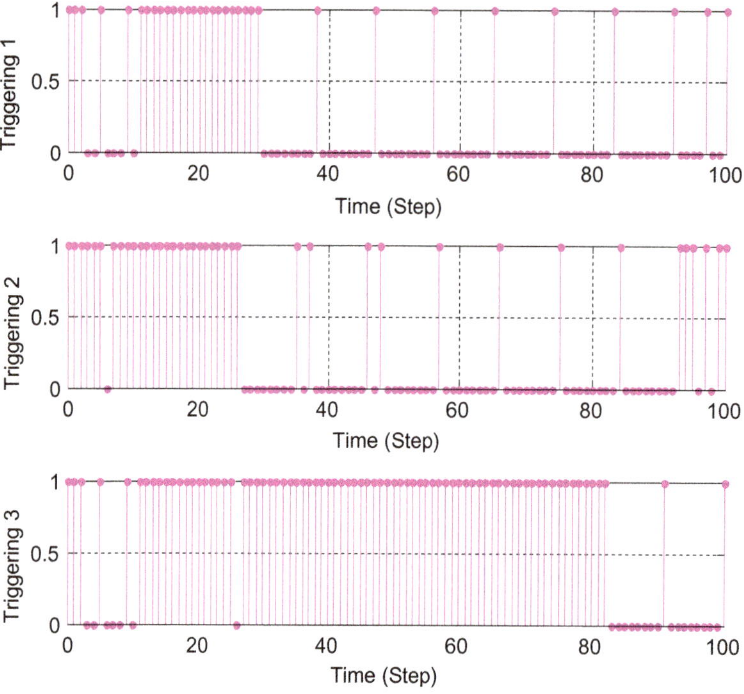


**Figure 3. Triggering instants.**

Notes: The value 1 means the local OCP is triggered at the corresponding time instant and the value 0 means the local OCP is not triggered.

**Funding**

This work was supported by National Natural Science Foundation (NNSF) from China [grant number 61374107], [grant number 61273073].

**Author details**

Ling Lu[1]

E-mail: 872298436@qq.com

Yuanyuan Zou[1]

E-mail: yyzou@ecust.edu.cn

Yugang Niu[1]

E-mail: acniuyg@ecust.edu.cn

[1] Key Laboratory of Advanced Control and Optimization for Chemical Process, East China University of Science & Technology, Ministry of Education, Shanghai 200237, China.

**References**

Alessio, A., Barcelli, D., & Bemporad, A. (2011). Decentralized model predictive control of dynamically coupled linear systems. *Journal of Process Control, 21*, 705–714. doi:10.1016/j.jprocont.2010.11.003

Dimarogonas, D., Frazzoli, E., & Johansson, K. (2012). Distributed event-triggered control for multi-agent systems. *IEEE Transactions on Automatic Control, 57*, 1291–1297. doi:10.1109/TAC.2011.2174666

Dong, H., Wang, Z., Alsaadi, F. E., & Ahmad, B. (2015). Event-triggered robust distributed state estimation for sensor networks with state-dependent noises. *International Journal of General Systems, 44*, 254–266. doi:10.1155/2015/261738

Dong, H., Wang, Z., Ding, S. X., & Gao, H. (2015). Event-based H-infinity filter design for a class of nonlinear time-varying systems with fading channels and multiplicative noises. *IEEE Transactions on Signal Processing, 63*, 3387–3395. doi:10.1109/TSP.2015.2422676

Eqtami, A., Dimarogonas, D., & Kyriakopoulos, K. (2011a). Novel event-triggered strategies for model predictive controllers. In *Proceedings of the 50th Conference Decision and Control* (pp. 3392–3397). doi:10.1109/CDC.2011.6161348

Eqtami, A., Dimarogonas, D., & Kyriakopoulos, K. (2011b). Event-triggered strategies for decentralized model predictive controllers. *Proceedings of 18th International Federation of Automatic Control* (pp. 10068–10073). doi:10.3182/20110828-6-IT-1002.03540

Eqtamin, A., Dimarogonas, D., & Kyriakopoulos, K. (2010). Event-triggered control for discrete-time systems. In *Proceedings of American Control Conference* (pp. 4719–4724). doi:10.1109/ACC.2010.5531089

Hua, C., Leng, J., & Guan, X. (2012). Decentralized MRAC for large-scale interconnected systems with time-varying delays and applications to chemical reactor systems. *Journal of Process Control, 22*, 1985–1996. doi:10.1016/j.jprocont.2012.08.009

Keviczky, T., Borrelli, F., & Balas, G. (2006). Decentralized receding horizon control for large scale dynamically decoupled systems. *Automatica, 42*, 2105–2115. doi:10.1016/j.automatica.2006.07.008

Lehmann, D., Henriksson, E., & Hohansson, K. (2013). Event-triggered model predictive control of discrete-time linear systems subjected to disturbances. In *Proceedings of European Control Conference* (pp. 1156–1161). Zürich: IEEE.

Li, H., & Shi, Y. (2014). Event-triggered robust model predictive control of continuous-time nonlinear systems. *Automatica, 50*, 1507–1513. doi:10.1016/j.automatica.2014.03.015

Liu, D., & Hao, F. (2013). Decentralized event-triggered control strategy in distributed networked systems with delays. *International Journal of Control, Automation, and Systems, 11*, 33–40. doi:10.1007/s12555-012-0094-1

Magni, L., & Scattolini, R. (2006). Stabilizing decentralized model predictive control of nonlinear systems. *Automatica, 42*, 1231–1236. doi:10.1016/j.automatica.2006.02.010

Mayne, D. (2014). Model predictive control: Recent developments and future promise. *Automatica, 50*, 2967–2986. doi:10.1016/j.automatica.2014.10.128

Mayne, D., Rawlings, J., Rao, C., & Scokaert, P. (2000). Constrained model predictive control: Stability and optimality. *Automatica, 36*, 789–814. doi:10.1016/S0005-1098(99)00214-9

Raimondo, D., Magni, L., & Scattolini, R. (2007). Decentralized MPC of nonlinear systems: An input-to-state stability approach. *International Journal of Robust and Nonlinear Control, 17*, 1651–1667. doi:10.1002/rnc.1214

Richards, A., & How, J. (2006). Robust stable model predictive control with constraint tightening. In *Proceedings of American Control Conference* (pp. 1557–1562). doi:10.1109/ACC.2006.1656440

Riverso, S., Farina, M., & Trecate, G. (2013). Plug and play decentralized model predictive control for linear systems. *IEEE Transactions on Automatic Control, 58*, 2608–2614. doi:10.1109/TAC.2013.2254641

Tran, T., & Ha, Q. (2014). Decentralized model predictive control for networks of linear systems with coupling delay. *Journal of Optimization Theory and Applications, 161*, 933–950. doi:10.1007/s10957-013-0379-4

Yan, X., Edwards, C., Spurgeon, S., & Bleijs, J. (2004). Decentralized sliding-mode control for multimachine power systems using only output information. *IEE Proceedings-Control Theory and Applications, 151*, 627–635. doi:10.1049/ip-cta:20040956

Yan, X., Lam, J., Li, H., & Chen, I. (2000). Decentralized control of nonlinear large-scale systems using dynamic output feedback. *Journal of Optimization Theory and Applications, 104*, 459–475. doi:10.1023/A:1004674032740

Zhang, J., & Liu, J. (2013). Distributed moving horizon state estimation for nonlinear systems with bounded uncertainties. *Journal of Process Control, 23*, 1281–1295. doi:10.1016/j.jprocont.201308.005

Zhang, X., Zhang, C., & Wang, Y. (2014). Decentralized output feedback stabilization for a class of large-scale feedforward nonlinear time-delay systems. *International Journal of Robust and Nonlinear Control, 24*, 2628–2639. doi:10.1002/rnc.3013

Zou, Y., & Niu, Y. (2013). Predictive control of constrained linear systems with multiple missing measurements. *Circuits, Systems, and Signal Processing, 32*, 615–630. doi:10.1007/s00034-012-9482-2

# 3

# Frequency interval balanced truncation of discrete-time bilinear systems

Ahmad Jazlan[1,2]*, Victor Sreeram[1], Hamid Reza Shaker[3] and Roberto Togneri[1]

*Corresponding author: Ahmad Jazlan, School of Electrical, Electronics and Computer Engineering, University of Western Australia, 35 Stirling Highway, Crawley, Perth, Western Australia 6009, Australia; Faculty of Engineering, Department of Mechatronics Engineering, International Islamic University Malaysia, Jalan Gombak, 53100 Kuala Lumpur, Malaysia

E-mail: ahmadjazlan@iium.edu.my

Reviewing editor: James Lam, University of Hong Kong, Hong Kong

**Abstract:** This paper presents the development of a new model reduction method for discrete-time bilinear systems based on the balanced truncation framework. In many model reduction applications, it is advantageous to analyze the characteristics of the system with emphasis on particular frequency intervals of interest. In order to analyze the degree of controllability and observability of discrete-time bilinear systems with emphasis on particular frequency intervals of interest, new generalized frequency interval controllability and observability gramians are introduced in this paper. These gramians are the solution to a pair of new generalized Lyapunov equations. The conditions for solvability of these new generalized Lyapunov equations are derived and a numerical solution method for solving these generalized Lyapunov equations is presented. Numerical examples which illustrate the usage of the new generalized frequency interval controllability and observability gramians as part of the balanced truncation framework are provided to demonstrate the performance of the proposed method.

**Subjects: Dynamical Control Systems; Non-Linear Systems; Systems & Control Engineering**

**Keywords: model reduction; bilinear systems; balanced truncation; frequency interval gramians; finite frequency interval**

## ABOUT THE AUTHORS

Ahmad Jazlan is currently a PhD student at the School of Electrical, Electronics and Computer Engineering, University of Western Australia.

Victor Sreeram is currenty a professor at the School of Electrical, Electronic, and Computer Engineering, University of Western Australia. He is on the editorial board of many journals including IET Control, Theory and Applications, Asian Journal of Control, and Smart Grid and Renewable Energy.

Hamid Reza Shaker is currently an associate professor at the Center for Energy Informatics, University of Southern Denmark.

Roberto Togneri is currently a professor at the School of Electrical, Electronic and Computer Engineering. He is currently an associate fditor for IEEE Signal Processing Magazine Lecture Notes and IEEE Transactions on Speech, Audio and Language Processing.

This research work is applicable to both engineering and mathematical problems which can be formulated as bilinear systems.

## PUBLIC INTEREST STATEMENT

Nonlinear mathematical models are commonly used to describe the processes in many branches of engineering. Bilinear systems are an important class of nonlinear systems which have well-established theories and are applicable to many practical applications. Mathematical models in the form of bilinear systems can be found in a variety of fields such as the mathematical models which describe the processes of electrical networks, hydraulic systems, heat transfer, and chemical processes. Many nonlinear systems can be modeled as bilinear systems with appropriate state feedback or can be approximated as bilinear systems by using the bilinearization process. The mathematical modeling of a large-scale bilinear system may result in a high-order bilinear model. To address the complexity associated with high-order models, we present a new model reduction technique for discrete-time bilinear systems.

## 1. Introduction

Model reduction which is of fundamental importance in many modeling and control applications deals with the approximation of a higher order model by a lower order model such that the input-output behavior of the original system is preserved to a required accuracy. The balanced truncation model reduction technique originally developed by Moore for continuous-time linear systems is one of the most widely applied model reduction techniques (Moore, 1981). In recent years, many variations to this original balanced truncation technique have been developed (Li, Yu, Gao, & Zhang, 2014; Minh, Battle, & Fossas, 2014; Opmeer & Reis, 2015; Zhang, Wu, Shi, & Zhao, 2015).

One of the further developments to the original balanced truncation technique was the work by Gawronski and Juang which involved the development of frequency interval controllability and observability gramians (Gawronski & Juang, 1990). The significance of emphasizing particular frequency intervals of interest in a variety of control engineering problems has led to extensive theoretical developments in robust control techniques which emphasize particular frequency intervals of interest which have been presented in (Ding, Du, & Li, 2015; Ding, Li, Du, & Xie, 2016; Du, Fan, & Ding, 2016; Imran & Ghafoor, 2015; Li & Yang, 2015; Li, Yin, & Gao, 2014; Li, Yu, & Gao, 2015).

In the context of discrete-time systems, digital systems are designed to work with signals with known frequency characteristics, therefore it is essential to have model reduction techniques which generate reduced-order models which function well with signals which have specified frequency characteristics. The works by Horta, Juang, and Longman (1993), Wang and Zilouchian (2000) and more recently by Imran and Ghafoor (2014) described the formulation of frequency interval gramians for discrete-time systems.

Bilinear systems are an important category of non-linear systems which have well-established theories (Al-Baiyat, Bettayeb, & Al-Saggaf, 1994; Dorissen, 1989; D'Alessandro, Isidori, & Ruberti, 1974; Shaker & Tahavori, 2014b, 2015). Many non-linear systems in various branches of engineering can be well represented by bilinear systems. Similar to the case of linear systems, the mathematical modeling process to obtain bilinear system models may result in obtaining high-order models. Fortunately, by formulating a state space model for these bilinear system models, the application of model reduction techniques becomes possible to reduce the order of these bilinear system models. The balanced truncation technique for continuous-time bilinear systems has been presented in Zhang & Lam (2002) whereas the balanced truncation technique for discrete-time bilinear systems has been presented in Zhang, Lam, Huang, and Yang (2003). More recently further developments have been carried out to the original balanced truncation technique for continuous-time bilinear systems in order to reduce the approximation error between the outputs of the original bilinear model and reduced-order bilinear model by incorporating time and frequency interval techniques (Shaker & Tahavori, 2014a, 2014c).

The contributions of this paper are as follows. Firstly, new generalized frequency interval controllability and observability gramians are defined for discrete-time bilinear systems. Secondly, it is shown that these frequency interval controllability and observability gramians are solutions to a pair of new generalized Lyapunov equations. Thirdly, conditions for solvability of these new generalized Lyapunov equation are proposed together with a numerical solution method for solving these new Lyapunov equations. Finally, numerical examples are provided to demonstrate the performance of the proposed method relative to existing techniques.

The notation used in this paper is as follows. $M^*$ refers to the transpose of the matrix $M$ if $M \in \mathbb{R}^{n \times m}$ and complex conjugate transpose if $M \in \mathbb{C}^{n \times m}$. The $\otimes$ symbol denotes a Kronecker product.

## 2. Preliminaries

### *2.1. Controllability and observability gramians of discrete-time linear systems*

Considering the following time-invariant and asymptotically stable discrete-time linear system $(A, B, C)$:

$$\begin{aligned} x(k+1) &= Ax(k) + Bu(k) \\ y(k) &= Cx(k) \end{aligned} \tag{1}$$

where $u \in \mathbb{R}^p$, $y \in \mathbb{R}^q$, $C \in \mathbb{R}^n$ are the input, output and states respectively. $A \in \mathbb{R}^{n \times n}$, $B \in \mathbb{R}^{n \times p}$, $C \in \mathbb{R}^{q \times n}$ are matrices with appropriate dimensions.

Definition 1 The discrete-time domain controllability and observability gramian definitions are given by:

$$P = \sum_{k=0}^{\infty} A^k BB^* (A^*)^k \tag{2}$$

$$Q = \sum_{k=0}^{\infty} (A^*)^k C^* C A^k \tag{3}$$

Remark 1 It is established that (2) and (3) satisfy the following Lyapunov equations:

$$APA^* - P + BB^* = 0 \tag{4}$$

$$A^*QA - Q + C^*C = 0 \tag{5}$$

Remark 2 By applying a direct application of Parseval's theorem to (2) and (3), the controllability and observability gramians in the frequency domain are given by:

$$P = \frac{1}{2\pi} \int_{-\pi}^{\pi} (e^{j\theta} I - A)^{-1} BB^* (e^{-j\theta} I - A^*)^{-1} d\theta \tag{6}$$

$$Q = \frac{1}{2\pi} \int_{-\pi}^{\pi} (e^{-j\theta} I - A^*)^{-1} C^* C (e^{j\theta} I - A)^{-1} d\theta \tag{7}$$

where $I$ is an identity matrix.

### *2.2. Frequency interval controllability and observability gramians of discrete-time linear systems*

Definition 2 The frequency interval controllability and observability gramians for discrete-time systems are defined as (Horta et al., 1993):

$$P_{cf} = \frac{1}{2\pi} \int_{\delta\theta} (e^{j\theta} I - A)^{-1} BB^* (e^{-j\theta} I - A^*)^{-1} d\theta \tag{8}$$

$$Q_{cf} = \frac{1}{2\pi} \int_{\delta\theta} (e^{-j\theta} I - A^*)^{-1} C^* C (e^{j\theta} I - A)^{-1} d\theta \tag{9}$$

where $\delta\theta = [\theta_1, \theta_2]$ is the frequency range of operation and $0 \le \theta_1 < \theta_2 \le \pi$. Due to the symmetry of the discrete Fourier transform, the integration is carried out throughout the frequency intervals $[\theta_1, \theta_2]$ and $[-\theta_2, -\theta_1]$ (Horta et al., 1993). Therefore the gramians $P_{cf}$ and $Q_{cf}$ in (8) and (9) will always be real.

Remark 3 It has been shown that the frequency interval controllability and observability gramians defined in (8) and (9) are the solutions to the following Lyapunov equations (Wang et al., 2000):

$$AP_{cf}A^* - P_{cf} + X_{cf} = 0 \tag{10}$$

$$A^*Q_{cf}A - Q_{cf} + Y_{cf} = 0 \tag{11}$$

where

$$X_{cf} = F_{cf}BB^* + BB^*F_{cf}^* \tag{12}$$

$$Y_{cf} = F_{cf}^*C^*C + C^*CF_{cf} \tag{13}$$

and

$$F_{cf} = \frac{-(\theta_2 - \theta_1)}{2\pi}I + \frac{1}{2\pi}\int_{\delta\theta}(I - Ae^{-j\theta})^{-1}d\theta \tag{14}$$

***2.3. Controllability and observability gramians of discrete-time bilinear systems***

Considering the following discrete-time bilinear system represented by:

$$x(k+1) = Ax(k) + \sum_{j=1}^{m} N_j x(k)u_j(k) + Bu(k) \tag{15}$$

$$y(k) = Cx(k) \tag{16}$$

where $x(k) \in \mathbb{R}^{nxn}$ is the state vector, $u(k) \in \mathbb{R}^{mxm}$ is the input vector and $u_j(k)$ is the corresponding $j$th element of $u(k)$, $y(k) \in \mathbb{R}^{qxq}$ is the output vector and $A$, $B$, $C$ and $N_j$ are matrices with suitable dimensions. This bilinear system is denoted as $(A, N_j, B, C)$.

The controllability gramian for this system is defined as (Zhang et al., 2003):

$$P = \sum_{i=1}^{\infty}\sum_{k_i=0}^{\infty}\cdots\sum_{k_1=0}^{\infty} P_i P_i^* \tag{17}$$

where

$$P_1(k_1) = A^{k_1}B$$

$$P_i(k_1, \ldots k_i) = A^{k_i}\begin{bmatrix} N_1P_{i-1} & N_2P_{i-1} \cdots N_mP_{i-1} \end{bmatrix}, \quad i \geq 2$$

whereas the observability gramian is defined as (Zhang et al., 2003):

$$Q = \sum_{i=1}^{\infty}\sum_{k_i=0}^{\infty}\cdots\sum_{k_1=0}^{\infty} Q_i^* Q_i \tag{18}$$

where

$$Q_1(k_1) = CA^{k_1}$$

$$Q_i(k_1, \ldots k_i) = \begin{bmatrix} Q_{i-1}N_1 \\ Q_{i-1}N_2 \\ \vdots \\ Q_{i-1}N_m \end{bmatrix}$$

The controllability and observability gramians defined in (17) and (18) are the solution to the following generalized Lyapunov equations (Zhang et al., 2003):

$$APA^* - P + \sum_{j=1}^{\infty} N_j P N_j^* + BB^* = 0 \tag{19}$$

$$A^*QA - Q + \sum_{j=1}^{\infty} N_j^* Q N + C^*C = 0 \tag{20}$$

The generalized Lyapunov equations corresponding to the controllability and observability gramians in (19) and (20) can be solved iteratively. The controllability gramian can be obtained by (Zhang et al., 2003):

$$P = \lim_{i \to \infty} \hat{P}_i \tag{21}$$

where

$$A\hat{P}_1 A^* - \hat{P}_i + BB^* = 0,$$
$$A\hat{P}_i A^* - \hat{P} + \sum_{j=1}^{\infty} N_j \hat{P}_{i-1} N_j^* + BB^* = 0, \quad i \geq 2 \tag{22}$$

whereas the observability gramian can be obtained by (Zhang et al., 2003)

$$Q = \lim_{i \to \infty} \hat{Q}_i \tag{23}$$

where

$$A^*\hat{Q}_1 A - \hat{Q}_i + C^*C = 0,$$
$$A^*\hat{Q}_i A - \hat{Q} + \sum_{j=1}^{\infty} N_j^* \hat{Q}_{i-1} N_j + C^*C = 0, \quad i \geq 2 \tag{24}$$

## 3. Main work

### *3.1. Frequency interval controllability and observability gramians of discrete-time bilinear systems*

For a particular discrete-time frequency interval $\Omega = [\gamma_1, \gamma_2]$, we define the frequency interval controllability and observability gramians as follows:

Definition 3 The generalized frequency interval controllability gramian for discrete-time bilinear systems is defined as:

$$\hat{P}(\theta) := \sum_{i=1}^{\infty} \frac{1}{(2\pi)^i} \int_{\delta\theta} \cdots \int_{\delta\theta} \hat{P}_i(\theta_1, ..., \theta_i) \hat{P}_i^*(\theta_1, ..., \theta_i) d\theta_1 \ldots d\theta_i \tag{25}$$

where $\delta\theta = [\gamma_1, \gamma_2]$ and

$$\hat{P}_1(\theta_1) = (e^{j\theta_1} I - A)^{-1} B$$
$$\vdots$$
$$\hat{P}_i(\theta_1, ..., \theta_i) = (e^{j\theta_i} I - A)^{-1} \begin{bmatrix} N_1 \hat{P}_{i-1} & N_2 \hat{P}_{i-1} \ldots N_m \hat{P}_{i-1} \end{bmatrix}$$

Similarly, the generalized frequency interval observability gramian is defined as:

$$\hat{Q}(\theta) := \sum_{i=1}^{\infty} \frac{1}{(2\pi)^i} \int_{\delta\theta} \cdots \int_{\delta\theta} \hat{Q}_i^*(\theta_1, ..., \theta_i) \hat{Q}_i(\theta_1, ..., \theta_i) d\theta_1 \ldots d\theta_i \tag{26}$$

where $\delta\theta = [\gamma_1, \gamma_2]$ and

$$\hat{Q}_1(\theta_1) = C(e^{-j\theta_1}I - A^*)^{-1}$$

$\vdots$

$$\hat{Q}_i(\theta_1, \ldots, \theta_i) = \begin{bmatrix} N_1\hat{Q}_{i-1} \\ N_2\hat{Q}_{i-1} \\ \vdots \\ N_m\hat{Q}_{i-1} \end{bmatrix}(e^{-j\theta_i}I - A^*)^{-1}$$

These gramians defined in (25) and (26) are the solution to a pair of new generalized Lyapunov equations which is presented in Theorem 1. Lemmas 1, 2 and 3 together with Theorem 1 presented in the following sections are interrelated such that Lemma 1 and Lemma 2 are required as part of proving Lemma 3, whereas Lemma 3 is required for proving Theorem 1.

LEMMA 1 *Let A be a square matrix which is also stable and let M be a matrix with the appropriate dimension. If X satisfies the following equation:*

$$X = \sum_{i=0}^{+\infty} A^i M(A^i)^* \tag{27}$$

It follows that $X$ is the solution to:

$$AXA^* - X + M = 0 \tag{28}$$

*Proof* Since $X = \sum_{i=0}^{+\infty} A^i M(A^i)^*$ and $A$ is stable, it follows that:

$$\begin{aligned} AXA^* - X &= A\sum_{i=0}^{+\infty} A^i M(A^i)^*(A^i)^* - \sum_{i=0}^{+\infty} A^i M(A^i)^* \\ &= \sum_{i=0}^{+\infty} A^{i+1} M(A^*)^{i+1} - \sum_{i=0}^{+\infty} A^i M(A^i)^* \end{aligned}$$

*Since* $\sum_{i=0}^{+\infty} A^{i+1}M(A^*)^{i+1} = \sum_{i=1}^{+\infty} A^i M(A^*)^i$

$$AXA^* - A = \sum_{i=1}^{+\infty} A^i M(A^*)^i - \sum_{i=0}^{+\infty} A^i M(A^*)^i = -M.$$

LEMMA 2 *Let A be a square matrix which is also stable and let R be a matrix with the appropriate dimension. If Y satisfies the following*

$$Y = \sum_{i=0}^{+\infty} (A^i)^* R(A^i) \tag{29}$$

It follows that $Y$ is the solution to

$$A^*YA - Y + R = 0 \tag{30}$$

*Proof* Similar to the proof of Lemma 1 and is therefore omitted for brevity.

LEMMA 3 *Let M and R be matrices with the appropriate dimensions and let A be stable, if $\hat{P}_{cf}$ and $\hat{Q}_{cf}$ satisfy:*

$$\hat{P}_{cf} = \frac{1}{2\pi}\int_{\delta\theta}(e^{j\theta}I - A)^{-1}M(e^{-j\theta}I - A^{*})^{-1}d\theta \tag{31}$$

$$\hat{Q}_{cf} = \frac{1}{2\pi}\int_{\delta\theta}(e^{-j\theta}I - A^{*})^{-1}R(e^{j\theta}I - A)^{-1}d\theta \tag{32}$$

then $\hat{P}_{cf}$ and $\hat{Q}_{cf}$ are the solution to the following generalized Lyapunov equations:

$$A\hat{P}_{cf}A^{*} - \hat{P}_{cf} = X_{cf} \tag{33}$$

$$A^{*}\hat{Q}_{cf}A - \hat{Q}_{cf} = Y_{cf} \tag{34}$$

where

$$X_{cf} = -FM - MF^{*} \tag{35}$$

$$Y_{cf} = -F^{*}R - RF \tag{36}$$

and

$$F = \frac{(\theta_1 - \theta_2)}{2\pi}I + \frac{1}{2\pi}\int_{-\theta_2}^{\theta_2}(I - Ae^{-j\theta})^{-1}d\theta \ldots \\ - \frac{1}{2\pi}\int_{-\theta_1}^{\theta_1}(I - Ae^{-j\theta})^{-1}d\theta \tag{37}$$

*Proof* In this part we will prove that (31) is the solution to (33). This proof is a further development of the proof of equation 4.1a in Wang and Zilouchian (2000). The proof that (32) is the solution to (34) can then be obtained similarly by using lemma 2 and therefore is omitted for brevity. Firstly (28) can be re-written as follows:

$$-(e^{j\theta}I - A)X(e^{-j\theta}I - A^{*}) + X(e^{-j\theta}I - A^{*})e^{j\theta}I \ldots \\ + (e^{j\theta}I - A)Xe^{-j\theta}I = M \tag{38}$$

Multiplying (38) from the left by $(e^{j\theta}I - A)^{-1}$ and from the right by $(e^{-j\theta}I - A^{*})^{-1}$ followed by integrating both sides by $\frac{1}{2\pi}\int_{\delta\theta}d\theta$ yields:

$$\begin{aligned} &\frac{1}{2\pi}\int_{\delta\theta}(e^{j\theta}I - A)^{-1}M(e^{-j\theta}I - A^{*})^{-1}d\theta \\ &\quad = -\frac{1}{2\pi}\int_{\delta\theta}Xd\theta + \frac{1}{2\pi}\int_{\delta\theta}(e^{j\theta}I - A)^{-1}Xe^{j\theta}d\theta + \ldots \\ &\frac{1}{2\pi}\int_{\delta\theta}Xe^{-j\theta}(e^{-j\theta}I - A^{*})^{-1}\theta \\ &\quad = -\frac{1}{2\pi}\int_{\delta\theta}Xd\theta + \left(\frac{1}{2\pi}\int_{\delta\theta}(I - e^{-j\theta}IA)^{-1}d\theta\right)X + \ldots \\ &X\left(\frac{1}{2\pi}\int_{\delta\theta}(I - e^{-j\theta}IA)^{-1}d\theta\right)^{*} \end{aligned} \tag{39}$$

Denoting $K_1 = \frac{1}{2\pi}\int_{\delta\theta}(I - e^{-j\theta}IA)^{-1}d\theta$, (39) can be re-written as:

$$\hat{P}_{cf} = -\frac{1}{2\pi}\int_{\delta\theta} Xd\theta + K_1X + XK_1^* \tag{40}$$

Substituting (40) into the left-hand side of (33) yields:

$$\begin{aligned} &A\left[-\frac{1}{2\pi}\int_{\delta\theta} Xd\theta + K_1X + XK_1^*\right]A^*\ldots \\ &-\left[-\frac{1}{2\pi}\int_{\delta\theta} Xd\theta + K_1X + XK_1^*\right] = X_{cf} \end{aligned} \tag{41}$$

It has been shown in Wang and Zilouchian (2000) that the property $AK_1 = K_1A$ and $AK_1^* = K_1^*A$ holds true. As a result (41) can be re-written as

$$\begin{aligned} X_{cf} = &\ -\frac{1}{2\pi}\int_{\delta\theta} 1d\theta I[AXA^* - X] + K_1[AXA^* - X] + [AXA^* - X]K_1^* \\ &= \left(\frac{1}{2\pi}\int_{\delta\theta} 1d\theta I\right)(M) - K_1M - MK_1^* \\ &= -\left(-\frac{1}{4\pi}\int_{\delta\theta} 1d\theta I + K_1\right)(M) - (M)\left(-\frac{1}{4\pi}\int_{\delta\theta} 1d\theta I + K_1\right)^* \end{aligned} \tag{42}$$

(42) is equivalent to the right-hand side of (35). By comparing both expressions we have

$$F = -\frac{1}{4\pi}\int_{\delta\theta} 1d\theta I + K_1 \tag{43}$$

Due to the symmetry of the discrete Fourier transform, the integrations are carried out throughout the frequency intervals $[\theta_1, \theta_2]$ and $[-\theta_2, -\theta_1]$ (Horta et al., 1993). Therefore we have

$$\begin{aligned} F = &\frac{(\theta_1 - \theta_2)}{2\pi}I + \frac{1}{2\pi}\int_{-\theta_2}^{\theta_2}(I - Ae^{-j\theta})^{-1}d\theta\ldots \\ &-\frac{1}{2\pi}\int_{-\theta_1}^{\theta_1}(I - Ae^{-j\theta})^{-1}d\theta \end{aligned}$$

Lemma 3 derived in the previous section is now applied as part of the proof of Theorem 1 as follows.

THEOREM 1 The frequency interval controllablity and observability gramians $\hat{P}(\theta)$ and $\hat{Q}(\theta)$ defined in (25) and (26) are the solutions to the following generalized Lyapunov equations:

$$\begin{aligned} &A\hat{P}(\theta)A^* - \hat{P}(\theta) + F\left(\sum_{j=1}^{m} N_j\hat{P}(\theta)N_j^*\right) + \ldots \\ &\quad\left(\sum_{j=1}^{m} N_j\hat{P}(\theta)N_j^*\right)F^* + FBB^* + BB^*F^* = 0 \end{aligned} \tag{44}$$

$$\begin{aligned} &A^*\hat{Q}(\theta)A - \hat{Q}(\theta) + F^*\left(\sum_{j=1}^{m} N_j^*\hat{Q}(\theta)N_j\right) + \ldots \\ &\quad\left(\sum_{j=1}^{m} N_j^*\hat{Q}(\theta)N_j\right)F + F^*C^*C + C^*CF = 0 \end{aligned} \tag{45}$$

*Proof* The proof that the frequency interval controllability gramian $\hat{P}(\theta)$ defined in (25) is the solution to the generalized Lyapunov equation in (44) is presented in this section. The proof that the frequency interval observability gramian $\hat{Q}(\theta)$ defined in (26) is the solution to the generalized Lyapunov equation in (45) can be obtained in a similar manner and therefore is omitted for brevity. Firstly let:

$$\tilde{P}_1(\theta) = \frac{1}{2\pi}\int_{\delta\theta} \hat{P}_1(\theta)\hat{P}_1^*(\theta)d\theta_1 \tag{46}$$

$$\vdots$$

$$\tilde{P}_i(\theta) = \frac{1}{2\pi}\int_{\delta\theta}\cdots\int_{\delta\theta} \hat{P}_i(\theta_1,\theta_2,...,\theta_i)\hat{P}_i^*(\theta_1,\theta_2,...,\theta_i)d\theta_1...d\theta_i \tag{47}$$

we have

$$\hat{P}(\theta) = \sum_{i=1}^{\infty}\tilde{P}_i(\theta) \tag{48}$$

Using Lemma 3 with $M = BB^*$, it is observed that $\tilde{P}_1(\theta)$ is the solution to

$$A\tilde{P}_1(\theta)A^* - \tilde{P}_1(\theta) + FBB^* + BB^*F^* = 0 \tag{49}$$

For $\tilde{P}_2(\theta)$ we have

$$\begin{aligned}\tilde{P}_2(\theta) &= \frac{1}{(2\pi)^i}\int_{\delta\theta}\int_{\delta\theta}\hat{P}_2(\theta_1,\theta_2)\hat{P}_2^*(\theta_1,\theta_2)d\theta_1 d\theta_2\\ &= \frac{1}{(2\pi)^i}\int_{\delta\theta}\int_{\delta\theta}(e^{j\theta}I - A)\left[N_1\hat{P}_1 \dots N_m\hat{P}_1\right]\times...\end{aligned}$$

$$\begin{bmatrix}\hat{P}_1^*N_1^*\\ \vdots\\ \hat{P}_1^*N_m^*\end{bmatrix}(e^{-j\theta}I - A^*)d\theta_1 d\theta_2 = \frac{1}{2\pi}\int_{\delta\theta}(e^{j\theta}I - A)^{-1}\times...$$

$$\left(\sum_{j=1}^{m}N_j\left(\frac{1}{2\pi}\int_{\delta\theta}\hat{P}_1(\theta)\hat{P}_1^*(\theta)d\theta_1\right)N_j^*\right)(e^{-j\theta}I - A^*)^{-1}d\theta_2$$

$$= \frac{1}{2\pi}\int_{\delta\theta}(e^{j\theta}I - A)^{-1}\left(\sum_{j=1}^{m}N_j\tilde{P}_1(\theta)N_j^*\right)(e^{-j\theta}I - A^*)^{-1}d\theta_2$$

Denoting $M = \sum_{j=1}^{m}N_j\tilde{P}_1(\theta)N_j^*$, Lemma 3 applies and as a result $\tilde{P}_2(\theta)$ will be the solution to:

$$A\tilde{P}_2(\theta)A^* - \tilde{P}_2(\theta) + F\left(\sum_{j=1}^{m}N_j\tilde{P}_1(\theta)N_j^*\right) + \left(\sum_{j=1}^{m}N_j\tilde{P}_1(\theta)N_j^*\right)F^* = 0 \tag{50}$$

Similarly, according to Lemma 3, $\tilde{P}_i(\theta)$ will be the solution to

$$A\tilde{P}_i(\theta)A^* - \tilde{P}_i(\theta) + F\left(\sum_{j=1}^{m}N_j\tilde{P}_{i-1}(\theta)N_j^*\right) + \left(\sum_{j=1}^{m}N_j\tilde{P}_{i-1}(\theta)N_j^*\right)F^* = 0 \tag{51}$$

Adding (51) to (49) and applying a summation to infinity as in the right-hand side of (48) yields

$$A\sum_{i=1}^{\infty}\tilde{P}_i(\theta)A^* - \sum_{i=1}^{\infty}\tilde{P}_i(\theta) + F\left(\sum_{j=1}^{m}N_j\sum_{i=2}^{\infty}\tilde{P}_{i-1}(\theta)N_j^*\right) + \ldots \\ \left(\sum_{j=1}^{m}N_j\sum_{i=2}^{\infty}\tilde{P}_{i-1}(\theta)N_j^*\right)F^* + FBB^* + BB^*F^* = 0 \tag{52}$$

Equivalently, we have

$$A\sum_{i=1}^{\infty}\tilde{P}_i(\theta)A^* - \sum_{i=1}^{\infty}\tilde{P}_i(\theta) + F\left(\sum_{j=1}^{m}N_j\sum_{i=1}^{\infty}\tilde{P}_i(\theta)N_j^*\right) + \ldots \\ \left(\sum_{j=1}^{m}N_j\sum_{i=1}^{\infty}\tilde{P}_i(\theta)N_j^*\right)F^* + FBB^* + BB^*F^* = 0 \tag{53}$$

Finally applying the property in (48) to (53)

$$A\hat{P}(\theta)A^* - \hat{P}(\theta) + F\left(\sum_{j=1}^{m}N_j\hat{P}(\theta)N_j^*\right) + \left(\sum_{j=1}^{m}N_j\hat{P}(\theta)N_j^*\right)F^* + FBB^* + BB^*F^* = 0$$

### *3.2. Conditions for solvability of the Lyapunov equations corresponding to frequency interval controllability and observability gramians*

In this section, the condition for solvability of the generalized Lyapunov equation in (44) which corresponds to the frequency interval controllability gramian defined in (25) is presented herewith in Theorem 2. The condition for solvability of the generalized Lyapunov equation in (45) which corresponds to the frequency interval observability gramian defined in (26) can be derived in a similar manner and is therefore omitted for brevity.

THEOREM 2 The generalized Lyapunov equation in (44) is solvable and has a unique solution if and only if

$$W = (A \otimes A) - (I \otimes I) + (N_j \otimes FN_j) + (FN_j \otimes N_j) \tag{54}$$

is non-singular.

*Proof* Let *vec*(.) be an operator which converts a matrix into a vector by stacking the columns of the matrix on top of each other. This operator has the following useful property [20]

$$vec(M_1, M_2, M_3) = (M_3^* \otimes M_1)vec(M_2) \tag{55}$$

Applying vec(.) on both sides of (44) together with the property in (54) yields

$$\left\{(A \otimes A) - (I \otimes I) + \left(\sum_{j=1}^{m}N_j \otimes FN_j\right) + \left(\sum_{j=1}^{m}FN_j \otimes N_j\right)\right\}vec(\hat{P}(\theta)) = -vec(FBB^* + BB^*F) \tag{56}$$

The generalized Lyapunov equation in (44) is solvable provided that this equation presented in (56) is solvable and has a unique solution. It follows that (56) is solvable and has a unique solution if and only if

$$W = (A \otimes A) - (I \otimes I) + (N_j \otimes FN_j) + (FN_j \otimes N_j)$$

is non-singular. □

### 3.3. Numerical solution method for the Lyapunov equations corresponding to the frequency interval controllability and observability gramians

The iterative procedure for solving bilinear Lyapunov equations in previous studies can also be applied to obtain the solution to the generalized Lyapunov equation in (44) - $\hat{P}(\theta)$ as follows (Shaker et al., 2014a, 2014c; Zhang et al., 2003; Zhang & Lam, 2002):

$$\hat{P}(\theta) = \lim_{i \to +\infty} \tilde{P}_i(\theta) \tag{57}$$

where

$$A\tilde{P}_1(\theta)A^* - \tilde{P}_1(\theta) + FBB^* + BB^*F^* = 0$$

$$A\tilde{P}_i(\theta)A^* - \tilde{P}_i(\theta) + F\left(\sum_{j=1}^{m} N_j\tilde{P}_{i-1}(\theta)N_j^*\right) + \dots \tag{58}$$

$$\left(\sum_{j=1}^{m} N_j\tilde{P}_{i-1}(\theta)N_j^*\right)F^* + FBB^* + BB^*F^* = 0, \quad i \geq 2 \tag{59}$$

This iterative procedure can also be applied to solve the generalized Lyapunov equation corresponding to the frequency interval observability gramian in (45).

### 3.4. Model reduction algorithm

The procedure for obtaining the reduced-order model is described as follows

Step 1: The frequency interval controllability and observability gramians are calculated by solving (44) and (45), respectively.

Step 2: Both of these frequency interval controllability and observability gramians obtained by solving (44) and (45) are simultaneously diagonalized by using a suitable transformation matrix denoted by $T$ such that

$$T\hat{P}(\theta)T^T = T^{-T}\hat{Q}(\theta)T^{-1}$$

Step 3: Transform and partition to get the realization

$$\bar{A} = T^{-1}AT = \begin{bmatrix} A_{11} & A_{12} \\ A_{21} & A_{22} \end{bmatrix}, \quad \bar{N} = T^{-1}NT = \begin{bmatrix} N_{11} & N_{12} \\ N_{21} & N_{22} \end{bmatrix}$$

$$\bar{B} = T^{-1}B = \begin{bmatrix} B_1 \\ B_2 \end{bmatrix}, \quad \bar{C} = CT = \begin{bmatrix} C_1 & C_2 \end{bmatrix}$$

Step 4: The reduced order model is given by $A_r = A_{11}, N_r = N_{11}, B_r = B_1, C_r = C_1$

## 4. Results and discussion

### 4.1. Numerical example and results

Considering the following fifth-order discrete-time bilinear system originally presented by Hinamoto and Maekawa (1984) which has also been used by Zhang et al. (2003).

Table 1. Exact values of $y(k)$ from Figure 1

| k | y(k) | | | | |
|---|---|---|---|---|---|
| | Original | Proposed | Zhang et al. (2003) | E1 | E2 |
| 20 | 99.35 | **100.5** | 110.4 | **1.150** | 11.050 |
| 21 | 103.1 | **104.2** | 114.6 | **1.1** | 11.5 |
| 22 | 106.6 | **107.2** | 117.7 | **0.6** | 11.1 |
| 23 | 109.7 | **109.4** | 119.9 | **0.3** | 10.2 |
| 24 | 112.6 | **110.9** | 120.8 | **1.7** | 8.2 |
| 25 | 115.2 | **111.6** | 120.6 | **3.6** | 5.4 |

$$x(k+1) = Ax(k) + \sum_{j=1}^{m} N_j x(k) u_j(k) + Bu(k) \qquad (60)$$

$$y(k) = Cx(k)$$

where

$$A = \begin{bmatrix} 0 & 0 & 0.024 & 0 & 0 \\ 1 & 0 & -0.26 & 0 & 0 \\ 0 & 1 & 0.9 & 0 & 0 \\ 0 & 0 & 0.2 & 0 & -0.06 \\ 0 & 0 & 0.15 & 1 & 0.5 \end{bmatrix}, \quad B = \begin{bmatrix} 0.8 \\ 0.6 \\ 0.4 \\ 0.2 \\ 0.5 \end{bmatrix}, \quad N = \begin{bmatrix} 0.1 & 0 & 0 & 0 & 0 \\ 0 & 0.2 & 0 & 0 & 0 \\ 0 & 0 & 0.3 & 0 & 0 \\ 0 & 0 & 0 & 0.4 & 0 \\ 0 & 0 & 0 & 0 & 0.5 \end{bmatrix}$$

$$C = \begin{bmatrix} 0.2 & 0.4 & 0.6 & 0.8 & 1.0 \end{bmatrix}$$

The proposed technique involves firstly obtaining the frequency interval controllability and observability gramians defined in Theorem 1 and subsequently using these gramians as part of the balanced truncation-based technique described in Section 3.4 (Moore, 1981; Zhang & Lam, 2002; Zhang et al., 2003). This fifth-order model $\{A, N, B, C\}$ is reduced to the following second-order model $\{A_{r1}, N_{r1}, B_{r1}, C_{r1}\}$ in the form of (60) by using the proposed technique for the frequency interval $\Omega = [0.04\pi, 0.3\pi]$

Similarly, by applying the proposed technique for the frequency interval $\Omega = [0, 0.1\pi]$ to this fifth-order model $\{A, N, B, C\}$, a third-order discrete-time bilinear system with the following system matrices $\{A_{r2}, N_{r2}, B_{r2}, C_{r2}\}$ in the form of (60) is obtained

$$A_{r1} = \begin{bmatrix} 0.8247 & -0.2446 \\ 0.2394 & 0.5495 \end{bmatrix}, \quad B_{r1} = \begin{bmatrix} 1.2986 \\ -0.7881 \end{bmatrix}, \quad N_{r1} = \begin{bmatrix} 0.3214 & 0.1276 \\ 0.0878 & 0.2862 \end{bmatrix}$$

$$C_{r1} = \begin{bmatrix} 1.2974 & 0.7789 \end{bmatrix}$$

Table 2. Exact values of $y(k)$ from Figure 2

| k | y(k) | | | | |
|---|---|---|---|---|---|
| | Original | Proposed | Zhang et al. (2003) | E1 | E2 |
| 20 | 99.35 | **100.8** | 96.49 | **1.45** | 2.86 |
| 21 | 103.1 | **104.5** | 99.1 | **1.4** | 4 |
| 22 | 106.6 | **107.7** | 101.2 | **1.1** | 5.4 |
| 23 | 109.7 | **110.5** | 102.9 | **0.8** | 6.8 |
| 24 | 112.6 | **112.9** | 104.3 | **0.3** | 8.3 |
| 25 | 115.2 | **114.9** | 105.3 | **0.3** | 9.9 |

$$A_{r2} = \begin{bmatrix} 0.7588 & -0.1963 & 0.0881 \\ 0.1932 & 0.4929 & 0.2810 \\ 0.0947 & -0.2836 & 0.2436 \end{bmatrix} \quad B_{r2} = \begin{bmatrix} 1.4647 \\ -1.0558 \\ -0.6156 \end{bmatrix}, \quad N_{r2} = \begin{bmatrix} 0.3260 & 0.1387 & 0.0108 \\ 0.0998 & 0.3008 & -0.1865 \\ -0.0108 & -0.0910 & 0.3204 \end{bmatrix}$$

$$C_{r2} = \begin{bmatrix} 1.4583 & 1.0710 & -0.5848 \end{bmatrix}$$

For comparison, we apply the method by Zhang et al. (2003) which yields the following second- and third-order discrete-time bilinear systems with the following system matrices $\{A_{r3}, N_{r3}, B_{r3}, C_{r3}\}$ and $\{A_{r4}, N_{r4}, B_{r4}, C_{r4}\}$ in the form of (60)

$$A_{r3} = \begin{bmatrix} 0.8032 & -0.2445 \\ 0.1692 & 0.6050 \end{bmatrix}, \quad B_{r3} = \begin{bmatrix} 1.3335 \\ -0.8717 \end{bmatrix},$$

$$N_{r3} = \begin{bmatrix} 0.3387 & 0.1301 \\ 0.1075 & 0.2851 \end{bmatrix}, \quad C_{r3} = \begin{bmatrix} 1.3615 & 0.7053 \end{bmatrix}$$

and

$$A_{r4} = \begin{bmatrix} 0.8032 & -0.2445 & 0.0373 \\ 0.1692 & 0.6050 & 0.3500 \\ 0.0732 & -0.3154 & 0.3499 \end{bmatrix}, \quad B_{r4} = \begin{bmatrix} 1.3335 \\ -0.8717 \\ -0.4052 \end{bmatrix}$$

$$N_{r4} = \begin{bmatrix} 0.3387 & 0.1301 & 0.0221 \\ 0.1075 & 0.2851 & -0.1982 \\ -0.0315 & -0.0857 & 0.3059 \end{bmatrix}, \quad C_{r4} = \begin{bmatrix} 1.3615 & 0.7053 & -0.2884 \end{bmatrix}$$

### *4.2. Discussion of results*

Figure 1 shows the step responses of the original fifth-order model, second-order model obtained using the proposed method for a frequency interval $\Omega = [0.04\pi, 0.3\pi]$ and a second-order model obtained using the method by Zhang et al. (2003). Table 1 shows the exact values for $y(k)$ for the discrete times $k = 20, 21, 22, 23, 24, 25$ from Figure 1. E1 and E2 denote the absolute error between the value of $y(k)$ of the original model and the value of $y(k)$ obtained using the proposed method and the method by Zhang et al. (2003), respectively.

On the other hand Figure 2 shows the step responses of the original fifth-order model, third-order model obtained using the proposed method for a frequency interval $\Omega = [0, 0.1\pi]$ and a third-order model obtained using the method by Zhang et al. (2003). Table 2 shows the exact values for $y(k)$ for the discrete times $k = 20, 21, 22, 23, 24, 25$ from Figure 2. E1 and E2 denote the absolute error between the value of $y(k)$ of the original model and the value of $y(k)$ obtained using the proposed method and the method by Zhang et al. (2003), respectively. From both

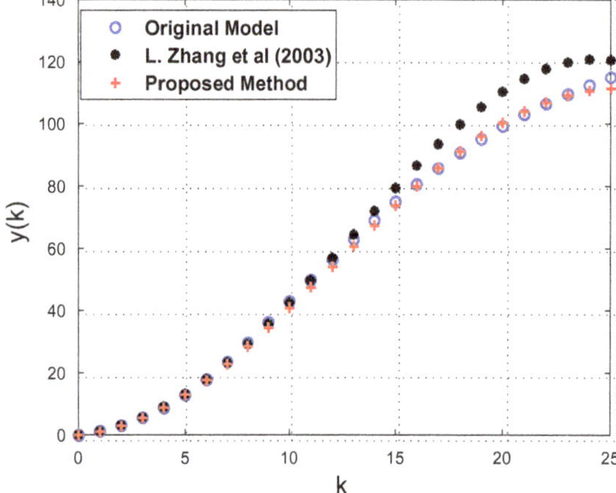


**Figure 1. Step responses of the original model, second-order reduced model obtained using the method by Zhang et al. (2003) and second-order reduced model obtained using the proposed technique for a frequency $\Omega = [0.04\pi, 0.3\pi]$.**

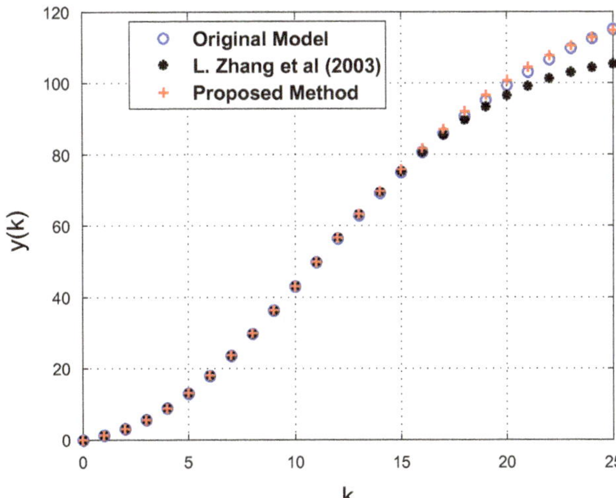


**Figure 2. Step responses of the original model, third-order reduced model obtained using the method by Zhang et al. (2003) and third-order reduced model obtained using the proposed technique for a frequency $\Omega = [\,0, 0.1\pi]$.**

Figure 1, Figure 2, Table 1, and Table 2 it is shown that applying the proposed technique yields a reduced-order model which is a closer approximation to the original model compared to the method by Zhang et al. (2003).

## 5. Conclusion

In conclusion, a new model reduction method for discrete time bilinear systems based on balanced truncation has been developed. The frequency interval controllability and observability gramians for discrete time bilinear systems are introduced and are shown to be solutions to a pair of new generalized Lyapunov equations. The conditions for solvability of these new Lyapunov equations are provided and the numerical solution method used to solve these equations is explained. Numerical results show that the proposed method yields reduced-order models which is a closer approximation to the original model as compared to existing techniques. The technique proposed in this paper is applicable to a variety of non-linear systems which can be formulated as bilinear systems.

**Funding**
The authors received no direct funding for this research.

**Author details**
Ahmad Jazlan[1,2]
E-mail: ahmadjazlan@iium.edu.my
Victor Sreeram[1]
E-mail: victor.sreeram@uwa.edu.au
Hamid Reza Shaker[3]
E-mail: hrsh@sdu.dk
Roberto Togneri[1]
E-mail: roberto.togneri@uwa.edu.au
[1] School of Electrical, Electronics and Computer Engineering, University of Western Australia, 35 Stirling Highway, Crawley, Perth, Western Australia 6009, Australia.
[2] Faculty of Engineering, Department of Mechatronics Engineering, International Islamic University Malaysia, Jalan Gombak, 53100 Kuala Lumpur, Malaysia.
[3] Center for Energy Informatics, University of Southern Denmark, Campusvej 55, DK-5230 Odense M, Denmark.

**References**

Al-Baiyat, S. A., Bettayeb, M., & Al-Saggaf, U. M. (1994). New model reduction scheme for bilinear systems. *International Journal of Systems Science, 25*, 1631–1642.

D'Alessandro, P., Isidori, A., & Ruberti, A. (1974). Realization and structure theory of bilinear dynamical systems. *SIAM Journal on Control, 12*, 517–535.

Ding, D., Du, X., & Li, X. (2015). Finite-frequency model reduction of two-dimensional digital filters. *IEEE Transactions on Automatic Control, 60*, 1624–1629.

Ding, D. W., Li, X. J., Du, X., & Xie, X. (2016). Finite-frequency model reduction of takagi-sugeno fuzzy systems. *IEEE Transactions on Fuzzy System, PP*(PP), 1–10.

Dorissen, H. (1989). Canonical forms for bilinear systems. *Systems and Control Letters, 13*, 153–160.

Du, X., Fan, F., & Ding, D. (2016). Finite-frequency model order reduction of discrete-time linear time-delayed systems. *Nonlinear Dynamics, X*(X), 1–12.

Gawronski, W., & Juang, J. (1990). Model reduction in limited time and frequency intervals. *International Journal of System Science, 21*, 349–376.

Hinamoto, T., & Maekawa, S. (1984). Approximation of polynomial state-affine discrete time systems. *IEEE Transactions on Circuits and Systems, 31*, 713–721.

Horta, L., Juang, J., & Longman, R. (1993). Discrete-time model reduction in limited frequency ranges. *Journal of Guidance, Control, and Dynamics, 16*, 1125–1130.

Imran, M., & Ghafoor, A. (2014). Stability preserving model reduction technique and error bounds using frequency limited gramians for discrete-time systems. *IEEE*

*Transactions on Circuits and Systems - II: Express Briefs, 61*, 716–720.

Imran, M., & Ghafoor, A. (2015). Model reduction of descriptor systems using frequency limited gramians. *Journal of the Franklin Institute, 352*, 33–51.

Li, X., & Yang, G. (2015). Adaptive control in finite frequency domain for uncertain linear systems. *Information Sciences, 314*, 14–27.

Li, X., Yin, S., & Gao, H. (2014). Passivity-preserving model reduction with finite frequency approximation performance. *Automatica, 50*, 2294–2303.

Li, X., Yu, C., & Gao, H. (2015). Frequency Limited $H_\infty$ Model Reduction for Positive Systems. *IEEE Transactions on Automatic Control, 60*, 1093–1098.

Li, X., Yu, C., Gao, H., & Zhang, L. (2014). *A New Approach to $h_\infty$ model reduction for positive systems.* Proceedings of the 19th IFAC World Congress. Cape Town.

Minh, H. B., Battle, C., & Fossas, E. (2014). A new estimation of the lower error bound in balanced truncation method. *Automatica, 50*, 2196–2198.

Moore, B. (1981). Principal component analysis in linear system: Controllability, observability, and model reduction. *IEEE Transactions on Automatic Control, AC-26*, 17–32.

Opmeer, M., & Reis, T. (2015). A lower bound for the balanced truncation error for mimo systems. *IEEE Transactions on Automatic Control, 60*, 2207–2212.

Shaker, H., & Tahavori, M. (2014a). Frequency interval model reduction of bilinear systems. *IEEE Transactions on Automatic Control, 59*, 1948–1953.

Shaker, H. & Tahavori, M. (2014b). *Generalized hankel interaction index array for control structure selection for discrete-time mimo bilinear processes and plants* (pp. 3149-3154). Proceedings of the 53rd IEEE Conference on Decision and Control (CDC). Los Angeles, CA.

Shaker, H., & Tahavori, M. (2014c). Time-interval model reduction of bilinear systems. *International Journal of Control, 87*, 1487–1495.

Shaker, H., & Tahavori, M. (2015). Control configuration selection for bilinear systems via generalised hankel interaction index array. *International Journal of Control, 88*, 30–37.

Wang, D., & Zilouchian, A. (2000). Model reduction of discrete linear systems via frequency-domain balanced structure. *IEEE Transactions on Circuits and Systems I: Fundamental Theory and Applications, 47*, 830–837.

Zhang, H., Wu, L., Shi, P., & Zhao, Y. (2015). Balanced truncation approach to model reduction of markovian jump time-varying delay systems. *Journal of the Franklin Institute, 352*, 4205–4224.

Zhang, L., & Lam, J. (2002). On $H_2$ model reduction of bilinear systems. *Automatica, 38*, 205–216.

Zhang, L., Lam, J., Huang, B., & Yang, G. H. (2003). On gramians and balanced truncation of discrete-time bilinear systems. *International Journal of Control, 76*, 414–427.

# 4

# Controllability of nonlocal second-order impulsive neutral stochastic functional integro-differential equations with delay and Poisson jumps

Diem Dang Huan[1,2*] and Hongjun Gao[3]

*Corresponding author: Diem Dang Huan, Faculty of Basic Sciences, Bacgiang Agriculture and Forestry University, Bacgiang 21000, Vietnam; Vietnam National University, Hanoi, 144 Xuan Thuy Street, Cau Giay, Hanoi 10000, Vietnam
E-mail: huandd@bafu.edu.vn
Reviewing editor: James Lam, University of Hong Kong, Hong Kong

**Abstract:** The current paper is concerned with the controllability of nonlocal second-order impulsive neutral stochastic functional integro-differential equations with infinite delay and Poisson jumps in Hilbert spaces. Using the theory of a strongly continuous cosine family of bounded linear operators, stochastic analysis theory and with the help of the Banach fixed point theorem, we derive a new set of sufficient conditions for the controllability of nonlocal second-order impulsive neutral stochastic functional integro-differential equations with infinite delay and Poisson jumps. Finally, an application to the stochastic nonlinear wave equation with infinite delay and Poisson jumps is given.

**Subjects: Non-Linear Systems; Probability Theory & Applications; Stochastic Models & Processes**

**Keywords: controllability; impulsive neutral stochastic integro-differential equations; Poisson jumps; cosine functions of operators; infinite delay; Banach fixed point theorem**

**2010 Mathematics Subject classifications: 34A37; 93B05; 93E03; 60H20; 34K50**

## 1. Introduction

As one of the fundamental concepts in mathematical control theory, controllability plays an important role both in deterministic and stochastic control problems such as stabilization of unstable systems by feedback control. It is well known that controllability of deterministic equation is widely

## ABOUT THE AUTHORS

Diem Dang Huan was born in Bacgiang, Vietnam, on 13 July 1980. He received his BS and MS degrees in Mathematics and Theory of Probability and Statistics from University of Science—Vietnam National University, Hanoi, in 2004 and 2008, respectively. From 2004 to August 2010, he has been employed at Bacgiang Agriculture and Forestry University. After he got a scholarship from the Vietnamese Government in August 2010, he started his PhD study in Applied Mathematic group in the Institute of Mathematics, School of Mathematical Science, in Nanjing Normal University, China. His research interests include stochastic functional differential equations, stochastic partial differential equations, and theory control of dynamical systems.

Hongjun Gao, professor, speciality in stochastic partial differential equations and its dynamics.

## PUBLIC INTEREST STATEMENT

Controllability plays an important role in the analysis and design of control systems. Roughly speaking, controllability generally means that it is possible to steer dynamical control system from an arbitrary initial state to an arbitrary final state using the set of admissible controls. It is well known that stochastic control theory is stochastic generalization of the classic control theory. In this paper, we study the controllability of nonlocal second-order impulsive neutral stochastic functional integro-differential equations with infinite delay and Poisson jumps and our results can complement the earlier publications in the existing literature.

used in many fields of science and technology, say, physics and engineering (e.g. see Ahmed, 2014a; Balachandran & Dauer, 2002; Coron, 2007; Curtain & Zwart, 1995; Zabczyk, 1992, and the references therein). Stochastic control theory is stochastic generalization of the classic control theory. The theory of controllability of differential equations in infinite dimensional spaces has been extensively studied in the literature, and the details can be found in various papers and monographs (Ahmed, 2014b; Astrom, 1970; Balachandran & Dauer, 2002; Karthikeyan & Balachandran, 2013; Yang, 2001; Zabczyk, 1991, and the references therein). Any control system is said to be controllable if every state corresponding to this process can be affected or controlled in a respective time by some control signals. If the system cannot be controlled completely, then different types of controllability can be defined such as approximate, null, local null, and local approximate null controllabilities. On this matter, we refer the reader to Ahmed (2014c), Chang (2007), Karthikeyan and Balachandran (2009), Ntouyas and ÓRegan (2009), Sakthivel, Mahmudov, and Lee (2009), and the references therein.

The theory of impulsive differential equations as much as neutral differential equations has been emerging as an important area of investigations in recent years, stimulated by their numerous applications to problems in physics, mechanics, electrical engineering, medicine biology, ecology, and so on. The impulsive differential systems can be used to model processes which are subject to abrupt changes, and which cannot be described by the classical differential systems (Lakshmikantham, Baĭnov, & Simeonov, 1989). Partial neutral integro-differential equation with infinite delay has been used for modeling the evolution of physical systems, in which the response of the system depends not only on the current state, but also on the past history of the system, for instance, for the description of heat conduction in materials with fading memory, we refer the reader to the papers of Gurtin and Pipkin (1968), Nunziato (1971), and the references therein related to this matter. Besides, noise or stochastic perturbation is unavoidable and omnipresent in nature as well as in man-made systems. Therefore, it is of great significance to import the stochastic effects into the investigation of impulsive neutral differential equations. As the generalization of the classic impulsive neutral differential equations, impulsive neutral stochastic integro-differential differential equations with infinite delays have attracted the researchers' great interest. On the existence and the controllability for these equations, we refer the reader to (e.g. see Chang, 2007; Chang, Anguraj, & Arjunan, 2008; Karthikeyan & Balachandran, 2009, 2013; Park, Balachandran, & Annapoorani, 2009; Park, Balasubramaniam, & Kumaresan, 2007; Shen & Sun, 2012; Yan & Yan, 2013, and the references therein).

Recently, Park, Balachandran, and Arthi (2009) investigated the controllability of impulsive neutral integro-differential systems with infinite delay in Banach spaces using Schauder-type fixed point theorem. Arthi and Balachandran (2012) established the controllability of damped second-order impulsive neutral functional differential systems with infinite delay by means of the Sadovskii fixed point theorem combined with a noncompact condition on the cosine family of operators. Very recently, also using Sadovskii's fixed point theorem, Muthukumar and Rajivganthi (2013) proved sufficient conditions for the approximate controllability of fractional order neutral stochastic integro-differential systems with nonlocal conditions and infinite delay.

By contrast, there has not been very much research on the controllability of second-order impulsive neutral stochastic functional differential equations with infinite delays, or in other words, the literature about controllability of second-order impulsive neutral stochastic functional differential equations with infinite delays is very scarce. To be more precise, Balasubramaniam and Muthukumar (2009) discussed on approximate controllability of second-order stochastic distributed implicit functional differential systems with infinite delay. Mahmudova and McKibben (2006) established the results concerning the global existence, uniqueness, approximate, and exact controllability of mild solutions for a class of abstract second-order damped McKean–Vlasov stochastic evolution equations in a real separable Hilbert space. More recently, using Holder's inequality, stochastic analysis, and fixed point strategy, Sakthivel, Ren, and Mahmudov (2010) considered sufficient conditions for the approximate controllability of nonlinear second-order stochastic infinite dimensional dynamical systems with impulsive effects. And Muthukumar and Balasubramaniam (2010) investigated sufficient conditions for the approximate controllability of a class of second-order damped McKean–Vlasov stochastic evolution equations in a real separable Hilbert space.

On the other hand, in recent years, stochastic partial differential equations with Poisson jumps have gained much attention since Poisson jumps not only exist widely, but also can be used to study many phenomena in real lives. Therefore, it is necessary to consider the Poisson jumps into the stochastic systems. For instance, Luo and Liu (2008) studied the existence and uniqueness of mild solutions to stochastic partial functional differential equations with Markovian switching and Poisson jumps using the Lyapunov–Razumikhin technique. Ren, Zhou, and Chen (2011) investigated the existence, uniqueness, and stability of mild solutions for a class of time-dependent stochastic evolution equations with Poisson jumps. More specifically, just recently, there is an article on the complete controllability of stochastic evolution equations with jumps in a separable Hilbert space discussed by Sakthivel and Ren (2011) and in reference Ren, Dai, and Sakthivel (2013), Ren et al. studied the approximate controllability of stochastic differential systems driven by Teugels martingales associated with a Lévy process. For more details about the stochastic partial differential equations with Poisson jumps, one can see a recent monograph of Peszat and Zabczyk (2007) as well as papers of Cao (2005), Marinelli & Rockner (2010), Rockner and Zhang (2007), and the references therein.

To the best of our knowledge, there is no work reported on nonlocal second-order impulsive neutral stochastic functional integro-differential equations with infinite delay and Poisson jumps. To close the gap, motivated by the above works, the purpose of this paper is to study the controllability of nonlocal second-order impulsive neutral stochastic functional integro-differential equations with infinite delay and Poisson jumps in Hilbert spaces. More precisely, we consider the following form:

$$\begin{cases} d\left[x'(t) - g\left(t, x_t, \int_0^t \sigma_1(t,s,x_s)ds\right)\right] = \left[Ax(t) + f\left(t, x_t, \int_0^t \sigma_2(t,s,x_s)ds\right) + Bu(t)\right]dt \\ \quad + \int_{-\infty}^t \sigma(t,s,x_s)dw(s) + \int_{\mathcal{U}} \gamma\left(t, x(t-), v\right)\widetilde{N}(dt,dv), \quad t_k \neq t \in J := [0,T], \\ \Delta x(t_k) = I_k^1(x_{t_k}), \quad k = \{1,\cdots,m\} =: \overline{1,m}, \\ \Delta x'(t_k) = I_k^2(x_{t_k}), \quad k = \overline{1,m}, \\ x'(0) = x_1 \in \mathbb{H}, \\ x(0) - q(x_{t_1}, x_{t_2}, \cdots, x_{t_n}) = x_0 = \varphi \in \mathcal{B}, \quad \text{for a.e. } s \in J_0 := (-\infty, 0], \end{cases} \tag{1.1}$$

where $0 < t_1 < t_2 < \cdots < t_n < T, n \in \mathbb{N}$; $x(\cdot)$ is a stochastic process taking values in a real separable Hilbert space $\mathbb{H}$; $A: D(A) \subset \mathbb{H} \to \mathbb{H}$ is the infinitesimal generator of a strongly continuous cosine family on $\mathbb{H}$. The history $x_t: J_0 \to \mathbb{H}$, $x_t(\theta) = x(t+\theta)$ for $t \geq 0$, belongs to the phase space $\mathcal{B}$, which will be described in Section 2. Assume that the mappings $f, g: J \times \mathcal{B} \times \mathbb{H} \to \mathbb{H}$, $\sigma: J \times J \times \mathcal{B} \to \mathcal{L}_2^0$, $\sigma_i: J \times J \times \mathcal{B} \to \mathbb{H}, i = 1, 2, I_k^1, I_k^2: \mathcal{B} \to \mathbb{H}, k = \overline{1,m}, q: \mathcal{B}^n \to \mathcal{B}$, and $\gamma: J \times \mathbb{H} \times \mathcal{U} \to \mathbb{H}$ are appropriate functions to be specified later. The control function $u(\cdot)$ takes values in $L^2(J,U)$ of admissible control functions for a separable Hilbert space $U$ and $B$ is a bounded linear operator from $U$ into $\mathbb{H}$. Furthermore, let $0 = t_0 < t_1 < \cdots < t_m < t_{m+1} = T$ be prefixed points, and $\Delta x(t_k) = x(t_k^+) - x(t_k^-)$ represents the jump of the function $x$ at time $t_k$ with $I_k$, determining the size of the jump, where $x(t_k^+)$ and $x(t_k^-)$ represent the right and left limits of $x(t)$ at $t = t_k$, respectively. Similarly $x'(t_k^+)$ and $x'(t_k^-)$ denote, respectively, the right and left limits of $x'(t)$ at $t_k$. Let $\varphi(t) \in \mathcal{L}_2(\Omega, \mathcal{B})$ and $x_1(t)$ be $\mathbb{H}$-valued $\mathcal{F}_t$-measurable random variables independent of the Wiener process $\{w(t)\}$ and the Poisson point process $p(\cdot)$ with a finite second moment.

The main techniques used in this paper include the Banach contraction principle and the theories of a strongly continuous cosine family of bounded linear operators.

The structure of this paper is as follows: in Section 2, we briefly present some basic notations, preliminaries, and assumptions. The main results in Section 3 are devoted to study the controllability for the system (1.1) with their proofs. An example is given in Section 4 to illustrate the theory. In Section 5, concluding remarks are given.

## 2. Preliminaries

In this section, we briefly recall some basic definitions and results for stochastic equations in infinite dimensions and cosine families of operators. For more details on this section, we refer the reader to Da Prato and Zabczyk (1992), Fattorini (1985), Protter (2004), and Travis and Webb (1978).

Let $(\mathbb{H}, \|\cdot\|_{\mathbb{H}}, \langle\cdot,\cdot\rangle_{\mathbb{H}})$ and $(\mathbb{K}, \|\cdot\|_{\mathbb{K}}, \langle\cdot,\cdot\rangle_{\mathbb{K}})$ denote two real separable Hilbert spaces, with their vectors, norms, and their inner products, respectively. We denote by $\mathcal{L}(\mathbb{K};\mathbb{H})$ the set of all linear bounded operators from $\mathbb{K}$ into $\mathbb{H}$, which is equipped with the usual operator norm $\|\cdot\|$. In this paper, we use the symbol $\|\cdot\|$ to denote norms of operators regardless of the spaces potentially involved when no confusion possibly arises. Let $(\Omega, \mathcal{F}, \mathbb{F} = \{\mathcal{F}_t\}_{t\geq 0}, \mathbf{P})$ be a complete filtered probability space satisfying the usual condition (i.e. it is right continuous and $\mathcal{F}_0$ contains all $\mathbf{P}$-null sets). Let $w = (w(t))_{t\geq 0}$ be a $Q$-Wiener process defined on the probability space $(\Omega, \mathcal{F}, \mathbb{F}, \mathbf{P})$ with the covariance operator $Q$ such that $Tr(Q) < \infty$. We assume that there exists a complete orthonormal system $\{e_k\}_{k\geq 1}$ in $\mathbb{K}$, a bounded sequence of nonnegative real numbers $\lambda_k$ such that $Qe_k = \lambda_k e_k$, $k = 1, 2, \dots,$ and a sequence of independent Brownian motions $\{\beta_k\}_{k\geq 1}$ such that

$$\langle w(t), e\rangle_{\mathbb{K}} = \sum_{k=1}^{\infty} \sqrt{\lambda_k}\langle e_k, e\rangle_{\mathbb{K}}\beta_k(t), \quad e \in \mathbb{K}, t \geq 0.$$

Let $\mathcal{L}_2^0 = \mathcal{L}_2(Q^{\frac{1}{2}}\mathbb{K};\mathbb{H})$ be the space of all Hilbert–Schmidt operators from $Q^{\frac{1}{2}}\mathbb{K}$ into $\mathbb{H}$ with the inner product $\langle\Psi, \phi\rangle_{\mathcal{L}_2^0} = Tr[\Psi Q\phi^*]$, where $\phi^*$ is the adjoint of the operator $\phi$. Let $p = p(t)$, $t \in D_p$ (the domain of $p(t)$) be a stationary $\mathcal{F}_t$-Poisson point process taking its value in a measurable space $(\mathcal{U}, \mathfrak{B}(\mathcal{U}))$ with a $\sigma$-finite intensity measure $\lambda(dv)$ by $N(dt, dv)$ the Poisson counting measure associated with $p$, that is,

$$N(t, \mathcal{U}) = \sum_{s\in D_p, s\leq t} \mathbb{I}_{\mathcal{U}}(p(s))$$

for any measurable set $\mathcal{U} \in \mathfrak{B}(\mathbb{K} - \{0\})$, which denotes the Borel $\sigma$-field of $(\mathbb{K} - \{0\})$. Let

$$\widetilde{N}(dt, dv) := N(dt, dv) - \lambda(dv)dt$$

be the compensated Poisson measure that is independent of $w(t)$. Denote by $\mathcal{P}^2(J \times \mathcal{U};\mathbb{H})$ the space of all predictable mappings $\gamma: J \times \mathcal{U} \to \mathbb{H}$ for which

$$\int_0^t \int_{\mathcal{U}} \mathbf{E}\|\gamma(t, v)\|_{\mathbb{H}}^2 \lambda(dv)dt < \infty.$$

We may then define the $\mathbb{H}$-valued stochastic integral $\int_0^t \int_{\mathcal{U}} \gamma(t, v)\widetilde{N}(dt, dv)$, which is a centered square-integrable martingale. For the construction of this kind of integral, we can refer to Protter (2004).

The collection of all strongly measurable, square-integrable $\mathbb{H}$-valued random variables, denoted by $\mathcal{L}_2(\Omega, \mathbb{H})$, is a Banach space equipped with norm $\|x\|_{\mathcal{L}_2} = \left(\mathbf{E}\|x\|^2\right)^{\frac{1}{2}}$. Let $C(J, \mathcal{L}_2(\Omega, \mathbb{H}))$ be the Banach space of all continuous maps from $J$ to $\mathcal{L}_2(\Omega, \mathbb{H})$, satisfying the condition $\sup_{t\in J} \mathbf{E}\|x(t)\|^2 < \infty$. An important subspace is given by $\mathcal{L}_2^0(\Omega, \mathbb{H}) = \{f \in \mathcal{L}_2(\Omega, \mathbb{H}): f \text{ is } \mathcal{F}_0\text{-measurable}\}$. Further, let $\mathcal{L}_2^{\mathbb{F}}(0, T;\mathbb{H}) = \left\{g: J \times \Omega \to \mathbb{H}: g(\cdot) \text{ is } \mathbb{F}\text{-progressively measurable and } \mathbf{E}\left(\int_J \|g(t)\|_{\mathbb{H}}^2 dt\right) < \infty\right\}$.

Next, to be able to access controllability for the system (1.1), we need to introduce the theory of cosine functions of operators and the second-order abstract Cauchy problem.

*Definition 2.1* (1) The one-parameter family $\{C(t)\}_{t\in\mathbb{R}} \subset \mathcal{L}(\mathbb{H})$ is said to be a strongly continuous cosine family if the following hold:

(i) $C(0) = I$, $I$ is the identity operators in $\mathbb{H}$;

(ii) $C(t)x$ is continuous in $t$ on $\mathbb{R}$ for any $x \in \mathbb{H}$; and

(iii) $C(t+s) + C(t-s) = 2C(t)C(s)$ for all $t, s \in \mathbb{R}$.

(2) The corresponding strongly continuous sine family $\{S(t)\}_{t\in\mathbb{R}} \subset \mathcal{L}(\mathbb{H})$, associated to the given strongly continuous cosine family $\{C(t)\}_{t\in\mathbb{R}} \subset \mathcal{L}(\mathbb{H})$ is defined by

$$S(t)x = \int_0^t C(s)xds, \quad t \in \mathbb{R}, x \in \mathbb{H}.$$

(3) The infinitesimal generator $A: \mathbb{H} \to \mathbb{H}$ of $\{C(t)\}_{t\in\mathbb{R}} \subset \mathcal{L}(\mathbb{H})$ is given by

$$Ax = \frac{d^2}{dt^2}C(t)x\Big|_{t=0}, \quad \text{for all } x \in D(A) = \{x \in \mathbb{H}: C(\cdot) \in C^2(\mathbb{R}, \mathbb{H})\}$$

It is well known that the infinitesimal generator $A$ is a closed, densely defined operator on $\mathbb{H}$, and the following properties hold (see Travis & Webb, 1978).

*Proposition 2.1* Suppose that $A$ is the infinitesimal generator of a cosine family of operators $\{C(t)\}_{t\in\mathbb{R}}$. Then, the following hold:

(i) There exist a pair of constants $M_A \geq 1$ and $\alpha \geq 0$ such that $\|C(t)\| \leq M_A e^{\alpha|t|}$, and hence $\|S(t)\| \leq M_A e^{\alpha|t|}$;

(ii) $A\int_s^r S(u)xdu = [C(r) - C(s)]x$, for all $0 \leq s \leq r < \infty$; and

(iii) There exists $N \geq 1$ such that $\|S(s) - S(r)\| \leq N\left|\int_s^r e^{\alpha|s|}ds\right|$, $0 \leq s \leq r < \infty$.

Thanks to the Proposition 2.1 and the uniform boundedness principle that we see a direct consequence that both $\{C(t)\}_{t\in J}$ and $\{S(t)\}_{t\in J}$ are uniformly bounded by $\widetilde{M} = M_A e^{\alpha|T|}$.

The existence of solutions for the second-order linear abstract Cauchy problem

$$\begin{cases} x''(t) = Ax(t) + h(t), & t \in J, \\ x(0) = z, \quad x'(0) = w, \end{cases} \tag{2.1}$$

where $h: J \to \mathbb{H}$ is an integrable function that has been discussed in Travis and Webb (1977). Similarly, the existence of solutions of the semilinear second-order abstract Cauchy problem has been treated in Travis and Webb (1978).

*Definition 2.2* The function $x(\cdot)$ given by

$$x(t) = C(t)z + S(t)w + \int_0^t S(t-s)h(s)ds, \quad t \in J,$$

is called a mild solution of (2.1), and that when $z \in \mathbb{H}$, $x(\cdot)$ is continuously differentiable and

$$x'(t) = AS(t)z + C(t)w + \int_0^t C(t-s)h(s)ds, \quad t \in J.$$

For additional details about cosine function theory, we refer the reader to Travis and Webb (1977, 1978).

Since the system (1.1) has impulsive effects, the phase space used in Balasubramaniam and Ntouyas (2006) and Park et al. (2007) cannot be applied to these systems. So, we need to introduce an abstract phase space $\mathcal{B}$, as follows:

Assume that $l:J_0 \to (0,+\infty)$ is a continuous function with $l_0 = \int_{J_0} l(t)dt < \infty$. For any $a > 0$, we define

$\mathcal{B}:= \Big\{ \psi:J_0 \to \mathbb{H}:(\mathbf{E}\|\psi(\theta)\|^2)^{\frac{1}{2}}$ is a bounded and measurable function on $[-a,0]$ and $\int_{J_0} l(s) \sup_{\theta\in[s,0]}(\mathbf{E}\|\psi(\theta)\|^2)^{\frac{1}{2}} ds < +\infty \Big\}$.

If $\mathcal{B}$ is endowed with the norm

$$\|\psi\|_{\mathcal{B}} = \int_{J_0} l(s) \sup_{\theta\in[s,0]} (\mathbf{E}\|\psi(\theta)\|^2)^{\frac{1}{2}} ds, \quad \forall \psi \in \mathcal{B},$$

then, it is clear that $(\mathcal{B}, \|\cdot\|_{\mathcal{B}})$ is a Banach space (Hino, Murakami, & Naito, 1991).

Let $J_T = (-\infty, T]$. We consider the space

$$\mathcal{B}_T: = \Big\{ x:J_T \to \mathbb{H} \text{ such that } x_k \in C(J_k, \mathbb{H}) \text{ and there exist } x(t_k^-) \text{ and } x(t_k^+) \text{ with}$$
$$x(t_k^-) = x(t_k^+), x(0) - q(x_{t_1}, x_{t_2}, \cdots, x_{t_n}) = \varphi \in \mathcal{B}, k = \overline{1,m} \Big\},$$

where $x_k$ is the restriction of $x$ to $J_k = (t_k, t_{k+1}], k = \overline{1,m}$. Set $\|\cdot\|_T$ be a seminorm in $\mathcal{B}_T$ defined by

$$\|x\|_T = \|\varphi\|_{\mathcal{B}} + \sup_{s\in J}(\mathbf{E}\|x(s)\|^2)^{\frac{1}{2}}, \quad x \in \mathcal{B}_T.$$

Now, we recall the following useful lemma that appeared in Chang (2007).

Lemma 2.1 *(Chang, 2007) Assume that $x \in \mathcal{B}_T$, then for $t \in J$, $x_t \in \mathcal{B}$. Moreover,*

$$l_0(\mathbf{E}\|x(t)\|^2)^{\frac{1}{2}} \le \|x_t\|_{\mathcal{B}} \le \|x_0\|_{\mathcal{B}} + l_0 \sup_{s\in[0,t]} (\mathbf{E}\|x(s)\|^2)^{\frac{1}{2}}.$$

*Next, we give the definition of mild solution for (1.1).*

*Definition 2.3* An $\mathcal{F}_t$-adapted càdlàg stochastic process $x:J_T \to \mathbb{H}$ is called a mild solution of (1.1) on $J_T$ if $x(0) - q(x_{t_1}, x_{t_2}, \cdots, x_{t_n}) = x_0 = \varphi \in \mathcal{B}$ and $x'(0) = x_1 \in \mathbb{H}$, satisfying $\varphi, x_1, q \in \mathcal{L}_2^0(\Omega, \mathbb{H})$; the functions $C(t-s)g(s, x_s, \int_0^s \sigma_1(s,\tau,x_\tau)d\tau)$ and $S(t-s)f(s, x_s, \int_0^s \sigma_2(s,\tau,x_\tau)d\tau)$ are integrable on $[0,T)$ such that the following conditions hold:

(i) $\{x_t:t \in J\}$ is a $\mathcal{B}$-valued stochastic process;

(ii) For arbitrary $t \in J$, $x(t)$ satisfies the following integral equation:

$$\begin{aligned} x(t) =& C(t)[\varphi(0) + q(x_{t_1}, x_{t_2}, \cdots, x_{t_n})(0)] + S(t)[x_1 - g(0, x_0, 0)] \\ &+ \int_0^t C(t-s)g(s, x_s, \int_0^s \sigma_1(s,\tau,x_\tau)d\tau)ds \\ &+ \int_0^t S(t-s)f(s, x_s, \int_0^s \sigma_2(s,\tau,x_\tau)d\tau)ds + \int_0^t S(t-s)Bu(s)ds \\ &+ \int_0^t S(t-s)\int_{-\infty}^s \sigma(s,\tau,x_\tau)dw(\tau)ds + \int_0^t S(t-s)\int_{\mathcal{V}} \gamma(t, x(t-), v)\widetilde{N}(dt,dv) \\ &+ \sum_{0<t_k<t} C(t-t_k)I_k^1(x_{t_k}) + \sum_{0<t_k<t} S(t-t_k)I_k^2(x_{t_k}); \text{ and} \end{aligned} \tag{2.2}$$

(iii) $\Delta x(t_k) = I_k^1(x_{t_k}), \Delta x'(t_k) = I_k^2(x_{t_k}), k = \overline{1,m}$.

*Definition 2.4* The system (1.1) is said to be controllable on the interval $J_T$, if for every initial stochastic process $\varphi \in \mathcal{B}$ defined on $J_0$, $x'(0) = x_1 \in \mathbb{H}$ and $y_1 \in \mathbb{H}$; there exists a stochastic control $u \in L^2(J, U)$ which is adapted to the filtration $\{\mathcal{F}_t\}_{t \in J}$ such that the solution $x(\cdot)$ of the system (1.1) satisfies $x(T) = y_1$, where $y_1$ and $T$ are the preassigned terminal state and time, respectively.

To prove our main results, we list the following basic assumptions of this paper.

**(H1)** There exists positive constants $M_C$, $M_S$, and $M_{\sigma_1}$ such that for all $t, s \in J, x, y \in \mathcal{B}$

$$\|C(t)\|^2 \le M_C, \quad \|S(t)\|^2 \le M_S;$$

$$\mathbf{E}\left\| \int_0^t [\sigma_1(t,s,x) - \sigma_1(t,s,y)]ds \right\|^2 \le M_{\sigma_1} \|x - y\|_{\mathcal{B}}^2.$$

**(H2)** The function $g: J \times \mathcal{B} \times \mathbb{H} \to \mathbb{H}$ is continuous and there exists a positive constant $M_g$ such that for all $t \in J, x_1, x_2 \in \mathcal{B}, y_1, y_2 \in \mathcal{L}_2(\Omega, \mathbb{H})$

$$\mathbf{E}\|g(t, x_1, y_1) - g(t, x_2, y_2)\|^2 \le M_g(\|x_1 - x_2\|_{\mathcal{B}}^2 + \mathbf{E}\|y_1 - y_2\|^2).$$

**(H3)** For each $(t, s) \in J \times J$, the function $\sigma_2: J \times J \times \mathcal{B} \to \mathbb{H}$ is continuous and there exists a positive constant $M_{\sigma_2}$ such that for all $t, s \in J, x, y \in \mathcal{B}$

$$\mathbf{E}\left\| \int_0^t [\sigma_2(t,s,x) - \sigma_2(t,s,y)]ds \right\|^2 \le M_{\sigma_2} \|x - y\|_{\mathcal{B}}^2.$$

**(H4)** The function $f: J \times \mathcal{B} \times \mathbb{H} \to \mathbb{H}$ is continuous and there exists a positive constant $M_f$ such that for all $t \in J, x_1, x_2 \in \mathcal{B}, y_1, y_2 \in \mathcal{L}_2(\Omega, \mathbb{H})$

$$\mathbf{E}\|f(t, x_1, y_1) - f(t, x_2, y_2)\|^2 \le M_f(\|x_1 - x_2|_{\mathcal{B}}^2 + \mathbf{E}\|y_1 - y_2\|^2).$$

**(H5)** The functions $I_k^1, I_k^2 \in C(\mathcal{B}, \mathbb{H})$, $k = \overline{1,m}$ and there exist positive constants $M_{I_k^1}, \overline{M}_{I_k^1}$, $M_{I_k^2}$, and $\overline{M}_{I_k^2}$ such that for all $x, y \in \mathcal{B}$

$$\mathbf{E}\|I_k^1(x)\|^2 \le M_{I_k^1}, \quad \mathbf{E}\|I_k^2(x)\|^2 \le M_{I_k^2};$$

$$\mathbf{E}\|I_k^1(x) - I_k^1(y)\|^2 \le \overline{M}_{I_k^1}\|x - y\|_{\mathcal{B}}^2, \quad \mathbf{E}\|I_k^2(x) - I_k^2(y)\|^2 \le \overline{M}_{I_k^2}\|x - y\|_{\mathcal{B}}^2.$$

**(H6)** For each $\varphi \in \mathcal{B}, h(t) = \lim_{c \to \infty} \int_{-c}^0 \sigma(t, s, \varphi)dw(s)$ exists and continuous. Further, there exists a positive constant $M_h$ such that

$$\mathbf{E}\|h(t)\|^2 \le M_h.$$

**(H7)** The function $\sigma: J \times J \times \mathcal{B} \to \mathcal{L}(\mathbb{K}, \mathbb{H})$ is continuous and there exists positive constants $M_\sigma, \overline{M}_\sigma$ such that for all $s, t \in J$ and $x, y \in \mathcal{B}$

$$\mathbf{E}\|\sigma(t,s,x)\|_{\mathcal{L}_2^0}^2 \le M_\sigma;$$

$$\mathbf{E}\|\sigma(t,s,x) - \sigma(t,s,y)\|_{\mathcal{L}_2^0}^2 \le \overline{M}_\sigma\|x - y\|_{\mathcal{B}}^2.$$

**(H8)** The function $q: \mathcal{B}^n \to \mathcal{B}$ is continuous and there exist positive constants $M_q, \overline{M}_q$ such that for all $x, y \in \mathcal{B}, t \in J_0$

$$\mathbf{E}\|q(x_{t_1}, x_{t_2}, \cdots, x_{t_n})(t)\|^2 \le M_q;$$

$$\mathbf{E}\|q(x_{t_1}, x_{t_2}, \cdots, x_{t_n})(t) - q(y_{t_1}, y_{t_2}, \cdots, y_{t_n})(t)\|^2 \leq \overline{M}_q \|x - y\|^2_{\mathcal{B}}.$$

(**H9**) The linear operator $W{:}L^2(J, U) \to L^2(\Omega, \mathbb{H})$ defined by

$$Wu = \int_J S(T-s)Bu(s)ds$$

has an induced inverse $W^{-1}$ which takes values in $L^2(J, U)/KerW$ (see Carimichel & Quinn, 1984) and there exist two positive constants $M_B$ and $M_W$ such that

$$\|B\|^2 \leq M_B \quad and \quad \|W^{-1}\|^2 \leq M_W.$$

(**H10**) The function $\gamma{:}J \times \mathbb{H} \times \mathcal{U} \to \mathbb{H}$ is a Borel measurable function and satisfies the Lipschitz continuity condition, the linear growth condition, and there exists positive constants $M_\gamma, \overline{M}_\gamma$ such that for any $x, y \in \mathcal{L}_2^{\mathbb{F}}(0, T; \mathbb{H}), t \in J$

$$\mathbf{E}\Big(\int_0^t \int_{\mathcal{U}} \|\gamma(t, x(s-), v)\|^2_{\mathbb{H}} \lambda(dv)ds\Big) \vee \mathbf{E}\Big(\int_0^t \int_{\mathcal{U}} \|\gamma(t, x(s-), v)\|^4_{\mathbb{H}} \lambda(dv)ds\Big)^{\frac{1}{2}}$$

$$\leq M_\gamma \mathbf{E} \int_0^t (1 + \|x(s)\|^2_{\mathbb{H}})ds;$$

$$\mathbf{E}\Big(\int_0^t \int_{\mathcal{U}} \|\gamma(t, x(s-), v) - \gamma(t, y(s-), v)\|^2_{\mathbb{H}} \lambda(dv)ds\Big)$$

$$\vee \mathbf{E}\Big(\int_0^t \int_{\mathcal{U}} \|\gamma(t, x(s-), v) - \gamma(t, y(s-), v)\|^4_{\mathbb{H}} \lambda(dv)ds\Big)^{\frac{1}{2}} \leq \overline{M}_\gamma \mathbf{E} \int_0^t \|x(s) - y(s)\|^2_{\mathbb{H}} ds.$$

## 3. Main results

In this section, we shall investigate the controllability of nonlocal second-order impulsive neutral stochastic functional integro-differential equations with infinite delay and Poisson jumps in Hilbert spaces.

The main result of this section is the following theorem.

THEOREM 3.1 Assume that the assumptions (**H1**) – –(**H10**) hold. If $\Xi < 1$ and $\Theta < 1$, then the system (1.1) is controllable on $J_T$, where

$$\Xi := 32\left(1 + 9T^2 M_B M_S M_W\right)\left\{2l_0^2 T^2 \left[M_C M_g (1 + 2M_{\sigma_1}) + M_S M_f (1 + 2M_{\sigma_2})\right] + TM_\gamma \widetilde{C}\right\},$$

$$\Theta := \left\{98 l_0^2 T^2 M_B M_C M_S M_W \overline{M}_q + 14 l_0^2 \left(1 + 7T^2 M_B M_S M_W\right)\left[T^2 M_C M_g \left(1 + M_{\sigma_1}\right)\right.\right.$$

$$\left.\left. + T^2 M_S M_f \left(1 + M_{\sigma_2}\right) + T^3 M_S \overline{M}_\sigma Tr(Q) + \frac{T\overline{M}_\gamma \widetilde{C}}{2l_0^2} + mM_C \sum_{k=1}^m \overline{M}_{I_k^1} + mM_S \sum_{k=1}^m \overline{M}_{I_k^2}\right]\right\}.$$

*Proof* Using the assumption (**H9**), for an arbitrary function $x(\cdot)$, we define the control process

$$
\begin{aligned}
u_x^T(t) =& W^{-1}\Big\{y_1 - C(T)[\varphi(0) + q(x_{t_1}, x_{t_2}, \cdots, x_{t_n})(0)] - S(T)[x_1 - g(0, x_0, 0)] \\
&- \int_0^T C(T-s)g(s, x_s, \int_0^s \sigma_1(s,\tau,x_\tau)d\tau)ds - \sum_{0<t_k<T} C(T-t_k)I_k^1(x_{t_k}) \\
&- \int_0^T S(T-s)f(s, x_s, \int_0^s \sigma_2(s,\tau,x_\tau)d\tau)ds - \sum_{0<t_k<T} S(T-t_k)I_k^2(x_{t_k}) \\
&- \int_0^T S(T-s)\Big[h(s) + \int_0^s \sigma(s,\tau,x_\tau)dw(\tau)\Big]ds \\
&- \int_0^T S(T-s)\int_{\mathcal{V}} \gamma\big(t, x(t-), v\big)\widetilde{N}(dt, dv)\Big\}(t).
\end{aligned} \tag{3.1}
$$

We transform (1.1) into a fixed point problem. Consider the operator $\Pi:\mathcal{B}_T \to \mathcal{B}_T$ defined by

$$
\begin{aligned}
\Pi x(t) =& \varphi(t) + q(x_{t_1}, x_{t_2}, \cdots, x_{t_n})(t), \quad t \in J_0; \\
\Pi x(t) =& C(t)[\varphi(0) + q(x_{t_1}, x_{t_2}, \cdots, x_{t_n})(0)] + S(t)[x_1 - g(0, x_0, 0)] \\
&+ \int_0^t C(t-s)g(s, x_s, \int_0^s \sigma_1(s,\tau,x_\tau)d\tau)ds \\
&+ \int_0^t S(t-s)f(s, x_s, \int_0^s \sigma_2(s,\tau,x_\tau)d\tau)ds + \int_0^t S(t-s)Bu_x^T(s)ds \\
&+ \int_0^t S(t-s)\Big[h(s) + \int_0^s \sigma(s,\tau,x_\tau)dw(\tau)\Big]ds \\
&+ \int_0^t S(t-s)\int_{\mathcal{V}} \gamma\big(t, x(t-), v\big)\widetilde{N}(dt, dv) \\
&+ \sum_{0<t_k<t} C(t-t_k)I_k^1(x_{t_k}) + \sum_{0<t_k<t} S(t-t_k)I_k^2(x_{t_k}), \text{ for a.e.} t \in J.
\end{aligned}
$$

In what follows, we shall show that using the control $u_x^T(\cdot)$, the operator $\Pi$ has a fixed point, which is then a mild solution for system (1.1).

Clearly, $\Pi x(T) = y_1$.

For $\varphi \in \mathcal{B}$, we defined $\widetilde{\varphi}$ by

$$
\widetilde{\varphi}(t) = \begin{cases} \varphi(t) + q(x_{t_1}, x_{t_2}, \cdots, x_{t_n})(t) & \text{if } t \in J_0, \\ C(t)\big[\varphi(0) + q(x_{t_1}, x_{t_2}, \cdots, x_{t_n})(0)\big] & \text{if } t \in J, \end{cases}
$$

then $\widetilde{\varphi} \in \mathcal{B}_T$.

Set $x(t) = z(t) + \widetilde{\varphi}(t)$, $t \in J_T$. It is easy to see that x satisfies (2.2) if and only if z satisfies $z_0 = 0$, $x'(0) = x_1 = z'(0) = z_1$ *and*

$$
\begin{aligned}
z(t) =& S(t)[z_1 - g(0, \widetilde{\varphi}_0, 0)] + \int_0^t C(t-s)g(s, z_s + \widetilde{\varphi}_s, \int_0^s \sigma_1(s,\tau,z_\tau + \widetilde{\varphi}_\tau)d\tau)ds \\
&+ \int_0^t S(t-s)f(s, z_s + \widetilde{\varphi}_s, \int_0^s \sigma_2(s,\tau,z_\tau + \widetilde{\varphi}_\tau)d\tau)ds + \int_0^t S(t-s)Bu_{z+\widetilde{\varphi}}^T(s)ds \\
&+ \int_0^t S(t-s)\Big[h(s) + \int_0^s \sigma(s,\tau,z_\tau + \widetilde{\varphi}_\tau)dw(\tau)\Big]ds \\
&+ \int_0^t S(t-s)\int_{\mathcal{V}} \gamma\big(t, z(t-) + \widetilde{\varphi}(t-), v\big)\widetilde{N}(dt, dv) \\
&+ \sum_{0<t_k<t} C(t-t_k)I_k^1(z_{t_k} + \widetilde{\varphi}_{t_k}) + \sum_{0<t_k<t} S(t-t_k)I_k^2(z_{t_k} + \widetilde{\varphi}_{t_k}), \quad t \in J,
\end{aligned}
$$

where $u^T_{z+\widetilde{\varphi}}(t)$ is obtained from (3.1) by replacing $x_t = z_t + \widetilde{\varphi}_t$.

*Let* $\mathcal{B}^0_T = \{y \in \mathcal{B}_T : y_0 = 0 \in \mathcal{B}\}$. *For any* $y \in \mathcal{B}^0_T$, *we have*

$$\|y\|_T = \|y_0\|_{\mathcal{B}} + \sup_{s\in J}(\mathbf{E}\|y(s)\|^2)^{\frac{1}{2}} = \sup_{s\in J}(\mathbf{E}\|y(s)\|^2)^{\frac{1}{2}},$$

and thus $(\mathcal{B}^0_T, \|\cdot\|_T)$ is a Banach space. Set

$$B_r = \{y \in \mathcal{B}^0_T : \|y\|^2_T \le r\} \quad \text{for some } r \ge 0,$$

then $B_r \subseteq \mathcal{B}^0_T$ is uniformly bounded, and for $u \in B_r$, by Lemma 2.1, we have

$$\begin{aligned}
\|z_t + \widetilde{\varphi}_t\|^2_{\mathcal{B}} &\le 2(\|z_t\|^2_{\mathcal{B}} + \|\widetilde{\varphi}_t\|^2_{\mathcal{B}}) \\
&\le 4\big(l_0^2 \sup_{s\in[0,t]}(\mathbf{E}\|z(s)\|^2 + \|z_0\|^2_{\mathcal{B}} + l_0^2 \sup_{s\in[0,t]}(\mathbf{E}\|\widetilde{\varphi}(s)\|^2 + \|\widetilde{\varphi}_0\|^2_{\mathcal{B}}\big) \\
&\le 4l_0^2\Big(r + 2M_C\big[\mathbf{E}\|\varphi(0)\|^2 + M_q\big]\Big) + 4\|\widetilde{\varphi}\|^2_{\mathcal{B}} \\
&:= r^\star.
\end{aligned} \tag{3.2}$$

Define the map $\overline{\Pi} : \mathcal{B}^0_T \to \mathcal{B}^0_T$ defined by $\overline{\Pi}z(t) = 0$, for $t \in J_0$ and

$$\begin{aligned}
\overline{\Pi}z(t) =& S(t)[z_1 - g(0, \widetilde{\varphi}_0, 0)] + \int_0^t C(t-s)g\Big(s, z_s + \widetilde{\varphi}_s, \int_0^s \sigma_1(s,\tau,z_\tau + \widetilde{\varphi}_\tau)d\tau\Big)ds \\
&+ \int_0^t S(t-s)f\Big(s, z_s + \widetilde{\varphi}_s, \int_0^s \sigma_2(s,\tau,z_\tau + \widetilde{\varphi}_\tau)d\tau\Big)ds + \int_0^t S(t-s)Bu^T_{z+\widetilde{\varphi}}(s)ds \\
&+ \int_0^t S(t-s)\Big[h(s) + \int_0^s \sigma(s,\tau,z_\tau + \widetilde{\varphi}_\tau)dw(\tau)\Big]ds \\
&+ \int_0^t S(t-s)\int_{\mathcal{V}} \gamma\big(t, z(t-) + \widetilde{\varphi}(t-), v\big)\widetilde{N}(dt, dv) \\
&+ \sum_{0<t_k<t} C(t-t_k)I^1_k(z_{t_k} + \widetilde{\varphi}_{t_k}) + \sum_{0<t_k<t} S(t-t_k)I^2_k(z_{t_k} + \widetilde{\varphi}_{t_k}), \quad t \in J.
\end{aligned}$$

Obviously, the operator $\Pi$ has a fixed point which is equivalent to prove that $\overline{\Pi}$ has a fixed point. Note that, by our assumptions, we infer that all the functions involved in the operator are continuous, therefore $\overline{\Pi}$ is continuous.

Let $z, \overline{z} \in \mathcal{B}^0_T$. From (3.1), by our assumptions, Hölder's inequality, the Doob martingale inequality, and the Burkholder–Davis–Gundy inequality for pure jump stochastic integral in Hilbert space (see Luo & Liu, 2008), Lemma 2.1, and in view of (3.2), for $t \in J$, we obtain the following estimates.

$$\begin{aligned}
&\mathbf{E}\|u^T_{z+\widetilde{\varphi}}(t)\|^2 \\
&\le 9M_W\Big\{\mathbf{E}\|y_1\|^2 + 2M_C(\mathbf{E}\|\varphi(0)\|^2 + M_q) + 2M_S\big[\mathbf{E}\|x_1\|^2 + 2(M_g\|\widetilde{\varphi}\|^2_{\mathcal{B}} + C_2)\big] \\
&\quad + 2T^2M_C\Big[M_g\big([1 + 2M_{\sigma_1}]r^\star + 2C_1\big) + C_2\Big] + 2T^2M_S\Big[M_f\big([1 + 2M_{\sigma_2}]r^\star + 2C_3\big) + C_4\Big] \\
&\quad + 2T^2M_S(M_h + TTr(Q)M_\sigma) + TM_\gamma\widetilde{C}(1 + \frac{r^\star}{l_0^2}) + mM_C\sum_{k=1}^m M_{I^1_k} + mM_S\sum_{k=1}^m M_{I^2_k}\Big\} := \ell,
\end{aligned}$$

and

$$
\begin{aligned}
&\mathbf{E}\|u^T_{z+\widetilde{\varphi}}(t) - u^T_{\bar{z}+\widetilde{\varphi}}(t)\|^2 \\
&\le 14 l_0^2 M_W \Big\{ M_C \overline{M}_q + T^2 M_C M_g(1+M_{\sigma_1}) + T^2 M_S M_f(1+M_{\sigma_2}) \\
&\quad + T^3 M_S \overline{M}_\sigma Tr(Q) + \frac{T\overline{M}_\gamma \widetilde{C}}{2l_0^2} + mM_C \sum_{k=1}^m \overline{M}_{I_k^1} + mM_S \sum_{k=1}^m \overline{M}_{I_k^2} \Big\} \sup_{s\in J} \mathbf{E}\|z(t) - \bar{z}(t)\|^2,
\end{aligned}
$$

where $\widetilde{C} > 0$ is a positive constant and

$$
\begin{aligned}
C_1 &:= T \sup_{(t,s)\in J\times J} \sigma_1^2(t,s,0), \quad C_2 := \sup_{t\in J} \|g(t,0,0)\|^2, \\
C_3 &:= T \sup_{(t,s)\in J\times J} \sigma_2^2(t,s,0), \quad C_4 := \sup_{t\in J} \|f(t,0,0)\|^2.
\end{aligned}
$$

LEMMA 3.1 *Under the assumptions of Theorem 3.1, there exists $r > 0$ such that $\overline{\Pi}(B_r) \subseteq B_r$.*

*Proof* If this property is false, then for each $r > 0$, there exists a function $z^r(\cdot) \in B_r$, but $\overline{\Pi}(z^r) \notin B_r$, i.e. $\|\overline{\Pi}(z^r)(t)\|^2 > r$ for some $t \in J$. However, by our assumptions, Hölder's inequality and the Burkholder–Davis–Gundy inequality, we have

$$
\begin{aligned}
r &< \mathbf{E}\|\overline{\Pi}(z^r)(t)\|^2 \\
&\le 8\Big[2M_S[\mathbf{E}\|x_1\|^2 + 2(M_g\|\widetilde{\varphi}\|_{\mathcal{B}}^2 + C_2)] + 2T^2 M_C\Big[M_g([1+2M_{\sigma_1}]r^\star + 2C_1) + C_2\Big] \\
&\quad + 2T^2 M_S\Big[M_f([1+2M_{\sigma_2}]r^\star + 2C_3) + C_4\Big] + T^2 M_S M_B \ell \\
&\quad + 2T^2 M_S(M_h + T Tr(Q) M_\sigma) + TM_\gamma \widetilde{C}(1 + \frac{r^\star}{l_0^2}) + mM_C \sum_{k=1}^m M_{I_k^1} + mM_S \sum_{k=1}^m M_{I_k^2}\Big], \\
&\le M^{\star\star} + 8(1 + 9T^2 M_B M_S M_W)\Big[2T^2\Big(M_C M_g(1+2M_{\sigma_1}) + M_S M_f(1+2M_{\sigma_2})\Big) + \frac{TM_\gamma \widetilde{C}}{l_0^2}\Big] r^\star,
\end{aligned}
\tag{3.3}
$$

where

$$
\begin{aligned}
&M^{\star\star} \\
&:= 72(1 + 9T^2 M_B M_S M_W)\Big[\mathbf{E}\|y_1\|^2 + 2M_C(\mathbf{E}\|\varphi(0)\|^2 + M_q)\Big] + 8(1 + 9T^2 M_B M_S M_W) \\
&\quad \times \Big[2M_S[\mathbf{E}\|x_1\|^2 + 2(M_g\|\widetilde{\varphi}\|_{\mathcal{B}}^2 + C_2)] + 2T^2 M_C(2M_g C_1 + C_2) + 2T^2 M_S(2M_f C_3 + C_4) \\
&\quad + 2T^2 M_S(M_h + T Tr(Q) M_\sigma) + TM_\gamma \widetilde{C} + mM_C \sum_{k=1}^m M_{I_k^1} + mM_S \sum_{k=1}^m M_{I_k^2}\Big].
\end{aligned}
$$

Dividing both sides of (3.3) by $r$ and noting that

$$
r^\star = 4l_0^2\Big(r + 2M_C[\mathbf{E}\|\varphi(0)\|^2 + M_q]\Big) + 4\|\widetilde{\varphi}\|_{\mathcal{B}}^2 \xrightarrow{r\to\infty} \infty
$$

and taking the limit as $r \to \infty$, we obtain

$$
1 \le \Xi
$$

which contradicts our assumption. Thus, for some positive number $r$, $\overline{\Pi}(B_r) \subseteq B_r$. This completes the proof of Lemma 3.1.

LEMMA 3.2 *Under the assumptions of Theorem 3.1, $\overline{\Pi}: \mathcal{B}_T^0 \to \mathcal{B}_T^0$ is a contraction mapping.*

*Proof* Let $z, \bar{z} \in \mathcal{B}_T^0$. Then, by our assumptions, Hölder's inequality, Burkholder–Davis–Gundy's inequality, Lemma 2.1, and since $\|z_0\|_{\mathcal{B}}^2 = 0$ and $\|\bar{z}_0\|_{\mathcal{B}}^2 = 0$, for each $t \in J$, we see that

$$\begin{aligned}
&\mathbf{E}\|(\overline{\Pi}z)(t) - (\overline{\Pi}\bar{z})(t)\|^2 \\
&\le 14l_0^2\Big\{T^2M_CM_g(1+M_{\sigma_1}) + T^2M_SM_f(1+M_{\sigma_2}) + T^3M_S\overline{M}_\sigma Tr(Q) + \frac{T\overline{M}_\gamma\widetilde{C}}{2l_0^2} \\
&\quad + mM_C\sum_{k=1}^m \overline{M}_{I_k^1} + mM_S\sum_{k=1}^m \overline{M}_{I_k^2}\Big\}\sup_{s\in J}\mathbf{E}\|z(t)-\bar{z}(t)\|^2 + 7T^2M_SM_B\mathbf{E}\|u_{z+\widetilde{\varphi}}^T(t) - u_{\bar{z}+\widetilde{\varphi}}^T(t)\|^2 \\
&\le \Big\{98l_0^2T^2M_BM_CM_SM_W\overline{M}_q + 14l_0^2(1+7T^2M_BM_SM_W) \\
&\quad \times\Big[T^2M_CM_g(1+M_{\sigma_1}) + T^2M_SM_f(1+M_{\sigma_2}) + T^3M_S\overline{M}_\sigma Tr(Q) + \frac{T\overline{M}_\gamma\widetilde{C}}{2l_0^2} \\
&\quad + mM_C\sum_{k=1}^m \overline{M}_{I_k^1} + mM_S\sum_{k=1}^m \overline{M}_{I_k^2}\Big]\Big\}\sup_{s\in J}\mathbf{E}\|z(t)-\bar{z}(t)\|^2.
\end{aligned}$$

Taking the supremum over t, we obtain

$$\|(\overline{\Pi}z) - (\overline{\Pi}\bar{z})\|_T^2 \le \Theta\|z-\bar{z}\|_T^2.$$

By our assumption, we conclude that $\overline{\Pi}$ is a contraction on $\mathcal{B}_T^0$. Thus, we have completed the proof of Lemma 3.2.

On the other hand, by Banach fixed point theorem, there exists a unique fixed point $x(\cdot) \in \mathcal{B}_T^0$ such that $(\Pi x)(t) = x(t)$. This fixed point is then the mild solution of the system (1.1). Clearly, $x(T) = (\Pi x)(T) = y_1$. Thus, the system (1.1) is controllable on $J_T$. The proof for Theorem 3.1 is thus complete.

Now, let us consider a special case for the system (1.1).

If $\gamma(t, x(t-), v) \equiv 0$, the system (1.1) becomes the following nonlocal second-order impulsive neutral stochastic functional integro-differential equations with infinite delay without Poisson jumps:

$$\begin{cases}
d\big[x'(t) - g(t, x_t, \int_0^t \sigma_1(t,s,x_s)ds)\big] = \big[Ax(t) + f(t, x_t, \int_0^t \sigma_2(t,s,x_s)ds) + Bu(t)\big]dt \\
\qquad + \int_{-\infty}^t \sigma(t,s,x_s)dw(s), \quad t_k \ne t \in J := [0,T], \\
\Delta x(t_k) = I_k^1(x_{t_k}), \quad k = \{1,\cdots,m\} =: \overline{1,m}, \\
\Delta x'(t_k) = I_k^2(x_{t_k}), \quad k = \overline{1,m}, \\
x'(0) = x_1 \in \mathbb{H}, \\
x(0) - q(x_{t_1}, x_{t_2}, \cdots, x_{t_n}) = x_0 = \varphi \in \mathcal{B}, \quad \text{for a.e. } s \in J_0 := (-\infty, 0],
\end{cases} \tag{3.4}$$

COROLLARY 3.1 *Assume that all assumptions of Theorem 3.1 hold except that* (**H11**) *and* $\Xi, \Theta$ *replaced by* $\widehat{\Xi}, \widehat{\Theta}$ *such that*

$$\widehat{\Xi} := 56l_0^2T^2(1+8T^2M_BM_SM_W)\big[M_CM_g(1+2M_{\sigma_1}) + M_SM_f(1+2M_{\sigma_2})\big],$$

*and*

$$\begin{aligned}
\widehat{\Theta} := &\Big\{72l_0^2T^2M_BM_CM_SM_W\overline{M}_q + 12l_0^2(1+6T^2M_BM_SM_W)\Big[T^2M_CM_g(1+M_{\sigma_1}) \\
&+ T^2M_SM_f(1+M_{\sigma_2}) + T^3M_S\overline{M}_\sigma Tr(Q) + mM_C\sum_{k=1}^m \overline{M}_{I_k^1} + mM_S\sum_{k=1}^m \overline{M}_{I_k^2}\Big]\Big\}.
\end{aligned}$$

*If* $\widehat{\Xi} < 1$ *and* $\widehat{\Theta} < 1$, *then the system (3.4) is controllable on* $J_T$.

## 4. Application

In this section, the established previous results are applied to study the controllability of the stochastic nonlinear wave equation with infinite delay and Poisson jumps. Specifically, we consider the following controllability of nonlocal second-order impulsive neutral stochastic functional integro-differential equations with infinite delay and Poisson jumps of the form:

$$\begin{cases} \frac{\partial}{\partial t}\left[\frac{\partial}{\partial t}y(t,\xi) - \int_{-\infty}^{t}\delta_1(t,\xi,s-t,)P_1(y(s,\xi))ds - \int_0^t\int_{-\infty}^{s}b_1(s-\tau)P_2(y(\tau,\xi))d\tau ds\right] \\ = \left[\frac{\partial^2}{\partial \xi^2}y(t,\xi) + \int_{-\infty}^{t}\delta_2(t,\xi,s-t,)G_1(y(s,\xi))ds + \int_0^t\int_{-\infty}^{s}b_2(s-\tau)G_2(y(\tau,\xi))d\tau ds \right. \\ \left. +b(\xi)u(t)\right]dt + \int_{-\infty}^{t}\delta(s-t)y(t,\xi)d\beta(s) + \int_{\mathcal{V}}y(t-,\xi)v\widetilde{N}(dt,dv), t_k \neq t \in J, \xi \in [0,\pi], \\ \Delta y(t_k)(\xi) = \int_{-\infty}^{t_k}\eta_k(t_k-s)y(s,\xi)ds, \quad k=\overline{1,m}, \quad \xi \in [0,\pi], \\ \Delta y'(t_k)(\xi) = \int_{-\infty}^{t_k}\rho_k(t_k-s)y(s,\xi)ds, \quad k=\overline{1,m}, \quad \xi \in [0,\pi], \\ y(t,0) = y(t,\pi) = 0, \quad t \in J, \\ \frac{\partial}{\partial t}y(0,\xi) = x_1(\xi), \quad \xi \in [0,\pi], \\ y(t,\xi) - \sum_{i=1}^{n}\int_0^{\pi}p_i(\xi,\zeta)y(t_i,\zeta)d\zeta = \varphi(t,\xi), \quad t \in J_0, \quad \xi \in [0,\pi], \end{cases} \tag{4.1}$$

where $\beta(t)$ is a standard one-dimensional Wiener process in $\mathbb{H}$, defined on a stochastic basis $(\Omega, \mathcal{F}, \mathbf{P})$; $\mathcal{V} = \{v \in \mathbb{R}: 0 < \|v\|_{\mathbb{R}} \leq a, a > 0\}$; $0 < t_1 < t_2 < \cdots < t_n < T$, $n \in \mathbb{N}$; $0 = t_0 < t_1 < \cdots < t_m < t_{m+1} < T$ are prefixed numbers, and $\varphi \in \mathcal{B}$.

Let $p = p(t), t \in D_p$ be a $\mathbb{K}$-valued $\sigma$-finite stationary Poisson point process (independent of $\beta(t)$) on a complete probability space with the usual condition $(\Omega, \mathcal{F}, (\mathcal{F}_t)_{t\geqslant 0}, \mathbf{P})$. Let $\widetilde{N}(ds,dv) := N(ds,dv) - \lambda(dv)ds$, with the characteristic measure $\lambda(dv)$ on $\mathcal{V} \in \mathfrak{B}(\mathbb{K} - \{0\})$. Assume that

$$\int_{\mathcal{V}} v^2\lambda(dv) < \infty \quad and \quad \int_{\mathcal{V}} v^4\lambda(dv) < \infty.$$

To rewrite (4.1) into the abstract from of (1.1), we consider the space $\mathbb{H} = L^2([0,\pi])$ with the norm $\|\cdot\|$. Let $e_n(\xi) := \sqrt{\frac{2}{\pi}}\sin n\xi$, $n = 1,2,3,\ldots$ denote the completed orthogonal basics in $\mathbb{H}$ and $\beta(t) = \sum_{n=1}^{\infty}\sqrt{\lambda_n}\beta_n(t)e_n, t \geq 0, \lambda_n > 0$, where $\{\beta_n(t)\}_{n\geq 0}$ are one-dimensional standard Brownian motions mutually independent on a usual complete probability space $(\Omega, \mathcal{F}, (\mathcal{F}_t)_{t\geqslant 0}, \mathbf{P})$.

Defined $A: \mathbb{H} \to \mathbb{H}$ by $A = \frac{\partial^2}{\partial \xi^2}$, with domain $D(A) = \mathbb{H}^2([0,\pi]) \cap \mathbb{H}_0^1([0,\pi])$, where

$$\mathbb{H}_0^1([0,\pi]) = \{w \in L^2([0,\pi]): \frac{\partial w}{\partial z} \in L^2([0,\pi]), w(0) = w(\pi) = 0\}$$

and

$$\mathbb{H}^2([0,\pi]) = \{w \in L^2([0,\pi]): \frac{\partial w}{\partial z}, \frac{\partial^2 w}{\partial z^2} \in L^2([0,\pi])\}.$$

Then,

$$Ax = -\sum_{n=1}^{\infty} n^2\langle x, e_n\rangle e_n, \quad x \in D(A), \tag{4.2}$$

(see Travis & Webb, 1987, Example 5.1). Using (4.2), one can easily verify that the operators $C(t)$ defined by

$$C(t)x = \sum_{n=1}^{\infty} cos(nt)\langle x, e_n\rangle e_n, \quad t \in \mathbb{R},$$

from a cosine function on $\mathbb{H}$, with associated sine function

$$S(t)x = \sum_{n=1}^{\infty} \frac{sin(nt)}{n}\langle x, e_n\rangle e_n, \quad t \in \mathbb{R}.$$

It is clear that (see Travis & Webb, 1977), for all $x \in \mathbb{H}, t \in \mathbb{R}, C(\cdot)x$ and $S(\cdot)x$ are periodic functions with $\|C(t)\| \leq 1$ and $\|S(t)\| \leq 1$. Thus, (**H1**) is true.

Now, we give a special $\mathcal{B}$-space. Let $l(s) = e^{2s}, s \leq 0$, then $l_0 = \int_{J_0} l(s)ds = \frac{1}{2}$ and define

$$\|\psi\|_{\mathcal{B}} = \int_{J_0} l(s) \sup_{\theta\in[s,0]} (\mathbf{E}\|\psi(\theta)\|^2)^{\frac{1}{2}} ds, \quad \forall \psi \in \mathcal{B}.$$

It follows from Hino et al. (1991) that $(\mathcal{B}, \|\cdot\|_{\mathcal{B}})$ is a Banach space. Hence, for $(t, \psi) \in J \times \mathcal{B}$, where $\psi(\theta)x = \psi(\theta, x), (\theta, x) \in J_0 \times [0, \pi]$. Let $y(t)(\xi) = y(t, \xi)$.

To study the system (4.1), we assume that the following conditions hold:

(i) Let $B \in \mathcal{L}(\mathbb{R}, \mathbb{H})$ be defined as

$$Bu(\xi) = b(\xi)u, \quad 0 \leq \xi \leq \pi, \quad u \in \mathbb{R}, \quad b(\xi) \in L^2([0, \pi]).$$

(ii) The linear operator $W: L^2(J, U) \to \mathbb{H}$ defined by

$$Wu = \int_J S(T-s)b(\xi)u(s)ds$$

is a bounded linear operator but not necessarily one-to-one. Let $KerW = \{u \in L^2(J, U): Wu = 0\}$ be null space of $W$ and $[KerW]^{\perp}$ be its orthogonal complement in $L^2(J, U)$. Let $W^*: [KerW]^{\perp} \to Range(W)$ be the restriction of $W$ to $[KerW]^{\perp}$, $W^*$ is necessarily one-to-one operator. The inverse mapping theorem says that $(W^*)^{-1}$ is bounded since $[KerW]^{\perp}$ and $Range(W)$ are Banach spaces. Since the inverse operator $W^{-1}$ is bounded and takes values in $L^2(J, U)/KerW$, the assumption (**H9**) is satisfied.

(iii) The functions $p_i: [0, \pi] \times 0, \pi] \to \mathbb{R}$ are $C^2$-functions, for each $i = \overline{1, n}$.

(iv) The functions $\eta_k, \rho_k \in C(\mathbb{R}, \mathbb{R})$ such that for $k = \overline{1, m}$,

$$\overline{M}_{I_k^1} = \int_{J_0} l(s)\eta_k^2(s)ds < \infty, \quad \overline{M}_{I_k^2} = \int_{J_0} l(s)\rho_k^2(s)ds < \infty.$$

We define the functions $g, f: J \times \mathcal{B} \times \mathbb{H} \to \mathbb{H}, \sigma: J \times J \times \mathcal{B} \to \mathcal{L}_2^0, \gamma: J \times \mathbb{H} \times \mathcal{U} \to \mathbb{H}$, and $I_k^1, I_k^2: \mathcal{B} \to \mathbb{H}$, $k = \overline{1, m}$ by

$$g(t, \psi, V_1\psi)(\xi) = \int_{J_0} \delta_1(t, \xi, \theta)P_1(\psi(\theta)(\xi))d\theta + V_1\psi(\xi),$$

$$f(t, \psi, V_2\psi)(\xi) = \int_{J_0} \delta_2(t, \xi, \theta)G_1(\psi(\theta)(\xi))d\theta + V_2\psi(\xi),$$

$$\sigma(t, s, \psi)(\xi) = \int_{J_0} \delta(\theta)\psi(\theta)(\xi)d\theta, \quad \gamma(t, \psi(\xi), v) = \psi(\xi)v,$$

$$I_k^1(t, \psi)(\xi) = \int_{J_0} \eta_k(-s)\psi(\theta)(\xi)ds, \quad k = \overline{1, m},$$

$$I_k^2(t, \psi)(\xi) = \int_{J_0} \rho_k(-s)\psi(\theta)(\xi)ds, \quad k = \overline{1, m},$$

where

$$V_1\psi(\xi) = \int_0^t \int_{J_0} b_1(s-\theta)P_2(\psi(\theta)(\xi))d\theta ds, \quad V_2\psi(\xi) = \int_0^t \int_{J_0} b_2(s-\theta)G_2(\psi(\theta)(\xi))d\theta ds.$$

Then, the system (4.1) can be written in the abstract form as the system (1.1). Further, we can impose some suitable conditions on the above-defined functions as those in the assumptions (**H1**) – –(**H10**). Therefore, by Theorem 3.1, we can conclude that the system (4.1) is controllable on $J_T$.

## 5. Conclusion

In this paper, we have studied the controllability for a class of nonlocal second-order impulsive neutral stochastic functional integro-differential equations with infinite delay and Poisson jumps in Hilbert spaces, which is new and allows us to develop the controllability of the second-order stochastic partial differential equations. Using the Banach fixed point theorem combined with theories of a strongly continuous cosine family of bounded linear operators, and stochastic analysis theory, the controllability of nonlocal second-order impulsive neutral stochastic functional integro-differential equations with infinite delay and Poisson jumps is obtained. In addition, an application is provided to illustrate the effectiveness of the controllability results obtained. The results in our paper extend and improve the corresponding ones announced by Arthi and Balachandran (2012), Balasubramaniam and Muthukumar (2009), Muthukumar and Rajivganthi (2013), Park, Balachandran, and Annapoorani (2009), Park, Balachandran, and Arthi (2009), Travis and Webb (1978), and some other results.

**Funding**

This work was supported by the China NSF [grant number 11171158].

**Author details**

Diem Dang Huan[1,2]

E-mail: huandd@bafu.edu.vn

Hongjun Gao[3]

E-mail: gaohj@njnu.edu.cn

[1] Faculty of Basic Sciences, Bacgiang Agriculture and Forestry University, Bacgiang, 21000 Vietnam.

[2] Vietnam National University, Hanoi, 144 Xuan Thuy Street, Cau Giay, Hanoi, 10000 Vietnam.

[3] School of Mathematical Science, Nanjing Normal University, Nanjing, 210023 P.R. China.

**References**

Ahmed, H. M. (2014a). Non-linear fractional integro-differential systems with non-local conditions. *IMA Journal of Mathematical Control*. doi:10.1093/imamci/dnu049

Ahmed, H. M. (2014b). Controllability of impulsive neutral stochastic differential equations with fractional Brownian motion. *IMA Journal of Mathematical Control*. doi:10.1093/imamci/dnu019

Ahmed, H. M. (2014c). Approximate controllability of impulsive neutral stochastic differential equations with fractional Brownian motion in a Hilbert space. *Advances in Difference Equations, 113*, 1–11.

Arthi, G., & Balachandran, K. (2012). Controllability of damped second-order impulsive neutral functional differential systems with infinite delay. *Journal of Optimization Theory and Applications, 152*, 799–813.

Astrom, K. J. (1970). *Introduction to stochastic control theory*. New York, NY: Academic Press.

Balachandran, K., & Dauer, J. P. (2002). Controllability of nonlinear stochastic systems in Banach spaces. *Journal of Optimization Theory and Applications, 115*, 7–28.

Balasubramaniam, P., & Dauer, J. P. (2002). Controllability of semilinear stochastic delay evolution equations in Hilbert spaces. *International Journal of Mathematics and Mathematical Sciences, 31*, 157–166.

Balasubramaniam, P., & Muthukumar, P. (2009). Approximate controllability of second-order stochastic distributed implicit functional differential systems with infinite delay. *Journal of Optimization Theory and Applications, 143*, 225–244.

Balasubramaniam, P., & Ntouyas, S. K. (2006). Controllability for neutral stochastic functional differential inclusions with infinite delay in abstract space. *Journal of Mathematical Analysis, 324*, 161–176.

Cao, G. (2005). Stochastic differential evolution equations in infinite dimensional (MS thesis), Huazhong University of Science and Technology, Wuhan.

Carimichel, N., & Quinn, M. D. (1984). *Fixed point methods in nonlinear control* (Lecture Notes in Control and Information Society, Vol. 75). Berlin: Springer.

Chang, Y. K. (2007). Controllability of impulsive functional differential systems with infinite delay in Banach spaces. *Chaos, Solitons & Fractals, 33*, 1601–1609.

Chang, Y. K., Anguraj, A., & Arjunan, M. M. (2008). Existence results for impulsive neutral functional differential equations with infinite delay. *Nonlinear Analysis: Hybrid Systems, 2*, 209–218.

Coron, J. M. (2007). *Control and nonlinearity, mathematical surveys and monographs* (Vol. 136). Providence, RI: American Mathematical Society.

Curtain, R. F., & Zwart, H. J. (1995). *An introduction to infinite dimensional linear systems theory*. Berlin: Springer.

Da Prato, G., & Zabczyk, J. (1992). *Stochastic equations in infinite dimensions* (Vol. 44). Cambridge: Cambridge University Press.

Fattorini, H. O. (1985). *Second order linear differential equations in Banach spaces* (North-Holland mathematics studies, Vol. 108). Amsterdam: North-Holland.

Gurtin, M. E., & Pipkin, A. C. (1968). A general theory of heat conduction with finite wave speed. *Archive for Rational Mechanics and Analysis, 31*, 113–126.

Hino, Y., Murakami, S., & Naito, T. (1991). *Functional differential equations with infinite delay* (Lecture notes in mathematics, Vol. 1473). Berlin: Springer-Verlag.

Karthikeyan, S., & Balachandran, K. (2009). Controllability of nonlinear stochastic neutral impulsive systems. *Nonlinear Analysis, 3*, 266–276.

Karthikeyan, S., & Balachandran, K. (2013). On controllability for a class of stochastic impulsive systems with delays in control. *International Journal of Systems Science, 44*, 67–76.

Lakshmikantham, V., Baĭnov, D., & Simeonov, P. S. (1989). *Theory of impulsive differential equations* (Series in modern applied mathematics, Vol. 6). Teaneck, NJ: World Scientific Publishing.

Luo, J., & Liu, K. (2008). Stability of infinite dimensional stochastic evolution with memory and Markovian jumps. *Stochastic Processes and their Applications, 118*, 864–895.

Mahmudova, N. I., & McKibben, M. A. (2006). Abstract second-order damped McKean--Vlasov stochastic evolution equations. *Stochastic Processes and their Applications, 24*, 303–328.

Marinelli, C., & Röckner, M. (2010). Well-posedness and asymptotic behavior for stochastic reaction--diffusion equations with multiplicative Poisson noise. *Electronic Journal of Probability, 15*, 1528–1555.

Muthukumar, P., & Balasubramaniam, P. (2010). Approximate controllability of second-order damped McKean--Vlasov stochastic evolution equations. *Computers & Mathematics with Applications, 60*, 2788–2796.

Muthukumar, P., & Rajivganthi, C. (2013). Approximate controllability of fractional order neutral stochastic integro-differential systems with nonlocal conditions and infinite delay. *Taiwanese Journal of Mathematics, 17*, 1693–1713.

Ntouyas, S. K. (2009). D. ÓRegan; Some remarks on controllability of evolution equations in Banach spaces, Electronic Journal of Qualitative Theory of. *Differential Equations, 29*, 1–6.

Nunziato, J. W. (1971). On heat conduction in materials with memory. *Quarterly of Applied Mathematics, 29*, 187–204.

Park, J. Y., Balachandran, K., & Annapoorani, N. (2009). Existence results for impulsive neutral functional integrodifferential equations with infinite delay. *Nonlinear Analysis, 71*, 3152–3162.

Park, J. Y., Balachandran, K., & Arthi, G. (2009). Controllability of impulsive neutral integrodifferential systems with infinite delay in Banach spaces. *Nonlinear Analysis: Hybrid Systems, 3*, 184–194.

Park, J. Y., Balasubramaniam, P., & Kumaresan, N. (2007). Controllability for neutral stochastic functional integrodifferential infinite delay systems in abstract space. *Numerical Functional Analysis and Optimization, 28*, 1369–1386.

Peszat, S., & Zabczyk, J. (2007.). *Stochastic partial differential equations with lévy noise*. Cambridge: Cambridge University Press.

Protter, P. E. (2004). *Stochastic integration and differential equations* (2nd ed.). New York, NY: Springer.

Ren, Y., Dai, H. L., & Sakthivel, R. (2013). Approximate controllability of stochastic differential systems driven by a lévy process. *International Journal of Control, 86*, 1158–1164.

Ren, Y., Zhou, Q., & Chen, L. (2011). Existence, uniqueness and stability of mild solutions for time-dependent stochastic evolution equations with Poisson jumps and infinite delay. *Journal of Optimization Theory and Applications, 149*, 315–331.

Röckner, M., & Zhang, T. (2007). Stochastic evolution equation of jump type: Existence, uniqueness and large deviation principles. *Potential Analysis, 26*, 255–279.

Sakthivel, R., Mahmudov, N. I., & Lee, S.-G. (2009). Controllability of non-linear impulsive stochastic systems. *International Journal of Control, 82*, 801–807.

Sakthivel, R., & Ren, Y. (2011). Complete controllability of stochastic evolution equations with jumps. *Reports on Mathematical Physics, 68*, 163–174.

Sakthivel, R., Ren, Y., & Mahmudov, N. I. (2010). Approximate controllability of second-order stochastic differential equations with impulsive effects. *Modern Physics Letters B, 24*, 1559–1572.

Shen, L. J., & Sun, J. T. (2012). Approximate controllability of stochastic impulsive functional systems with infinite delay. *Automatica, 48*, 2705–2709.

Travis, C. C., & Webb, G. F. (1977). Compactness, regularity, and uniform continuity properties of strongly continuous cosine families. *Houston Journal of Mathematics, 3*, 555–567.

Travis, C. C., & Webb, G. F. (1978). Cosine families and abstract nonlinear second order differential equations. *Acta Mathematica Academiae Scientiarum Hungaricae, 32*, 76–96.

Travis, C. C., & Webb, G. F. (1987). Second oder differential equations in Banach spaces. In *Proceedinds International Symposium on Nonlinear Equations in Abstract Spaces* (pp. 331–361). New York, NY: Academic Press.

Yan, Z. M., & Yan, X. X. (2013). Existence of solutions for impulsive partial stochastic neutral integrodifferential equations with state-dependent delay. *Collectanea Mathematica, 64*, 235–250.

Yang, T. (2001). *Impulsive systems and control: Theory and applications*. Berlin: Springer.

Zabczyk, J. (1991). Controllability of stochastic linear systems. *Systems & Control Letters, 1*, 25–31.

Zabczyk, J. (1992). *Mathematical control theory*. Basel: Birkhauser.

# 5

# Position control under simultaneous limited torque and speed of a torque-driven nonlinear rotational mechanism

Adriana Salinas[1*], Rafael Kelly[1] and Javier Moreno-Valenzuela[2]

*Corresponding author: Adriana Salinas, Laboratorio de Robótica, Departamento de Electrónica y Telecomunicaciones, CICESE, Carretera Ensenada-Tijuana No. 3918, Zona Playitas, Ensenada 22860, Baja California, Mexico

E-mail: asalinas@cicese.edu.mx

Reviewing editor: James Lam, University of Hong Kong, Hong Kong

**Abstract:** This paper deals with shaft displacement regulation (position control) of a torque-driven nonlinear gravity unbalanced mechanism under both prescribed bounded torque and speed. This is a novel control objective formulation for mechanisms. A nonlinear dynamic controller to resolve this control formulation is proposed. This controller aims to take care of the actuator/plant by keeping them within a safe operating torque–speed zone. An experimental study complements the proposed theory. Potential applications of the proposed approach are in safe control of robots.

**Subjects: Automation Control; Control Engineering; Dynamical Control Systems; Mechatronics; Robotics & Cybernetics**

**Keywords: control; actuators; mechanism; bound**

## 1. Introduction

From an automatic control point of view, it is recognized that most physical actuators (electric motors included) have limited capabilities (see e.g. Åström & Murray, 2008; Dorf & Bishop, 2008; Tarbouriech, Garcia, Gomes da Silva Jr., & Queinnec, 2011). In order to ensure proper safe operation of actuators like electric motors, they must operate inside prescribed "safe operating zones"—delimited by their torque-speed curve—where motors may be constrained to "medium/high" torque but "low" speed (Hughes & Drury, 2013). When this kind of actuators is applied to handle mechanisms such as robots (see Figure 1), the control system should take care of retaining the actuators within their safe prescribed torque and speed limits.

## ABOUT THE AUTHORS

Adriana Salinas is currently a doctoral student at the Centro de Investigación Científica y de Educación Superior de Ensenada (CICESE), México. Her doctoral subject is focused on regulation of torque-driven mechanical systems equipped with torque-constrained actuators.

Rafael Kelly is currently a professor at the Centro de Investigación Científica y de Educación Superior de Ensenada (CICESE), México. His research interests include adaptive control systems, robot control, vision systems, multi agents systems, and neural networks.

Javier Moreno-Valenzuela is currently a professor at the Instituto Politécnico Nacional-CITEDI, México. His main research interests are control of electromechanical systems, robotics, and intelligent systems.

## PUBLIC INTEREST STATEMENT

This paper proposes a solution to position control of a gravity unbalanced shaft driven by a torque actuator by respecting prescribed safe desired torque and speed bounds. This is a first step toward the more challenging self hardware-protecting safe control of robot manipulators.

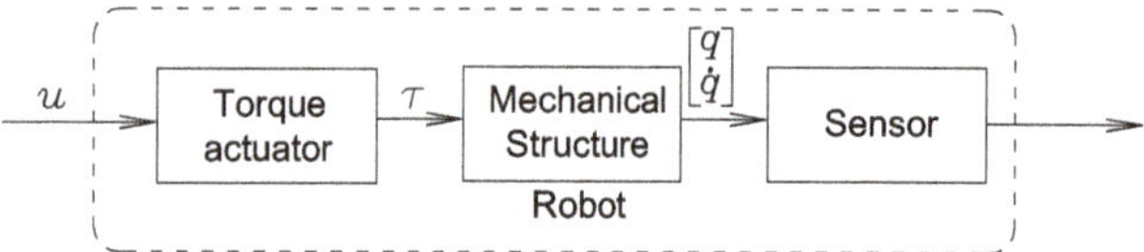


**Figure 1. Actuator applied to handle mechanism.**

The position control of robot manipulators (also called regulation or point-to-point robot control) can be defined as *"commanding the robot to go from an initial configuration to a desired final configuration without regard to the intermediate path to be followed by the end effector"* (Kelly, Santibánez, & Berghuis, 1997) and it has been the subject of several researches.

Regarding the position control of robot manipulators, studies which consider some type of restrictions in the actuators have been reported, such as Kelly et al. (1997) where torque limits are considered (without taking into account velocity constraints) and Ngo and Mahony (2006) where joint velocity constraints and bounded torque were taken into account. In the latter paper, torque is bounded in norm but no specific values of torque limits were stated in the control objective. The position control for other classes of robots such as nonholonomic mobile robots subject to input limits has been addressed (see for example Chen, 2014). Other studies where the system to be controlled presents input constraints have been reported in the literature such as Su, Chen, Wang, and Lam (2014), where the problem of consensus for a multiagent system subject to input saturations was addressed.

As a first approach toward the safe control of robot manipulators required that the torque actuators remain operating within their prescribed torque and speed limits, this paper considers the position control of a torque-driven nonlinear rotational mechanism.

The torque-driven nonlinear rotational mechanism is a rotational mechanical system whose input is provided by a torque actuator. The rotational mechanical system is modeled by a nonlinear differential equation (a physical example of such system is a pendulum, while a brushed DC motor is assumed to be a torque actuator (see Figure 2)).

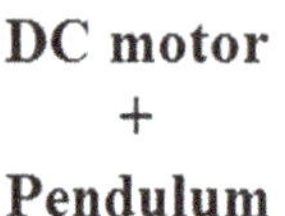

**Figure 2. Laboratory system: torque-driven gravity unbalanced device located at Instituto Politécnico Nacional-CITEDI.**

The control of linear and nonlinear servomechanisms is an active research topic. For example, in Kelly (1987) an adaptive control for DC motors is proposed for position control, in Kelly and Moreno-Valenzuela (2001) a study on shaft position regulation for DC motors was made using Proportional-Integral-Derivative controllers. Adaptive controller for the speed control of a servomechanism is proposed by Chen, and Cheng (2012). In those papers, neither input torque nor speed limits were considered.

For recent approaches to servomechanisms control under constrained input, see for example Aghili (2013), where a study on torque control taking into account voltage and current limits is presented; Guzmán-Guemez and Moreno-Valenzuela (2013) and Moreno-Valenzuela and Guzman-Guemez (2016) where the voltage control is addressed by considering limitation of the duty cycle percentage.

The contribution of this paper is to address the position control of a torque-driven mechanism but by respecting prescribed simultaneous torque and speed bounds of the actuator. To the best of the authors' knowledge, this is the first time that the issue of keeping the overall control system operating within both torque and speed bounds is addressed.

This paper is organized as follows: In Section 2, plant model and control objective are presented. The proposed controller is stated in Section 3. The analysis is given in Section 4. An experimental study to complement the proposed theory is provided in Section 5. Finally, the conclusion is presented in Section 6.

## 2. Problem statement

### *2.1. Plant model: gravity unbalanced shaft driven by a torque actuator*

This paper considers an "unbalanced rotational device driven by a torque actuator" (see in Figure 2 a picture of a laboratory setup) modeled by the normalized nonlinear second-order ordinary differential equation:

$$J\ddot{q} + f_v\dot{q} + mgl\sin(q) = \tau, \tag{1}$$

where $J > 0$ is the rotor-shaft-load moment of inertia, $m > 0$ is its mass assumed to be concentrated at a distance $l > 0$ from the rotation axis (gravity unbalanced load like a hanging pendulum shaped one), $f_v \geq 0$ is the viscous friction coefficient, and $g > 0$ is the gravity acceleration; variables $q$, $\dot{q}$ and $\ddot{q}$ are, the shaft angular position, speed and acceleration, respectively, and $\tau$ is the "input"— torque— provided by a motor-electronic driver actuator setup in torque-mode.

Plant model (Equation 1) may be seen as a particular simple case of more complex and challenging systems like torque-driven robot manipulators (Kelly, Santibánez, & Loría, 2005).

With regard to the system (Equation 1) to be controlled, in this paper, the following assumptions are made:

*Assumption 1:* The plant model (Equation 1) related parameters: $J, f_v$, and $mgl$ are known.

*Assumption 2:* Shaft position $q$ and speed $\dot{q}$ measurements are available from system (Equation 1).

*Assumption 3:* The "safe operating zone" of the motor-electronic driver actuator is defined by the region delimited by the torque-speed characteristic curve corresponding to "medium / high" absolute value torque $|\tau|$ bounded by $\tau_{max} > 0$, and "small" speed $|\dot{q}|$ bounded by $Vel_{max} > 0$. This is shown by the green left bottom rectangle in Figure 3.

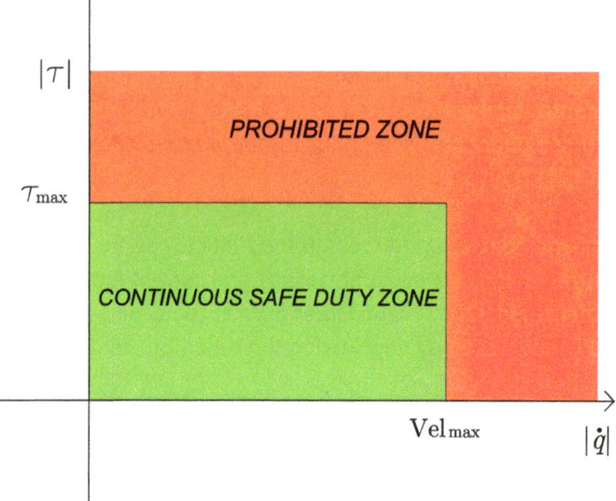


**Figure 3. Idealized conservative hard 'torquespeed' characteristic curve of motor-electronic driver: Safe operating and Prohibited hazardous.**

### *2.2. Actuator model*

In this paper, we use the following ideal identity actuator model:

$$\tau = u. \tag{2}$$

No torque limit is assumed for the actuator.

### *2.3. Control objective*

The control objective is shaft displacement regulation by respecting prescribed simultaneous torque and speed limits:

$$\lim_{t\to\infty} q(t) = q_d, \tag{3}$$

and

$$|\dot{q}(t)| \leq \mathrm{Vel}_{\max}\ \forall t \geq 0, \tag{4}$$

and

$$|\tau(t)| \leq \tau_{\max}\ \forall t \geq 0, \tag{5}$$

where $q_d$ is the known desired shaft displacement (arbitrary but constant), $\mathrm{Vel}_{\max} > 0$ is the known safe operational speed limit of the actuator/controlled system, and $\tau_{\max} > 0$ is the known maximum safe torque from the actuator.

Although no torque limit was assumed for the actuator, as a safety matter, a safe desired bound $\tau_{\max} < \infty$ is given to maintain the actuator/controlled system into the "safe operating zone."

According to Assumption 1, knowledge of some plant model parameters is required. If the plant model parameters are unknown, it would be necessary to explore alternative control methods, such as adaptive control (e.g. Kelly, 1987; Loría, Kelly, & Teel, 2005; Slotine & Li, 1988).

*Assumption 4:* The prescribed desired bounds $\tau_{\max}$ and $\mathrm{Vel}_{\max}$ are supposed to satisfy the following condition:

$$\tau_{\max} > f_v \mathrm{Vel}_{\max} + mgl. \tag{6}$$

Figure 3 shows a plot of $|\tau|$ versus $|\dot{q}|$. The red outer zone represents the prohibited unsafe operating area; meanwhile the "safe operating zone" is represented by the green left bottom inner rectangle. In this "safe zone," speed and torque constraints (Equations 4 and 5) hold simultaneously.

## 3. Proposed controller

This paper proposes the following control law:

$$\begin{aligned} u = {} & J\gamma_1 \mathrm{sech}^2(k_s\xi)k_s \tanh(\tilde{q}) \\ & - J\gamma_2 \mathrm{sech}^2\left(k_p\tilde{q}\right)k_p\dot{q} \\ & + f_v\dot{q} + mgl\sin(q), \\ \dot{\xi} = {} & \tanh(\tilde{q}), \end{aligned} \tag{7}$$

where the position error $\tilde{q}$ is defines as

$$\tilde{q} \triangleq q_d - q. \tag{8}$$

Controller parameters $\gamma_1$ and $\gamma_2$ are chosen to satisfy the following conditions:

$$\begin{aligned} & \gamma_1 < \gamma_2 < 0, \\ & \gamma_1 + \gamma_2 \leq \mathrm{Vel}_{max}. \end{aligned} \tag{9}$$

The initial velocity $\dot{q}(0)$ is assumed to be "small" in the sense:

$$|\dot{q}(0)| < \gamma_1 - \gamma_2, \tag{10}$$

which in virtue of (Inequalities 9) trivially satisfies (Equation 4) at $t = 0$.

The remaining controller parameters $k_s$ and $k_p$ are positive and are chosen to satisfy the tuning inequalities:

$$k_p < \frac{\tau_{max} - (f_v \mathrm{Vel}_{max} + mgl)}{J\gamma_2 \mathrm{Vel}_{max}}, \tag{11}$$

$$k_s \leq \frac{\tau_{max} - (f_v \mathrm{Vel}_{max} + mgl) - J\gamma_2 k_p \mathrm{Vel}_{max}}{J\gamma_1}. \tag{12}$$

Note that the right-hand side of (Inequality 11) is positive due to $\tau_{max}$ was assumed to satisfy condition (Inequality 6) and the right-hand side of (Inequality 12) is positive due to $k_p$ satisfies condition (Inequality 11).

And finally the initial condition $\xi(0)$ considered here as a parameter of the controller is obtained through the formula:

$$\xi(0) = \frac{1}{k_s} \mathrm{arctanh}\left(\frac{\dot{q}(0) - \gamma_2 \tanh\left(k_p\tilde{q}(0)\right)}{\gamma_1}\right). \tag{13}$$

which is well defined because $\dot{q}(0)$ satisfies (Inequality 10).

Inequality 10 ensures that the speed initial condition satisfies the control objective (Equation 4) at $t = 0$. However, given the (Condition 9) the value $|\dot{q}(0)|$ is restricted to $|\dot{q}(0)| < \mathrm{Vel}_{max}$, i.e., the initial velocities $\dot{q}(0) = -\mathrm{Vel}_{max}$ and $\dot{q}(0) = \mathrm{Vel}_{max}$ are not permissible values. In other words, the system (Equation 1) should not be started at the prescribed speed limits. The larger the difference $\gamma_1 - \gamma_2$, the larger the values of $|\dot{q}(0)|$ that are allowed but always under restriction $|\dot{q}(0)| < \mathrm{Vel}_{max}$.

To choose the parameters of the controller in (Equation 7), it is required to satisfy conditions (Inequalities 9, 11 and 12) and (Equation 13). The fulfillment of these conditions guarantees the achievement of the control objective in (Equation 3) and (Conditions 3 and 5) provided that initial speed is small enough in the sense of (Condition 10).

## 4. Analysis

The remaining of this section is devoted to prove that the proposed controller in (Equation 7) achieves the control objective in (Equation 3) and (Conditions 3 and 5).

### 4.1. *Speed boundedness*

Using torque actuator model in (Equation 2), control action $u$ in (Equation 7), and replacing into the plant model (Equation 1) yields:

$$\frac{d}{dt}\dot{q} = \gamma_1 \mathrm{sech}^2(k_s\xi)k_s \tanh(\tilde{q}) - \gamma_2 \mathrm{sech}^2\left(k_p\tilde{q}\right)k_p\dot{q}, \tag{14}$$

whose integral produces:

$$\begin{aligned}\dot{q} &= \gamma_1 \tanh(k_s\xi) + \gamma_2 \tanh(k_p\tilde{q}) \\ &\quad + \left[\dot{q}(0) - \gamma_1 \tanh(k_s\xi(0)) - \gamma_2 \tanh(k_p\tilde{q}(0))\right].\end{aligned} \tag{15}$$

By replacing (Equation 13) into (Equation 15), it becomes:

$$\dot{q} = \gamma_1 \tanh(k_s\xi) + \gamma_2 \tanh(k_p\tilde{q}). \tag{16}$$

Note that controller in (Equation 7) is very convenient, such that the value of the velocity $\dot{q}$ can be determined by (Equation 16). The resultant velocity depends on the position error $\tilde{q}$ and the integral of a nonlinear function of the position error. The obtained (Equation 16) is useful in the proof of the three parts of the control objective in (Equation 3) and (Conditions 4 and 5).

From (Equation 16), the velocity $\dot{q}$, can be bounded as follows:

$$\begin{aligned}|\dot{q}| &= |\gamma_1 \tanh(k_s\xi) + \gamma_2 \tanh(k_p\tilde{q})|, \\ &\leq |\gamma_1 \tanh(k_s\xi)| + |\gamma_2 \tanh(k_p\tilde{q})|, \\ &= \gamma_1 |\tanh(k_s\xi)| + \gamma_2 |\tanh(k_p\tilde{q})|, \\ &\leq \gamma_1 + \gamma_2, \\ &\leq \mathrm{Vel}_{max},\end{aligned} \tag{17}$$

where (Condition 9) on parameters $\gamma_1$ and $\gamma_2$ have been used.

Note that (Inequality 17) means that the desired speed boundedness (Condition 4) is achieved.

### 4.2. *Torque boundedness*

In this part of the paper, it is demonstrated that the requested torque $u$ to the actuator by the proposed controller in (Equation 7) is within the prescribed bound (Condition 5).

The requested torque $u$ to the actuator by the controller in (Equation 7) is bounded by:

$$
\begin{aligned}
|u| &= |J\gamma_1 \operatorname{sech}^2(k_s\xi)k_s \tanh(\tilde{q}) \\
&\quad - J\gamma_2 \operatorname{sech}^2\left(k_p\tilde{q}\right)k_p\dot{q} \\
&\quad + f_v\dot{q} + mgl\sin(q)|, \\
&\leq |J\gamma_1 \operatorname{sech}^2(k_s\xi)k_s \tanh(\tilde{q})| \\
&\quad + |J\gamma_2 \operatorname{sech}^2\left(k_p\tilde{q}\right)k_p\dot{q}| \\
&\quad + |f_v\dot{q}| + |mgl\sin(q)|, \\
&\leq |J\gamma_1 k_s||\operatorname{sech}^2(k_s\xi)\tanh(\tilde{q})| \\
&\quad + |J\gamma_2 k_p||\operatorname{sech}^2\left(k_p\tilde{q}\right)|\,|\dot{q}| \\
&\quad + |f_v||\dot{q}| + |mgl||\sin(q)|, \\
&= J\gamma_1 k_s|\operatorname{sech}^2(k_s\xi)\tanh(\tilde{q})| \\
&\quad + J\gamma_2 k_p|\operatorname{sech}^2\left(k_p\tilde{q}\right)|\,|\dot{q}| \\
&\quad + f_v|\dot{q}| + mgl|\sin(q)|, \\
&\leq J\gamma_1 k_s|\operatorname{sech}^2(k_s\xi)\tanh(\tilde{q})| \\
&\quad + J\gamma_2 k_p|\operatorname{sech}^2\left(k_p\tilde{q}\right)|\mathrm{Vel}_{max} \\
&\quad + f_v\mathrm{Vel}_{max} + mgl|\sin(q)|, \\
&\leq J\gamma_1 k_s + J\gamma_2 k_p\mathrm{Vel}_{max} + f_v\mathrm{Vel}_{max} + mgl, \\
&\leq \tau_{max},
\end{aligned}
\tag{18}
$$

where (Inequalities 9, 12 and 17) have been used.

Since the control action $u$ satisfies (Inequality 18), it follows from the actuator model (Equation 2) that

$$|\tau| \leq \tau_{max}. \tag{19}$$

The obtained (Inequality 19) means that the desired torque boundedness objective (Equation 5) is guaranteed.

### 4.3. Shaft displacement regulation

Taking into account (Equations 7, 8 and 16), a closed-loop system state representation in terms of $\tilde{q}$ and $\xi$ can be written as follows:

$$\frac{d}{dt}\begin{bmatrix} \tilde{q} \\ \xi \end{bmatrix} = \begin{bmatrix} -\gamma_1 \tanh(k_s\xi) - \gamma_2 \tanh(k_p\tilde{q}) \\ \tanh(\tilde{q}) \end{bmatrix}, \tag{20}$$

which is a nonlinear autonomous dynamic system having one unique equilibrium at $[\tilde{q}\ \xi]^T = \mathbf{0} \in R^2$.

Let us consider the following continuously differentiable function:

$$V(\tilde{q}, \xi) = \frac{k_s}{\gamma_1}\ln[\cosh(\tilde{q})] + \ln[\cosh(k_s\xi)]. \tag{21}$$

Furthermore, function $V(\tilde{q}, \xi)$ above is also globally positive definite and radially unbounded. So, $V(\tilde{q}, \xi)$ qualifies as a Lyapunov function candidate. Figure 4 depicts a typical shape of $V(\tilde{q}, \xi)$.

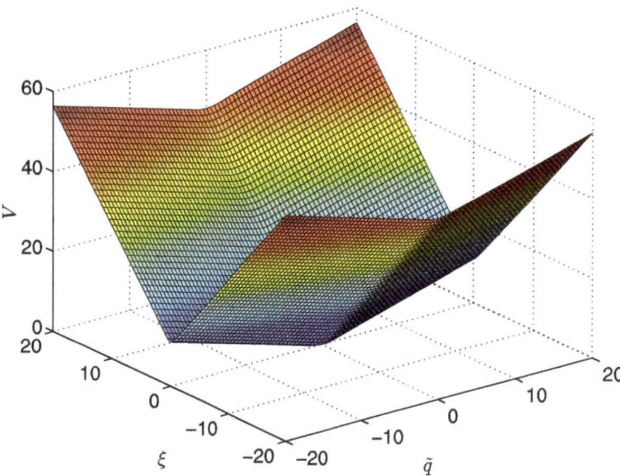


**Figure 4. Function** $V(\tilde{q},\xi)$ **in (Equation 21) with** $k_s = 2.3$ **and** $\gamma_1 = 4$.

The time derivative of $V(\tilde{q},\xi)$ in (Equation 21) is:

$$\begin{aligned}\dot{V}(\tilde{q},\xi) &= \frac{k_s}{\gamma_1}\tanh(\tilde{q})\dot{\tilde{q}} + \tanh(k_s\xi)k_s\dot{\xi},\\ &= \frac{k_s}{\gamma_1}\tanh(\tilde{q})[-\gamma_1\tanh(k_s\xi) - \gamma_2\tanh(k_p\tilde{q})]\\ &\quad + \tanh(k_s\xi)k_s\tanh(\tilde{q}),\\ &= -\frac{k_s}{\gamma_1}\gamma_2\tanh(\tilde{q})\tanh(k_p\tilde{q}).\end{aligned} \tag{22}$$

Since $k_s$, $\gamma_1$, and $\gamma_2$ are all positive constants, then $\dot{V}(\tilde{q},\xi)$ is a globally negative semidefinite function:

$$\dot{V}(\tilde{q},\xi) \leq 0 \quad \text{for all } \tilde{q},\xi \in R. \tag{23}$$

In virtue of the Lyapunov's direct method (see e.g. Kelly et al., 2005), this implies that the equilibrium $[\tilde{q}\ \xi]^T = \mathbf{0} \in R^2$ of the closed-loop system (Equation 20) is stable in the Lyapunov's sense, and it can be shown that both state variables $\tilde{q}$ and $\xi$ are bounded.

In order to prove achievement of the shaft displacement regulation objective (Equation 3), Barbalat's Lemma (Slotine & Li, 1991) shall be invoked.

Taking again the time derivative of $\dot{V}$ in (Equation 22), it follows that

$$\begin{aligned}\ddot{V}(\tilde{q},\xi) &= \frac{k_s\gamma_2}{\gamma_1}[\text{sech}^2(\tilde{q})\tanh(k_p\tilde{q})\\ &\quad + k_p\tanh(\tilde{q})\text{sech}^2(k_p\tilde{q})]\dot{q},\\ &\leq \frac{k_s\gamma_2}{\gamma_1}[1+k_p]\text{Vel}_{max},\end{aligned} \tag{24}$$

which allows the conclusion that $\ddot{V}$ is a bounded function. By invoking the Barbalat's Lemma (Slotine & Li, 1991) it leads to $\dot{V} \to 0$ as $t \to \infty$. Therefore, from Equation (22) it follows that the position error $\tilde{q}$ tends to zero:

$$\lim_{t\to\infty} \tilde{q}(t) = 0, \tag{25}$$

which is equivalent to the shaft displacement regulation objective (Equation 3).

## 5. Experimental study

A pendulum-like torque-driven gravity unbalanced mechanism located at Instituto Politécnico Nacional-CITEDI was used to illustrate through experiments the theoretical results. A diagram of the system is depicted in Figure 5.

The measurement data are obtained with the DAQ *Sensoray 626* which has input ports for optical encoders, bidirectional input/output digital ports and analog input/output ports. This DAQ allows interacting with *Matlab/Simulink* and performs real-time experiments through *Real-Time Windows Target* libraries. The control algorithm in (Equation 7) was executed at a sampling period of $1\times10^{-3}$ [s].

As actuator, a brushed direct current motor model 14207S-008 from *Pittman* was used with output torque limit

$$\tau_{max} = 0.35[\text{Nm}].$$

The value of the motor speed limit $Vel_{max}$ was set to

$$Vel_{max} = 5[\text{rad/s}].$$

The motor constant is

$$k_m = 0.0706\ [\text{Nm/A}].$$

Position measurement is carried out by an optical encoder with resolution of 2000 pulses per revolution (ppr). The motor is powered by the servoamplifier *Advanced Motion Controls* model *30A20AC* configured in current mode with relation

$$i = k_{sa}v,$$

where $v \in R$ is differential voltage input of the servoamplifier, $i \in R$ is the delivered current to the load and

$$k_{sa} = 1\ [\text{A/V}]$$

is the servoamplifier gain. Finally, the delivered torque is given by the relation

$$\tau = k_m i = k_m k_{sa} v = 0.0706\ v. \quad (26)$$

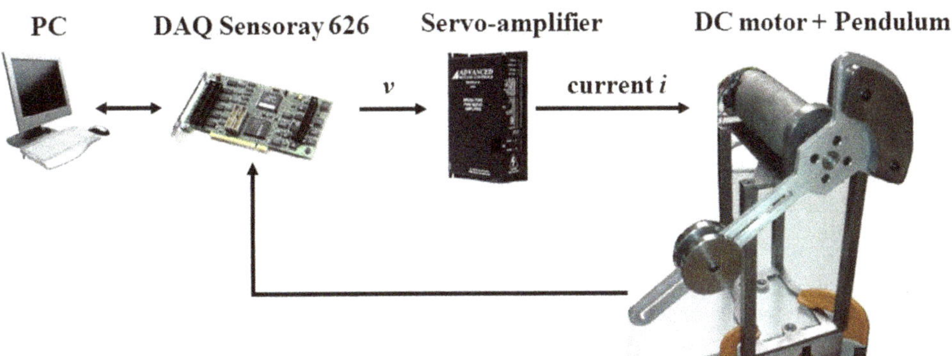


**Figure 5. Setup located at Instituto Politécnico Nacional-CITEDI.**

Considering actuator model (Equation 2) and Equation (26), the requested torque $u$ and the voltage $v$ are related by the following equation:

$$v = \frac{u}{0.0706}. \tag{27}$$

Estimate of the plant parameters (obtained by off-line least squares method) are:

$$J = 0.01204,$$
$$f_v = 0.004,$$
$$mgl = 0.1569.$$

To select the controller parameters, first choose $\gamma_1$ and $\gamma_2$ in order to satisfy (Condition 9). Choose $k_p$ according to (Condition 11). Then choose a large value of $k_s$ (which will be restricted by the previous selection of controller parameters $\gamma_1$, $\gamma_2$ and $k_p$, and by the plant parameters, and by the given speed and torque limits) that satisfy (Condition 12), and finally use (Equation 13) to compute $\xi(0)$.

Two experiments were performed. Parameters of the controller in (Equation 7) and desired joint positions of the experimental study were set as in Table 1.

The initial conditions of the system model (Equation 1) were:

$$q(0) = 0, \quad \dot{q}(0) = 0. \tag{28}$$

The controller parameters $\gamma_1$ and $\gamma_2$ selected as in Table 1, ensure that (Condition 9) are satisfied. On the other hand, the controller parameters $k_p$ and $k_s$, hold the tuning rules (Inequalities 11 and 12). The null initial speed $|\dot{q}(0)|$ in (Equation 28) satisfies (Inequality 10). Thus, all tuning rules: (Conditions 9, 11 and 12) and (Equation 13) are satisfied, so according to the closed-loop system analysis, it is expected that the control objective in Equations (3–5) are achieved.

Results of Experiment 1 are presented in Figures 6–9.

Figure 6 shows the shaft position $q(t)$ in blue continuous line and its desired position $q_d = 2$ in horizontal red dashed line.

Figure 7 depicts the estimation of shaft velocity $\dot{q}(t)$. Figure 8 shows the computed control action $u(t)$. Finally, Figure 9 shows $u(t)$ vs. $\dot{q}(t)$ evolving inside the "safe zone" as specified by the control objectives (Conditions 4 and 5).

**Table 1. Parameters of the controller in (7) and desired joint positions of the experimental study**

| | Experiment 1 | Experiment 2 |
|---|---|---|
| $q_d$ | 2 | 5 |
| $\gamma_1$ | 4 | 4 |
| $\gamma_2$ | 1 | 1 |
| $k_p$ | 1 | 2.5 |
| $k_s$ | 2.3 | 0.4 |
| $\xi(0)$ | -0.1069 | -0.6385 |

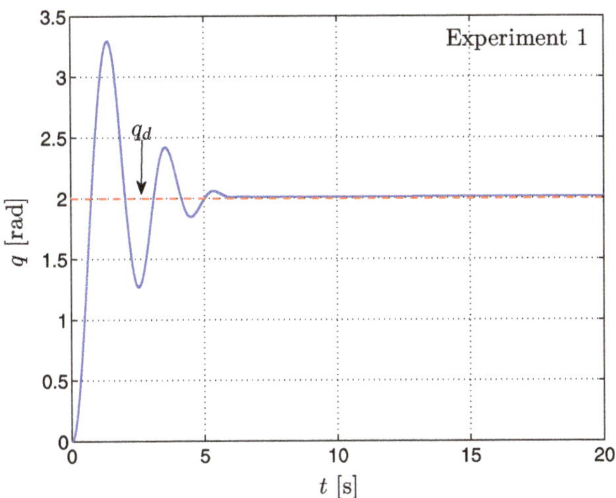


**Figure 6. Experiment 1. Shaft position *q(t)*. The desired value $q_d = 2$ is represented by the horizontal red dashed line.**

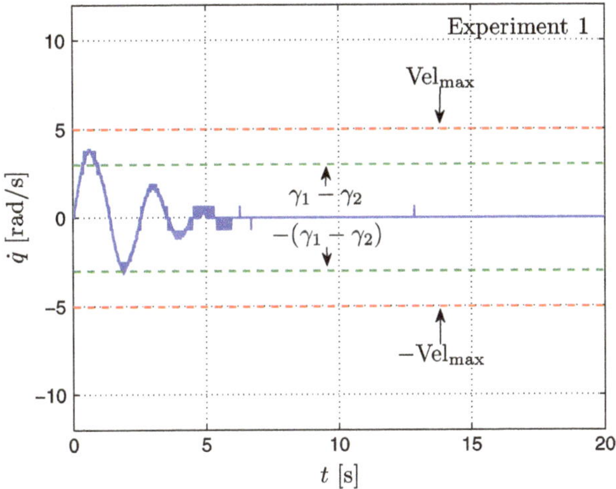


**Figure 7. Experiment 1. Estimation of shaft velocity $q(t)$. The limit values $\mathrm{Vel}_{max}$ and $-\mathrm{Vel}_{max}$ are in red dashed lines and the values of $\gamma_1 - \gamma_2$ and $-(\gamma_1 - \gamma_2)$ in green dashed lines.**

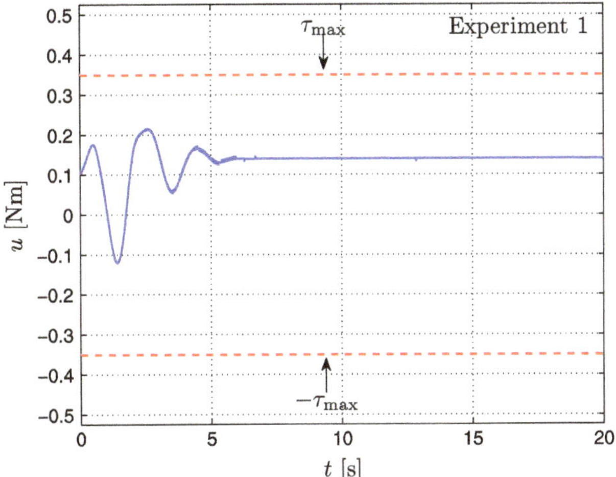


**Figure 8. Experiment 1. Computed control action *u(t)*. The limit values of $\tau_{max}$ and $-\tau_{max}$ are in horizontal red dashed lines.**

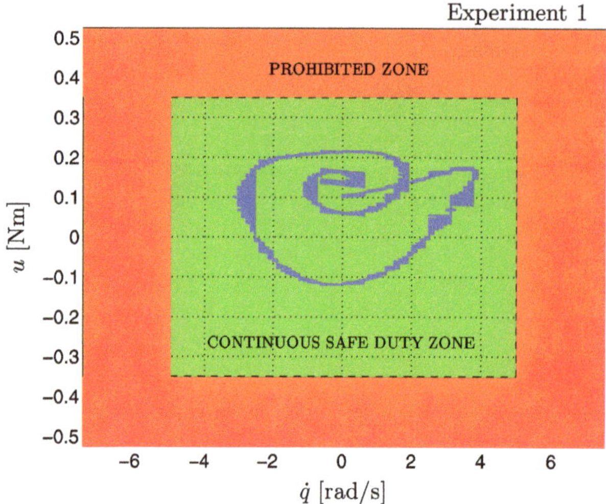


**Figure 9. Experiment 1. Plot of computed *u(t)* versus $\dot{q}(t)$ within the inner "safe operating zone".**

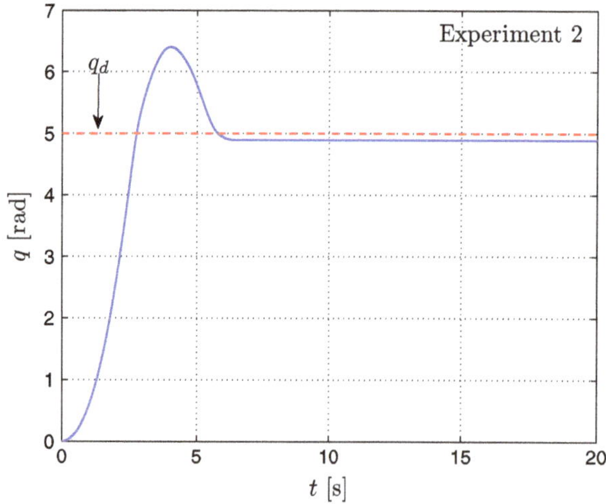


**Figure 10. Experiment 2. Shaft position *q(t)*. The desired value $q_d = 5$ is represented by the horizontal red dashed line.**

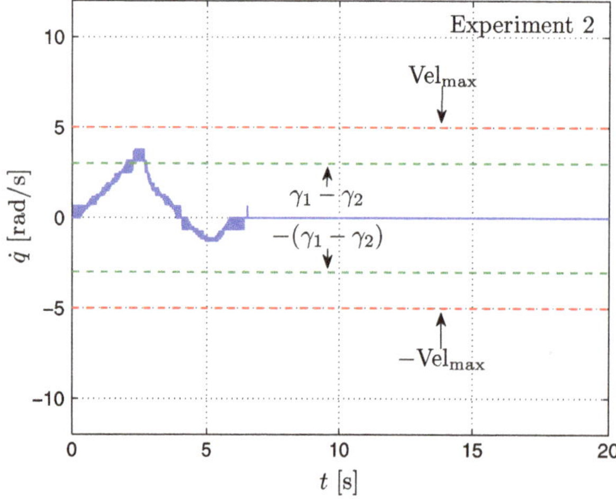


**Figure 11. Experiment 2. Estimation of shaft velocity $\dot{q}(t)$. The limit values $\mathrm{Vel}_{max}$ and $-\mathrm{Vel}_{max}$ are in red dashed lines and the values of $\gamma_1 - \gamma_2$ and $-(\gamma_1 - \gamma_2)$ in green dashed lines.**

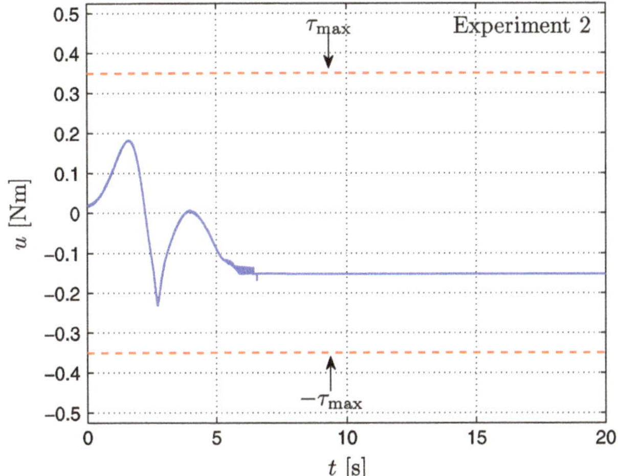


**Figure 12. Experiment 2. Computed control action *u(t)*. The limit values of $\tau_{max}$ and $-\tau_{max}$ are in horizontal red dashed lines.**

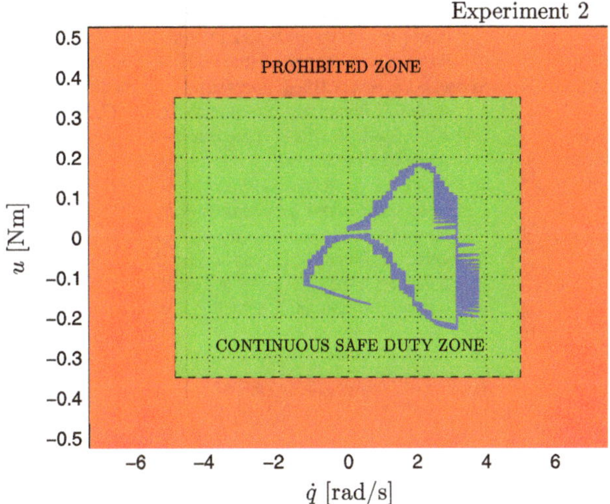


**Figure 13. Experiment 2. Plot of computed *u(t)* versus $\dot{q}(t)$ within the inner "safe operating zone".**

Results of Experiment 2 are presented in Figures 10–13, with the shaft position *q(t)*, estimation of velocity $\dot{q}(t)$, computed control action *u(t)* and computed *u(t)* versus $\dot{q}(t)$ in Figures 10, 11, 12, and 13, respectively. Note from Figure 13 that *u(t)* vs. $\dot{q}(t)$ evolve inside the "safe zone".

Steady-state errors of position *q(t)* in Figures 6 and 10 are observed, which may be due to: (1) unmodelled static friction present at the shaft-bearing of the DC motor actuator; (2) the implementation of the continuous controller in (Equation 7) as a discrete controller for the experimental study; (3) the velocity $\dot{q}$ is not being directly measured, instead of that, a velocity estimation based in position numerical differentiation is being used; and (4) the quantization noise in the digital-to-analog converter.

## 6. Conclusion

The control objective of shaft position regulation with simultaneous desired bounds of both torque and speed of an electro-mechanical system has been introduced in this paper. A nonlinear dynamic controller for the case of a torque-driven gravity nonlinear unbalanced device model was proposed. This research is a first step toward the formulation and design of "self-protecting" control system for more general, complex and challenging systems like robot manipulators where protection of motor-actuators is mandatory.

**Funding**

The authors would like to thank the CONACyT for the financial support for this research [grant number 166654], [grant number 176587].

**Author details**

Adriana Salinas[1]

E-mail: asalinas@cicese.edu.mx

Rafael Kelly[1]

E-mail: rkelly@cicese.edu.mx

Javier Moreno-Valenzuela[2]

E-mail: moreno@citedi.mx

[1] Laboratorio de Robótica, Departamento de Electrónica y Telecomunicaciones, CICESE, Carretera Ensenada-Tijuana No. 3918, Zona Playitas, Ensenada 22860, Baja California, Mexico.

[2] Instituto Politécnico Nacional-CITEDI, Department of Systems and Control, Ave. Instituto Politécnico Nacional No. 1310, Nueva Tijuana, Tijuana 22435, Baja California, Mexico.

**References**

Aghili, F. (2013). Optimal and Fault-Tolerant torque control of servo motors subject to voltage and current limits. *IEEE Transactions on Control Systems Technology, 21*, 1440–1448.

Åström, K. J., & Murray, R. M. (2008). *Feedback systems: An introduction for scientists and engineers*. Princeton, NJ: Princeton University Press.

Chen, C.-Y., & Cheng, M.-Y. (2012). Adaptive disturbance compensation and load torque estimation for speed control of a servomechanism. *International Journal of Machine Tools & Manufacture, 59*, 6–15.

Chen, H. (2014). Robust stabilization for a class of dynamic feedback uncertain nonholonomic mobile robots with input saturation. *International Journal of Control, Automation and Systems, 12*, 1216–1224.

Dorf, R. C., & Bishop, R. H. (2008). *Modern control systems* (11th ed.). Upper Saddle River, NJ: Prentice Hall.

Guzmán-Guemez, J., & Moreno-Valenzuela, J. (2013). Saturated control of boost DC-to-DC power converter. *Electronics Letters, 49*, 613–615.

Hughes, A., & Drury, B. (2013). *Electric motors and drives: Fundamentals, types and applications* (4th ed.). Oxford: Elsevier.

Kelly, R. (1987). A linear-state feedback plus adaptive feed-forward control for DC servomotors. *IEEE Transactions on Industrial Electronics, IE-34*, 153–157.

Kelly, R., & Moreno-Valenzuela, J. (2001). Learning PID structures in an introductory course of automatic control. *IEEE Transactions on Education, 44*, 373–376.

Kelly, R., Santibánez, V., & Berghuis, H. (1997). Point-to-point robot control under actuator constraints. *Control Engineering Practice, 5*, 1555–1562.

Kelly, R., Santibánez, V., & Loría, A. (2005). *Control of robot manipulators in joint space*. London: Springer-Verlag.

Loría, A., Kelly, R., & Teel, A. R. (2005). Uniform parametric convergence in the adaptive control of mechanical systems. *European Journal of Control, 11*, 87–100.

Moreno-Valenzuela, J., & Guzman-Guemez, J. (2016). Experimental evaluations of voltage regulators for a saturated boost DC-to-DC power converter. *Transactions of the Institute of Measurement and Control, 38*, 327–337.

Ngo, K. B., & Mahony, R. (2006). Bounded torque control for robot manipulators subject to joint velocity constraints. In *Proceedings of the 2006 IEEE International Conference on Robotics and Automation* (pp. 7-12). Orlando, FL.

Slotine, J.-J. E., & Li, W. (1988). Adaptive manipulator control: A case study. *IEEE Transactions on Automatic Control, 33*, 995–1003.

Slotine, J.-J. E., & Li, W. (1991). Applied nonlinear control. New Jersey, USA: Prentice Hall.

Su, H., Chen, M. Z. Q., Wang, X., & Lam, J. (2014). Semiglobal observer-based leader-following consensus with input saturation. *IEEE Transactions on Industrial Electronics, 61*, 2842–2850.

Tarbouriech, S., Garcia, G., Gomes da Silva, Jr., J. M., & Queinnec, I. (2011). *Stability and stabilization of linear systems with saturating actuators*. London: Springer.

# 6

# MRFT-based design of robust and adaptive controllers for gas loop of oil–gas separator

Hamdati Al Shehhi[1] and Igor Boiko[1]*

*Corresponding author: Igor Boiko, The Petroleum Institute, Abu Dhabi, UAE
E-mail: i.boiko@ieee.org
Reviewing editor: James Lam, University of Hong Kong, Hong Kong

**Abstract:** The modified relay feedback test (MRFT), which was recently proposed as a continuous oscillation method for identification of the process parameters and controller tuning, is used for the design of a robust and an adaptive Proportional-Integral (PI) controller for a gas loop in the oil–gas separator. The gas normally found in the separator is the natural gas (mostly methane) which is contained in crude oil coming from the reservoir. The robust and adaptive PI controllers are developed from analysis of 64 operating modes corresponding to certain ranges of the gas inflow and liquid-level values. It is shown through the developed model and simulations that these operating modes have significant effect on the dynamics of the gas loop. Dynamic properties of the process in each mode are studied through MRFT. The controllers are designed in order to maintain the pressure during the change of the operating conditions. Performance of the designed control system is studied by simulations.

**Subjects: Control Engineering; Dynamical Control Systems; Process Control - Chemical Engineering**

**Keywords: oil–gas separator control; controller tuning; discontinuous control**

### 1. Introduction

Controller tuning methodology based on continuous cycling can be traced back to the closed-loop Ziegler–Nichols (Z–N) method (1942). This method involves the excitation of continuous oscillation of the process variable. Recently, a few other continuous cycling tests were proposed (Boiko, 2008; Hang, Astom, & Wang, 2002; Kaya & Atherton, 2001; Majhi & Atherton, 1999; Wang, Lee, & Lin, 2003; Yu, 1998, 1999). Probably, the best-known test of this type is the Åström–Hägglund's (1984) relay

ABOUT THE AUTHORS

Hamdati Al Shehhi is a process control engineer at ZADCO, Abu Dhabi and a part-time MSc student at the Petroleum Institute, Abu Dhabi. Igor Boiko (PhD, DSc) is a professor at the Electrical Engineering department of the Petroleum Institute, Abu Dhabi. He leads a group that conducts research in the areas of industrial control applications and control theory. He authored monographs "Discontinuous control systems" (2009) and "Non-parametric tuning of PID controllers" (2013), many journal and conference publications. The group currently conducts research in process-control areas related to oil and gas production, separation, and treatment, as well as theoretical research in PID tuning, discontinuous and sliding-mode control.

PUBLIC INTEREST STATEMENT

In oil production, when the fluids from the oil reservoir reach the surface, they usually contain a mixture of gas, oil, and water. The first surface production step is to separate oil from gas and water that come from the wells, which is done by separators. Control of separators is important to the quality of oil and gas produced. A method of controller tuning for the gas pressure control in a separator is presented in the research. The approach proposed involves the modified relay feedback test. Two different approaches to controller design, robust and adaptive, are developed.

feedback test (RFT). It is worth noting that this test leads to the same ultimate gain and ultimate frequency as Z–N test if the model of the oscillations is based on the describing function (DF) method (Atherton, 1975).

In RFT, the excitation of the oscillations always occurs at the frequency $\omega_n$ corresponding to −180° of the phase response of the process. However, it is known that the controller introduction in such system would change the frequency $\omega_n$, so that the frequencies of the test oscillations and the phase crossover frequency of the open-loop system are different. The modified relay feedback test (MRFT), that introduced a coordinated selection of test parameters and tuning rules, was proposed by Boiko (2012) to eliminate the noted drawback of RFT. The MRFT can be considered as a further logical development of the ideas used in the Ziegler–Nichols and Åström–Hägglund tests. The approach proposed in MRFT was also termed as the holistic approach to the test and tuning. The main advantage of this test is that it can provide the desired value of the gain or phase margin exactly in non-parametric tuning (under the assumption that the DF method provides an exact model). In the present research, we aim to design a tuning procedure for the PID controller of the pressure loop of the oil–gas separator, using MRFT.

In oil production, when the fluids from the oil reservoir reach the surface, they usually contain a mixture of gas, oil, and water. The first surface production step is to separate oil from gas and water that comes from the wells, which is done by separators. The separation process involves equilibrium considerations or phase changes of these components. The equilibrium separation is done by four basic mechanisms as listed below (Roussean, 1987):

- Gravity separation which depends mainly on the density differences of the fluid components (water, crude oil, and associated gas) and occurs by reducing the velocity of a stream so that terminal velocity of particles due to gravity exceeds the velocity of bulk flow.
- Centrifugal force is similar to gravity separation, but it uses geometrical effects and dispersed phase concentrations for high-speed mechanical separation.
- Separation by impaction where the fluid will deflect around the body, while the particle having greater inertia will impact the body allowing the opportunity for separation.
- Electrically induced charge separation to attract the particles using fixed bodies such as plates with an opposite electrical charge.

We will further consider only gravity separators. There are two designs available: horizontal and the vertical separators, which are suitable to separate the fluids in oil and gas production facilities using the gravity separation techniques. The horizontal type is used when the amount of liquid is higher than the amount of gas, while the vertical type is used in the opposite case.

The horizontal-type liquid–gas separator is shown in Figure 1 and works as follows. First, the fluids flow into the separator and hit an inlet diverter, causing a sudden change in momentum. The initial separation of liquid and vapor occurs at the inlet diameter. The gravity force causes the liquid droplets which have a higher density to fall out of the gas stream (with a lower density) to the bottom of the vessel, where it is collected. Then, the liquid leaves the vessel through the level control valve (LCV). The gas passes through a mist extractor that removes very small droplets of liquid which are not easily separated by the gravity force (Roussean, 1987). The separation process is accomplished in one or more stages of decreasing pressure. There exist an optimum number of the stages required to separate the fluids. However, increasing the number of stages would normally increase the cost: due to necessity of additional separator, piping, controls, and space. According to Stewart and Arnold (2009), only one initial separation stage is necessary for the considered industrial case study, as the pressure of the gas in the plant is maintained at around 80 PSIG. Due to the relatively low pressure of fluid coming to the separator and the low water content (around 7%), the horizontal-type two-phase separator is used for initial separation within the considered platform production facilities.

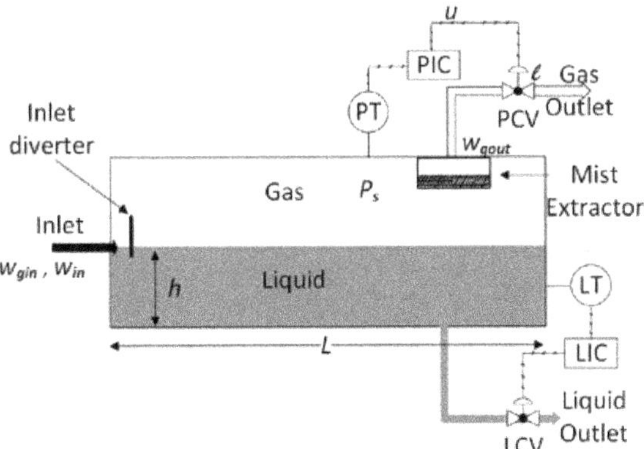


**Figure 1. The liquid–gas horizontal separator.**

In the considered case study, the separator is a horizontal-type two-phase separator. The gas is separated from the liquid with the gas and liquid being discharged. The gas is coming with crude oil from the wells. The separated gas is supplied to another plant for further treatment and use as a fuel gas for the gas turbine and high-pressure pumps. The separated oil is pumped to the storage and sales facilities. Therefore, two different loops are used to control the liquid and the gas in the vessel.

The model of the gas loop is considered as the process in this system which is to be controlled by the Proportional-Integral (PI) controller. The aim of the PI controller is to maintain the pressure equal to the set point despite various disturbances coming from the inflow stream and the gas pressure manifold. Maintaining the pressure equal to the set point is important not only to protect the vessel from the over-pressurization. When the separator pressure grows above the set-point of 80 PSIG, depressurization has to be done by sending the gas to flaring, which leads to a negative environmental impact. Thus, a pressure controller with a good performance is necessary for this application.

Conventional control used in the separator gas pressure loop is a PI control (Liptak, 2005), which also agrees with authors' industrial experience. To the best of the authors' knowledge, there are no publications suggesting other controls. This probably happens due to the fact that the basic model of the process (without actuator-valve model), which can be found in Sayda and Taylor (2007), can be written as a linear first-order differential equation. If a control is designed for this model only, then any proportional control with sufficiently high gain would reveal excellent performance. This design and model would not, however, match the reality: the actuator-valve dynamics has a significant effect on overall pressure loop dynamics, and cannot be neglected. A new model that involves all main contributors to the gas loop dynamics, such as gas-phase dynamics, valve dynamics, and gas flow between the separator and the gas manifold, is developed in the present paper.

This paper is organized as follows. First, the process model of the separator gas pressure is developed on the basis of the conservation laws. The models of the actuator and valve are presented too. Then MRFT is described. After that, a Simulink model of the industrial separator considered in the case study is introduced; MRFT is performed on this model; and with PI, optimal tuning rules optimized through ITAE criterion for the pressure loop are obtained. The robust and adaptive controllers are designed involving 64 operating modes. Finally, the simulation results showing performance of the system are presented and analyzed.

## 2. Process model

As shown in Figure 1, the typical industrial liquid–gas separator has two outlets to be controlled. The gas loop controls the gas outlet, where the pressure controller with indicating function (PIC) compares the set point with the process variable (gas pressure) which is measured by the pressure transmitter. The PIC sends the control signal (command) to the pressure control valve to operate

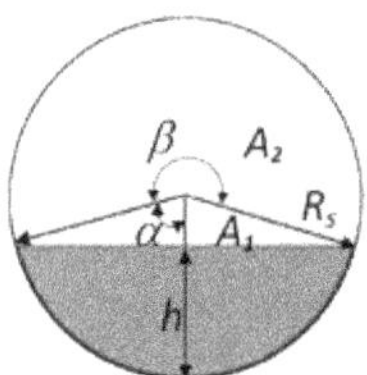


**Figure 2. Cross-sectional view of the separator at $h \leq R_s$.**

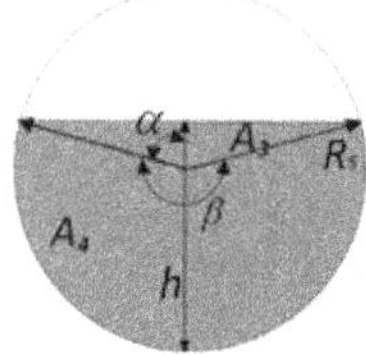


**Figure 3. Cross-sectional view of the separator at $h > R_s$.**

accordingly. The level loop controls the liquid level in the separator where the level controller with indicating function senses the change in the liquid level and controls the LCV accordingly. In order to design the model that represents the gas pressure loop for the separator, the following equations of gas pressure process given by Figure 1 are considered. The ideal gas law is used to represent gas pressure in the vessel:

$$p_s V_g = \frac{m_g}{\mu} R T_1 \tag{1}$$

where $p_s$ is the gas pressure inside the separator which is the process variable to be controlled, $V_g$ is the volume of the gas in the separator, $R$ is the universal gas constant, $T_1$ is the temperature of fluid inside the separator $\mu$ is the molecular weight, $m_g$ is the mass of the gas.

The difference between the total volume of the vessel and the volume of crude oil is equal to the volume of the natural gas inside the vessel, which is given by the following equation:

$$V_g = V_{max} - A_o L \tag{2}$$

where $V_{max}$ is the total volume of the vessel, $L$ is the length of the vessel, and $A_o$ is the part of the vertical cross-sectional area of the vessel filled with crude oil.

There are two cases for the cross-sectional area of crude oil which depends on the ratio between the level of the crude oil and the separator radius $R_s$:

Case 1: the level of the crude oil is less than or equal to the radius of the separator $h \leq R_s$ (Figure 2).

The cross-sectional area of crude oil is computed as the difference between the circle area, the area of the triangle $A_1$, and the area of the sector $A_2$:

$$A_o = \pi R_s^2 - A_1 - A_2 \tag{3}$$

where:

$$A_1 = (R_s - h)\sqrt{R_s^2 - (R_s - h)^2}$$

$$\cos(\alpha) = \frac{R_s - h}{R_s}$$

$$\alpha = \arccos\left(1 - \frac{h}{R_s}\right)$$

$$\beta = 2\pi - 2\alpha = 2\pi - 2\arccos\left(1 - \frac{h}{R_s}\right)$$

Then,

$$A_2 = \pi R_s^{\,2}\frac{\beta}{2\pi} = R_s^{\,2}\left(\pi - \arccos\left(1 - \frac{h}{R_s}\right)\right)$$

Case 2: the level of the crude oil is higher than the radius of the separator: $h > R_s$ (Figure 3).

The cross-sectional area of crude oil for Case 2 is computed as the sum of the triangle area $A_3$ and of the sector area $A_4$:

$$A_o = A_3 + A_4 \tag{4}$$

where:

$$A_3 = (h - R_s)\sqrt{R_s^{\,2} - (h - R_s)^2}$$

$$\cos(\alpha) = \frac{h - R_s}{R_s}$$

$$\alpha = \arccos\left(\frac{h}{R_s} - 1\right)$$

$$\beta = 2\pi - 2\alpha = 2\pi - 2\arccos\left(\frac{h}{R_s} - 1\right)$$

$$A_4 = \pi R_s^{\,2}\frac{\beta}{2\pi} = R_s^{\,2}\left(\pi - \arccos\left(\frac{h}{R_s} - 1\right)\right)$$

As shown in Figure 1, the separator has inlet and outlet lines (pipes). The control valve, installed in the outlet line, controls the gas outflow. The inlet flow is uncontrolled and acts as a disturbance applied to the system. The pressure process has also relation with rate of change for the mass of gas as given by the following mass balance equation:

$$\dot{m}_g = w_{gin} - w_{gout} \tag{5}$$

where $w_{gin}$ and $w_{gout}$ are the mass gas flow rates for the inlet and the outlet of the separator.

The outlet mass gas flow rate from the separator through the control valve is given by the St. Venant and Wantzel's equation (Beater, 2007):

$$w_{gout} = A_v(\ell)C_d p_s\sqrt{\frac{\gamma}{R_{spec}T_1}\left(\frac{2}{\gamma+1}\right)^{\left(\frac{\gamma+1}{\gamma-1}\right)}}\,\psi\left(\frac{p_{cin}}{p_s}\right) \tag{6}$$

where $C_d$ is the discharge coefficient. $\psi\left(\frac{p_{cin}}{p_s}\right)$ is the flow function given by:

$$\Psi\left(\frac{p_{cin}}{p_s}\right) = \begin{cases} 1 & \text{if} \quad \frac{p_{cin}}{p_s} \leq \beta_c \\ \sqrt{\frac{2}{\gamma-1}\left(\frac{\gamma+1}{2}\right)^{\frac{\gamma+1}{\gamma-1}}}\sqrt{\left(\frac{p_{cin}}{p_s}\right)^{\frac{2}{\gamma}} - \left(\frac{p_{cin}}{p_s}\right)^{\frac{\gamma+1}{\gamma}}} & \text{if } \frac{p_{cin}}{p_s} < \beta_c \end{cases}$$

$\beta_c = \left(\frac{2}{\gamma+1}\right)^{\frac{\gamma}{\gamma-1}}$ is the critical pressure ratio, $A_v$ is the function characterizing valve orifice pass area which related to the valve travel ($\ell$), $A_v(\ell) \propto \ell$ for the linear control valve (such as the globe valve), $P_{cin}$ is the discharge or manifold pressure of the control valve, $R_{spec}$ is the specific gas constant of the methane, and $\gamma$ is the isometric constant of methane.

The PI controller has the following equation:

$$u = (r_p - p_s)K_c + \frac{K_c}{T_i}\int_0^t \left(r_p - p_s(\tau)\right) d\tau \tag{7}$$

where $K_c$ is the proportional gain, $T_i$ is the integral time constant, $r_p$ is the set point pressure of the PI controller.

Finally, the relation between the valve travel and the controller commands is represented by the first-order plus dead time model (Boiko, 2013; Sayedain & Boiko, 2011):

$$\frac{l(s)}{u(s)} = k_v \frac{e^{-\tau_v s}}{T_v s + 1} \tag{8}$$

where $\tau_v$ is the dead time, $T_v$ is the time constant, and $k_v$ is the valve gain.

## 3. MRFT and holistic test and tuning

The MRFT is used in this research as a method of controller tuning for the pressure loop. The test signal that it generates (through the feedback principle) is similar to the one of a relay controller, which is discontinuous. The control output $u(t)$ of the MRFT is given as follows by Boiko (2012):

$$u(t) = \begin{cases} h \text{ if } e(t) \geq b_1 \text{or } \left(e(t) > -b_2 \text{ and } u(t-) = h\right) \\ -h \text{ if } e(t) \leq -b_2 \text{ or } \left(e(t) < b_1 \text{ and } u(t-) = -h\right) \end{cases} \tag{9}$$

where $h$ is the amplitude of the relay, $b_1 = \beta e_{max}$, $b_2 = -\beta e_{min}$, $e_{min} < 0$, $e_{max} > 0$ are the last "singular" point of the error sinusoid signal corresponding to the last maximum and minimum values of $e(t)$ after crossing the zero level, $\beta$ is a parameter of MRFT and of the tuning rules, and $u(t-) = \lim_{\epsilon\to 0,\, \epsilon<0} u(t-\epsilon)$ is the control value at the time immediately preceding the current time $t$ (Boiko, 2013).

The system under MRFT can be represented by the block diagram (Figure 4), in which, however, relay hysteresis value $b$ is not constant but variable—as defined above. Process model under the test is shown in Figure 4 in either the state space form or as the transfer function $W_p(s)$.

### 3.1. Describing function of MRFT

The DF method is applied in Boiko (2012) to analyze the periodic motion that occurs during the test. MRFT acts as hysteretic relay, in a sense, with unknown value of the hysteresis though. The DF of the hysteretic relay having hysteresis value $b$ is given by (Atherton, 1975):

$$N(a) = \frac{4h}{\pi a}\sqrt{1-\left(\frac{b}{a}\right)^2} - j\frac{4hb}{\pi a^2}, \quad a > b \tag{10}$$

where $a$ is the amplitude of the input sinusoidal signal.

But the system has unknown hysteretic value, which depends on the amplitude value: $b = \beta a$. Therefore, the DF formula can be written as follows (Boiko, 2012):

$$N(a) = \frac{4h}{\pi a}\left(\sqrt{1-\beta^2} - j\beta\right)$$

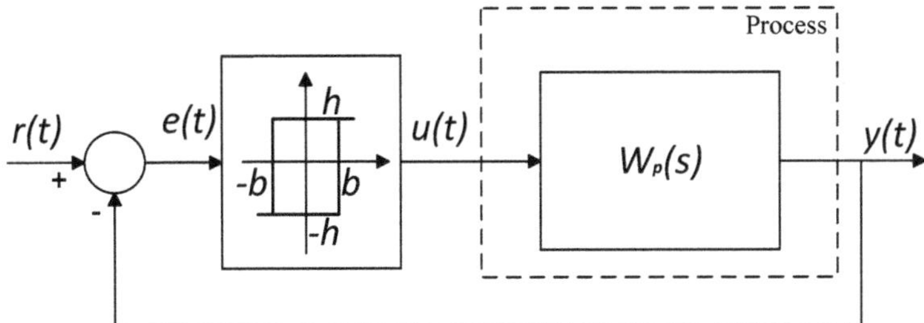


**Figure 4. Modified relay feedback test.**

The following harmonic balance equation can be used to find the parameters of oscillations during the MRFT.

$$W_p(j\Omega_0) = -\frac{1}{N(a_0)} = \frac{\pi a_0}{4h}\left(\sqrt{1-\beta^2}+j\beta\right) \tag{11}$$

where $a_0$ and $\Omega_0$ are the amplitude and the frequency of the periodic motions.

MRFT allows for the exact design of the gain margin (assuming that the DF method provides an exact model). Since the amplitude of the oscillation $a_0$ is measured from the test, the process gain at frequency $\Omega_0$ can be obtained as follows:

$$\left|W_p(j\Omega_0)\right| = \frac{\pi a_0}{4h} \tag{12}$$

which after introduction of the controller becomes the process gain at the critical frequency.

### *3.2. MRFT optimal design and tuning*

For a gas pressure process, from the point of view of possible parameter variation, it is important to find the gain margin of the system (Boiko, 2013). It was shown in Boiko (2013) that if the tuning rules have the format of:

$K_c = c_1 \frac{4h}{\pi a_0}$ for the proportional gain,

$T_i = c_2 \frac{2\pi}{\Omega_0}$ for the integral time constant, and

$T_d = c_3 \frac{2\pi}{\Omega_0}$ for the derivative time constant,

and the following equality constraint is satisfied:

$$\gamma_m c_1 \sqrt{1+\left(2\pi c_3 - \frac{1}{2\pi c_2}\right)^2} = 1 \tag{13}$$

where $\gamma_m$ is the required gain margin, and the designed controller will provide the desired gain margin exactly. To provide specified gain margin, MRFT must be carried out with parameter:

$$\beta = -\sin\varphi_c(\Omega_0) = -\sin\arctan\left(2\pi c_3 - \frac{1}{2\pi c_2}\right) = -\frac{2\pi c_3 - \frac{1}{2\pi c_2}}{\sqrt{1-\left(2\pi c_3 - \frac{1}{2\pi c_2}\right)^2}}$$

Yet, there is an issue of optimal combination of the coefficients that define tuning rules because there is only one Equation 13 and three variables. Optimal values of the PID parameters could be found by applying the tuning rules and optimization methods.

Among the optimization criteria, the most suitable are criteria that are measured from the output response of the system to the step change. The use of criteria that depend on the time domain rather than the frequency domain is preferred (Boiko, 2013). Therefore, the following popular time domain criteria can be used.

Integral absolute error (IAE):

$$Q_{\mathrm{IAE}} = \int_0^{\infty} |e(t)|\, dt$$

Integral time absolute error (ITAE):

$$Q_{\mathrm{ITAE}} = \int_0^{\infty} t\,|e(t)|\, dt \tag{14}$$

where $t$ is the time and $e(t)$ is the error.

**4. Separator model and controller design**

The Matlab and Simulink are used to simulate the system dynamics. The process and equipment data related to each part were collected in order to build the model of the process. Control valve data sheets were used to identify the actuator-valve characteristic and to calculate the required parameters: $A_v\ C_d$ = 0.0054921 m$^2$, with the dynamic model being $W_v(s) = \frac{e^{-0.5s}}{0.5s+1}$. The initial conditions of the model values are calculated such as the outlet gas pressure from the valve, gas mass, and inlet gas flow rate. It is considered that the control valve is linear and can open from 0 to 100% so that $\ell \in [0,1]$. The total volume of the two-phase separator is 100 m$^3$. However, the level of the liquid is specified by two limiting values, and the gas volume can be found from Equation 2. Gas properties such as the density, molar mass, and specific gas constant are those of methane, as it is the main ingredient of the gas contained in the separator, with 78% from the gas content.

From Equation 1, by considering the constant volume of gas (we disregard the dynamics of volume change because it is slow and addresses the variable volume in a quasi-static manner), the pressure rate of change:

$$\dot{p}_s = \frac{\dot{m}_g}{V_g \mu} R T_1$$

The modes of operation that affect the dynamics of the gas loop depend on the level and gas inflow for the separator, where

- Level of the liquid in the separator:
  $L \in$ [515 mm, 1825 mm]
- Gas inflow to the separator:
  $w_{in} \in$ [2.73 kg/s, 9.92 kg/s]

We split the above ranges of level and gas inflow into seven segments each and consider (7 + 1) × (7 + 1) = 64 operating modes of the separator in the controller design. The MRFT technique is applied for the 64 modes of operation. The model is run in Matlab-Simulink; first, the MRFT method is applied for each mode individually, and the critical amplitude $a_u$ and critical frequency $\Omega_u$ are calculated and presented in Table 1. The MRFT is carried out in the incremental way described in Boiko (2013), so that the produced control is the increment to the control value is a steady state corresponding to a certain operating mode. The tuning of the controller parameters is done by specifying the gain margin of *three* using the optimal tuning rules for gas pressure loops given in Boiko (2013), as the values of the constants used in the tuning rules: $c_1$ = 0.331, $c_2$ = 0.216, and the parameter of MRFT $\beta$ = 0.132. Gain margin of three is considered a good option for providing "moderately fast" closed-loop response—as per Boiko (2013). From Table 1, the parameters $k_c$ and $T_i$

for each mode are calculated using the tuning rules and are shown in Table 2. It can be noted that the value of the $k_c$ is very small since our control variable (gas pressure) is in Pascal as per SI standards.

**Table 1. The critical amplitude au (×10²) and critical frequency $\Omega_u$ of the system**

| Flow, kg/s | Level, m | | | | | | | |
|---|---|---|---|---|---|---|---|---|
| | **0.515** | **0.702** | **0.889** | **1.076** | **1.264** | **1.451** | **1.638** | **1.825** |
| **Critical amplitude au (×10²)** | | | | | | | | |
| 2.73 | 8.19 | 8.90 | 9.84 | 11.1 | 13.0 | 15.1 | 17.4 | 22.1 |
| 3.76 | 8.19 | 8.90 | 9.84 | 11.1 | 12.8 | 14.7 | 17.4 | 22.1 |
| 4.78 | 8.19 | 8.91 | 9.85 | 10.6 | 12.2 | 14.4 | 17.4 | 21.1 |
| 5.811 | 7.83 | 8.51 | 9.41 | 10.6 | 12.2 | 14.4 | 16.6 | 21.2 |
| 6.84 | 7.83 | 8.51 | 9.41 | 10.6 | 12.2 | 14.0 | 16.6 | 20.1 |
| 7.87 | 7.83 | 8.52 | 9.42 | 10.6 | 11.6 | 13.7 | 16.3 | 20.1 |
| 8.893 | 7.83 | 8.52 | 9.42 | 10.1 | 11.6 | 13.7 | 16.3 | 19.1 |
| 9.92 | 7.83 | 8.52 | 8.98 | 10.1 | 11.6 | 13.4 | 16.3 | 19.1 |
| **Critical frequency $\Omega_u$** | | | | | | | | |
| 2.73 | 0.805 | 0.805 | 0.805 | 0.805 | 0.805 | 0.805 | 0.829 | 0.829 |
| 3.76 | 0.805 | 0.805 | 0.805 | 0.805 | 0.805 | 0.813 | 0.829 | 0.829 |
| 4.78 | 0.805 | 0.805 | 0.805 | 0.826 | 0.826 | 0.826 | 0.829 | 0.852 |
| 5.811 | 0.826 | 0.826 | 0.826 | 0.826 | 0.826 | 0.826 | 0.852 | 0.857 |
| 6.84 | 0.826 | 0.826 | 0.826 | 0.826 | 0.826 | 0.826 | 0.852 | 0.875 |
| 7.87 | 0.826 | 0.826 | 0.826 | 0.826 | 0.849 | 0.849 | 0.863 | 0.875 |
| 8.893 | 0.826 | 0.826 | 0.826 | 0.849 | 0.849 | 0.849 | 0.863 | 0.897 |
| 9.92 | 0.826 | 0.829 | 0.849 | 0.849 | 0.849 | 0.863 | 0.872 | 0.897 |

**Table 2. The parameters $K_c$ (×10⁻⁵) and $T_i$ for gain margin 3**

| Flow, kg/s | Level, m | | | | | | | |
|---|---|---|---|---|---|---|---|---|
| | **0.515** | **0.702** | **0.889** | **1.076** | **1.264** | **1.451** | **1.638** | **1.825** |
| **$K_c$ parameter** | | | | | | | | |
| 2.73 | 2.06 | 1.90 | 1.71 | 1.52 | 1.30 | 1.11 | 0.968 | 0.764 |
| 3.76 | 2.06 | 1.89 | 1.71 | 1.52 | 1.32 | 1.15 | 0.968 | 0.763 |
| 4.78 | 2.06 | 1.89 | 1.71 | 1.59 | 1.38 | 1.18 | 0.967 | 0.800 |
| 5.811 | 2.15 | 1.98 | 1.79 | 1.59 | 1.38 | 1.18 | 1.01 | 0.797 |
| 6.84 | 2.15 | 1.98 | 1.79 | 1.59 | 1.38 | 1.20 | 1.01 | 0.840 |
| 7.87 | 2.15 | 1.98 | 1.79 | 1.59 | 1.45 | 1.23 | 1.04 | 0.839 |
| 8.893 | 2.15 | 1.98 | 1.79 | 1.67 | 1.45 | 1.23 | 1.04 | 0.883 |
| 9.92 | 2.15 | 1.98 | 1.88 | 1.67 | 1.45 | 1.26 | 1.04 | 0.884 |
| **$T_i$ parameter** | | | | | | | | |
| 2.73 | 9.485 | 9.485 | 9.485 | 9.485 | 9.485 | 9.485 | 9.211 | 9.211 |
| 3.76 | 9.485 | 9.485 | 9.485 | 9.485 | 9.485 | 9.394 | 9.211 | 9.211 |
| 4.78 | 9.485 | 9.485 | 9.485 | 9.242 | 9.242 | 9.242 | 9.211 | 8.968 |
| 5.811 | 9.242 | 9.242 | 9.242 | 9.242 | 9.242 | 9.242 | 8.968 | 8.907 |
| 6.84 | 9.242 | 9.242 | 9.242 | 9.242 | 9.242 | 9.242 | 8.968 | 8.725 |
| 7.87 | 9.242 | 9.242 | 9.242 | 9.242 | 8.998 | 8.998 | 8.846 | 8.725 |
| 8.893 | 9.242 | 9.242 | 9.242 | 8.998 | 8.998 | 8.998 | 8.846 | 8.512 |
| 9.92 | 9.242 | 9.211 | 8.998 | 8.998 | 8.998 | 8.846 | 8.755 | 8.512 |

### 4.1. Robust controller

From Table 2, the robust controller can be designed by choosing the parameters $K_c$ and $T_i$ that provide the least aggressive action and, therefore, will be suitable for all modes. In this design, the performance of the control system is guaranteed regardless of changes in the plant dynamics. However, none of the modes provides a combination of lowest gain and highest time constant. The robust controller values are selected from Table 2 at the point of (1.451 m, 2.73 kg/s), which provides the highest value of 9.485 for $T_i$ and sufficiently low value of $K_c$ being $1.1 \times 10^{-5}$. The overall performance is then verified by simulations.

### 4.2. Adaptive controller

Another method of approaching the issue of variable parameters of the process is the use of a controller having variable parameters. This approach constitutes the adaptive control. In using this approach, we utilize the data of Table 2 to design a gain schedule for the controller. Table 2 serves as a look-up table for finding the controller gain for a given mode, which is characterized by $L$ and $w_{in}$. This approach is straightforward but inconvenient in realization, as it requires storing look-up tables and using interpolation. A more convenient way would be to obtain a suitable approximation for the values of Table 2 as functions of $L$ and $w_{in}$.

The 3D graphs for the proportional gain $K_c$ and integral time constant $T_i$ as functions of $L$ and $w_{in}$ are given in Figures 5 and 6, respectively. One can notice that the dependence of $K_c$ on $L$ and $w_{in}$ is nearly linear, so that linear approximation would be appropriate. The dependence of $T_i$ on $L$ and $w_{in}$ has more complex character, which requires higher order approximations. Both approximating functions were found through minimization of mean square error technique. The dependence of $K_c$ on $L$ and $w_{in}$ is given by:

$K_c(L,w_{in}) = 0.0000254 - 0.00001001L + 0.000000179w_{in}$. And the approximating dependence of $T_i$ on $L$ and $w_{in}$ is given by third-order surface as follows:

$$T_i(L,w_{in}) = 9.733 - 0.5909L - 0.005925w_{in} + 0.7804L^2 + 0.003486Lw_{in} - 0.009414w_{in}^2 \\ - 0.3219L^3 - 0.00205L^2 w_{in} - 0.003218Lw_{in}^2 + 0.0007893w_{in}^3$$

### 5. Simulation

The dynamics of the system with the robust and adaptive (gain scheduled) controllers was tested by the Matlab simulations under two different operating modes which are the normal (inflow rate $w_{gin}$ = 6.3 kg/s and level $h$ = 1.17 m) and the "minimum" (inflow rate $w_{gin}$ = 2.73 kg/s and level

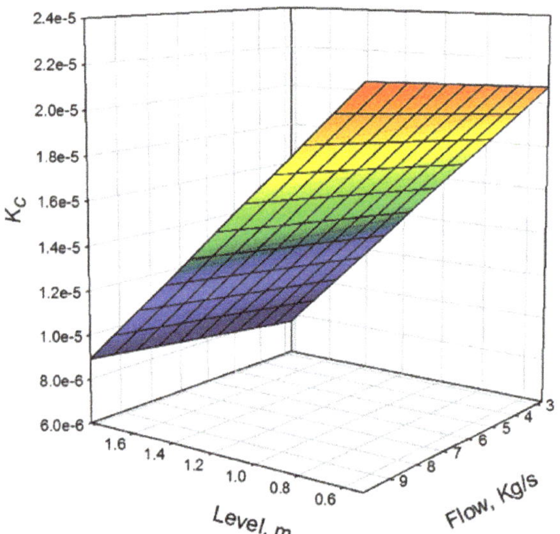


**Figure 5. 3D graph of $K_c$ dependence with the surface approximation.**

$h = 0.515$ m) operating modes. The Simulink diagram of the system is presented in Figure 7. Simulation is done using the fourth-order Runge–Kutta algorithm with the step size of 0.01s. Representative testing results are shown in Figures 8–27. The trends show the pressure loop performance under the simultaneous action of the gas inflow and level and step change of these disturbances, and the

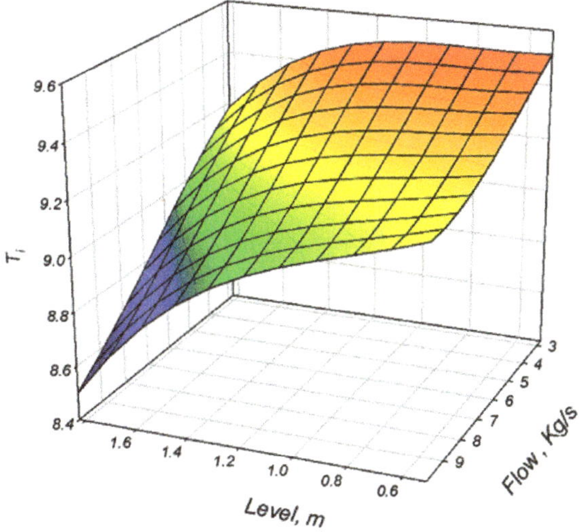


**Figure 6. 3D graph of $T_i$ dependence with the surface approximation.**

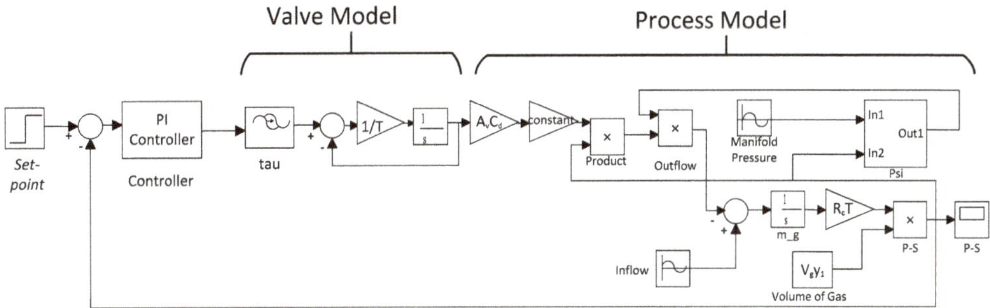


**Figure 7. The Simulink gas loop model.**

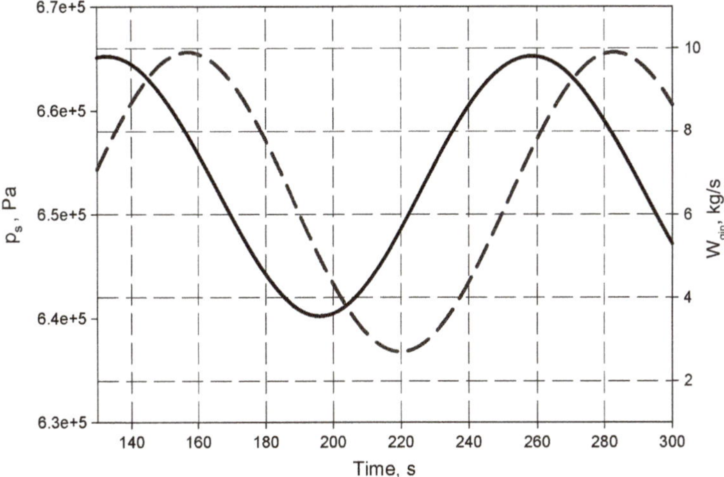


**Figure 8. Separator pressure (solid line) and sinusoidal inflow change (dashed line) with robust PI controller.**

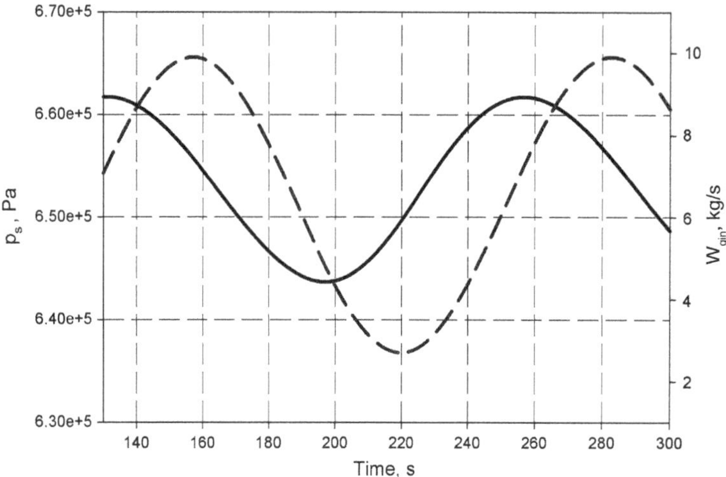


**Figure 9. Separator pressure (solid line) and sinusoidal inflow change (dashed line) with adaptive PI controller.**

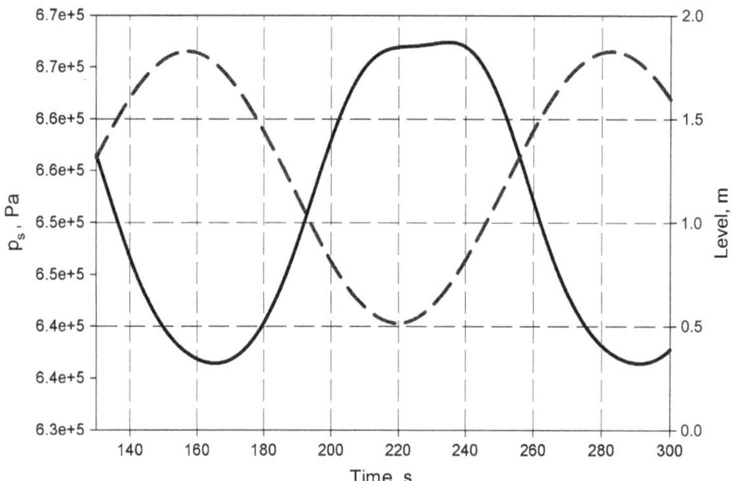


**Figure 10. Separator pressure (solid line) and sinusoidal level change (dashed line) with robust PI controller.**

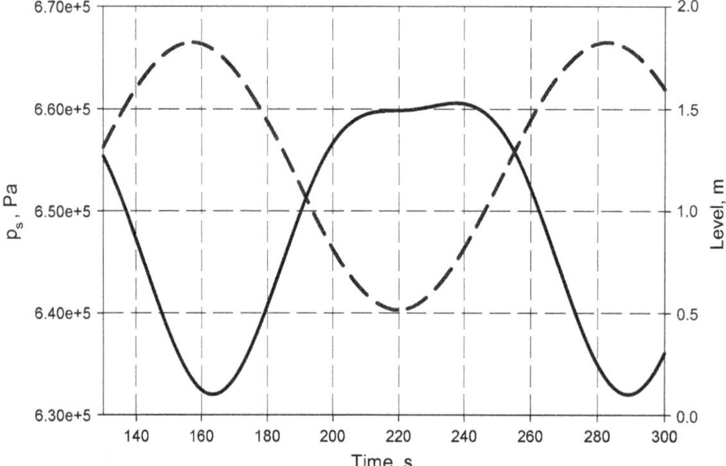


**Figure 11. Separator pressure (solid line) and sinusoidal level change (dashed line) with adaptive PI controller.**

manifold pressure and the pressure set point. The performance of the robust PI controller is demonstrated by the loop reaction to the disturbances and set point combinations, which is a sinusoidal change of the inflow and level with a step change of the separator pressure set point (Figures 8, 10, 12, 14, 16, 18, 20, 22, 24, and 26). The performance of the adaptive PI controller is demonstrated by the loop reaction to the three disturbances and set point change, which are the sinusoidal changes of

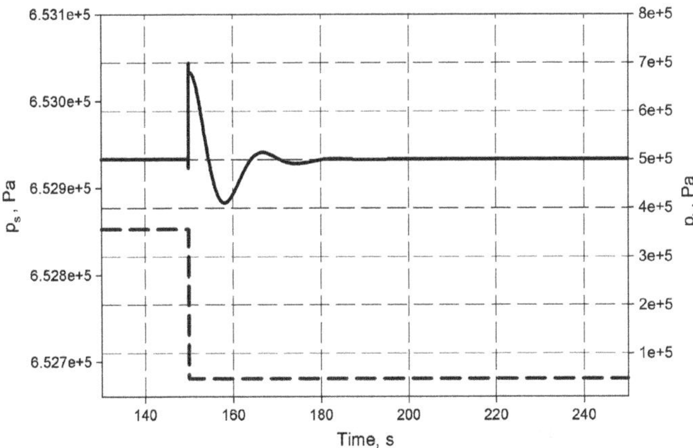


**Figure 12. Separator pressure (solid line) and step down change of manifold pressure (dashed line) for robust PI controller; in normal mode.**

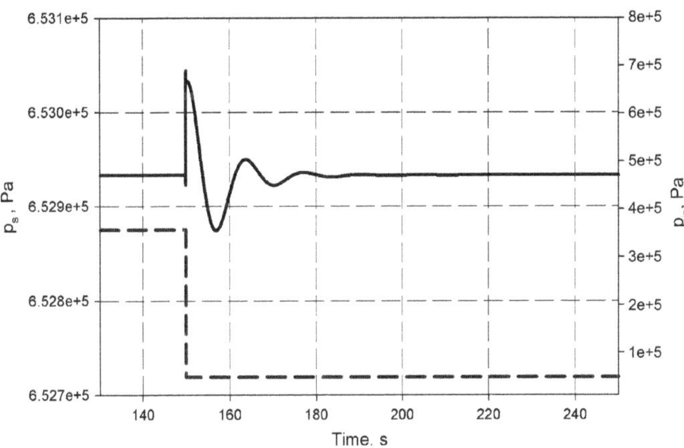


**Figure 13. Separator pressure (solid line) and step down change of manifold pressure (dashed line) with adaptive PI controller; in normal mode.**

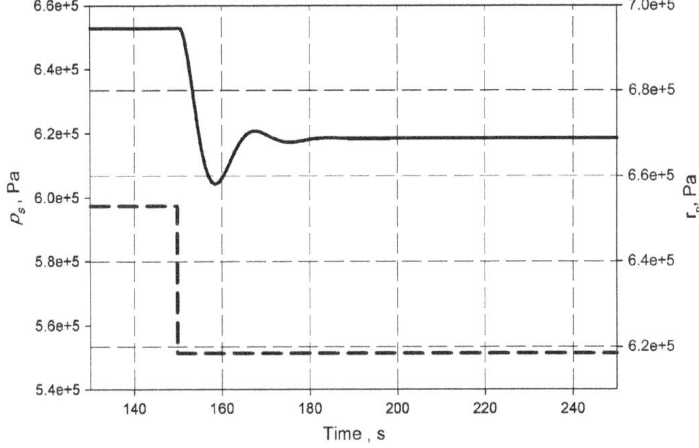


**Figure 14. Separator pressure (solid line) and step down change of set-point pressure (dashed line) with robust PI controller; in normal mode.**

the inflow and level and a step change of the separator pressure set point, the inflow and level and the manifold pressure (Figures 9, 11, 13, 15, 17, 19, 21, 23, 25, and 27). IAE index values for the step tests are presented in Table 3. IAE index was selected as a metric because it was the basis for the development of the tuning rules in Boiko (2013) that are used in this research. One can see from the

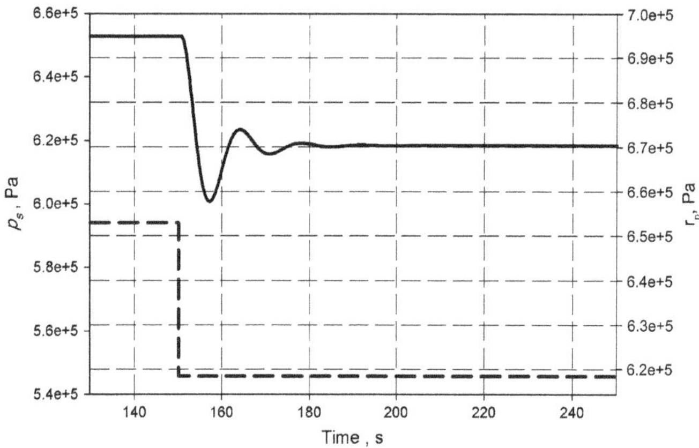


**Figure 15. Separator pressure (solid line) and step down change of set-point pressure (dashed line) with adaptive PI controller; in normal mode.**

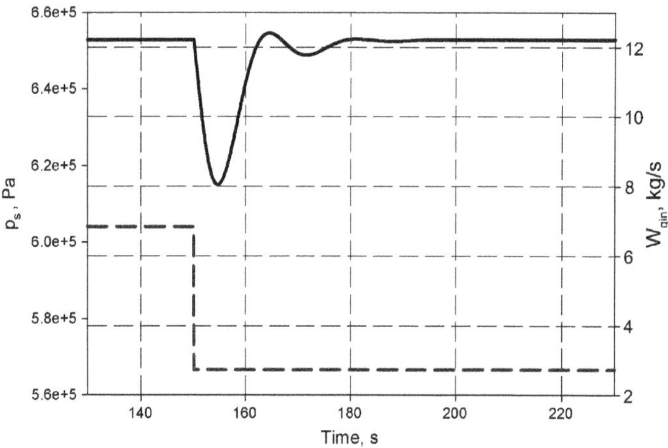


**Figure 16. Separator pressure (solid line) and step down inflow change (dashed line) with robust PI controller.**

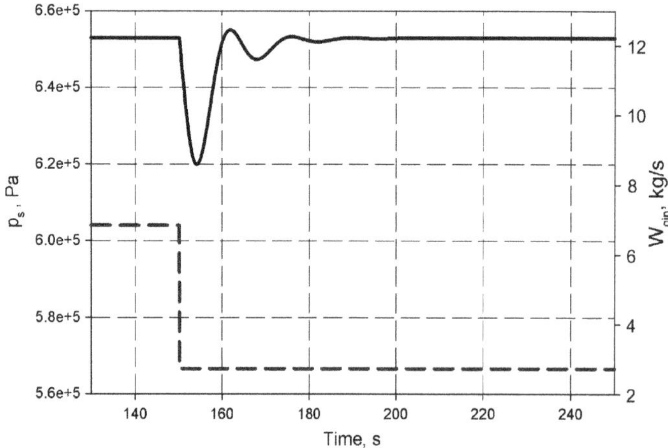


**Figure 17. Separator pressure (solid line) and step down inflow change (dashed line) with adaptive PI controller.**

trends of the separator pressure in Figures 12–27, and Table 3 that the performance of the robust controller is slightly higher, which is revealed as smaller pressure fluctuations. Also, the trends of the separator pressure with adaptive controller in Figures 9 and 11 are shifted down which means that

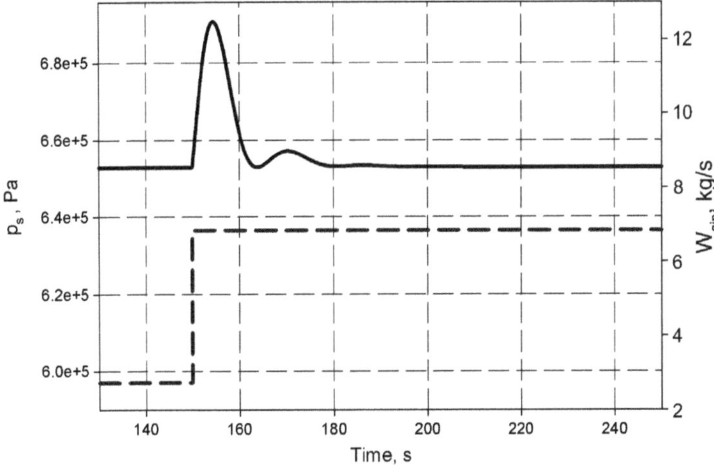


**Figure 18. Separator pressure (solid line) and step up change of inflow (dashed line) with robust PI controller.**

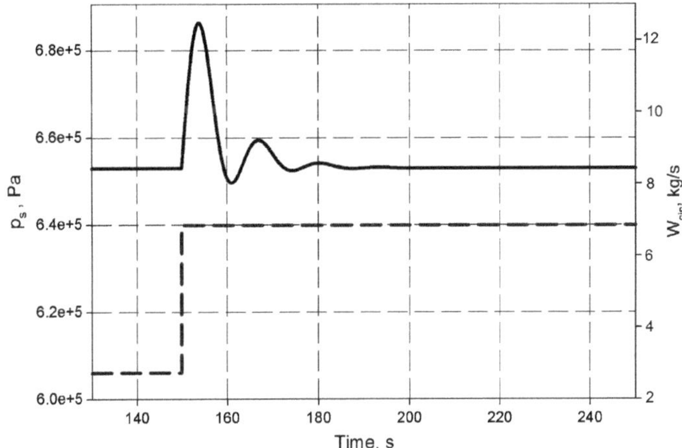


**Figure 19. Separator pressure (solid line) and step up change of inflow (dashed line) with adaptive PI controller.**

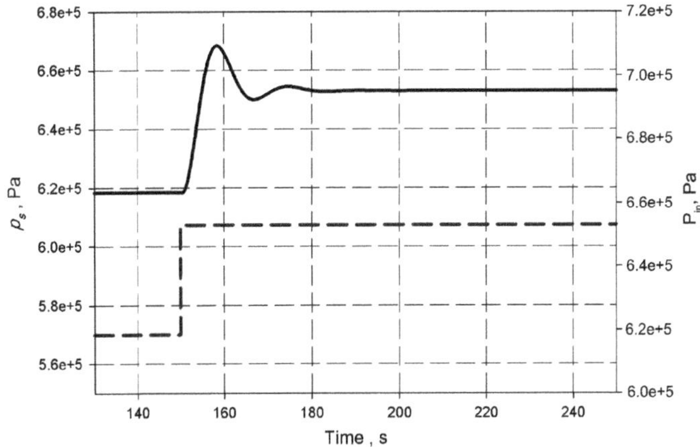


**Figure 20. Separator pressure (solid line) and step up change of set-point pressure (dashed line) with robust PI controller; in the normal mode.**

during the simultaneous action of the gas inflow and level, the separator pressure with adaptive controller does not go to high value as with the use of the robust controller in Figures 8 and 10. One of the main objectives to design the gas loop controller was to protect the vessel from overpressurization. As mention earlier, when the separator pressure grows above the set point of 80 PSIG,

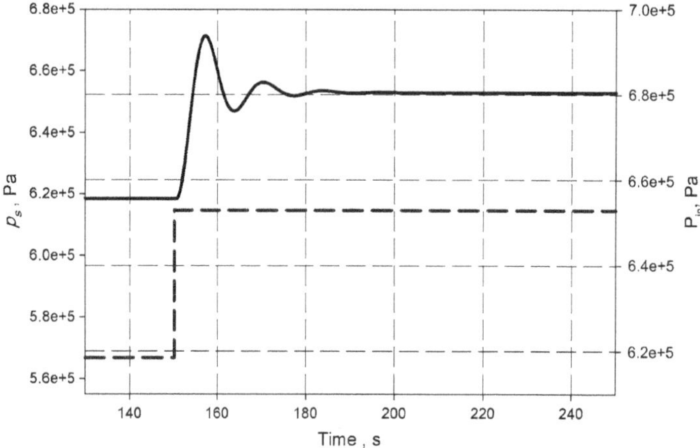


**Figure 21. Separator pressure (solid line) and step up change of set-point pressure (dashed line) with adaptive PI controller; in the normal mode.**

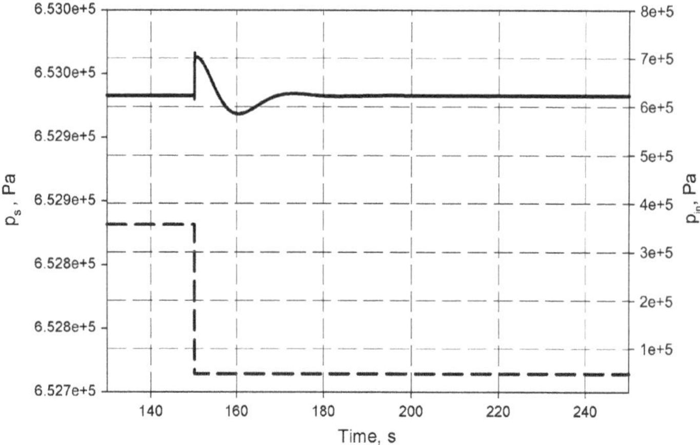


**Figure 22. Separator pressure (solid line) and step down change of manifold pressure (dashed line) with robust PI controller in the "minimum" mode.**

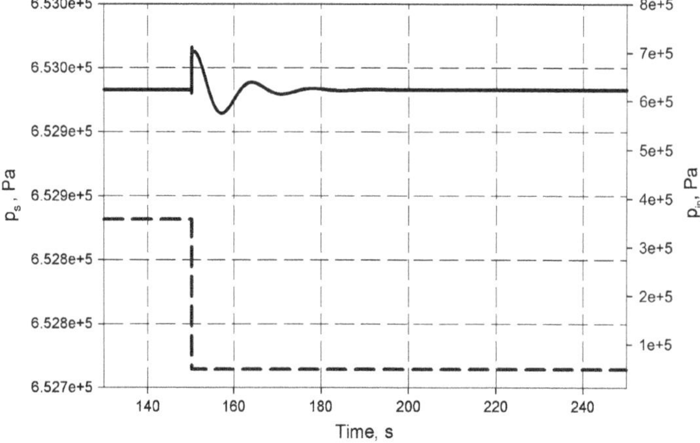


**Figure 23. Separator pressure (solid line) and step change of manifold pressure (dashed line) with adaptive PI controller in the "minimum" mode.**

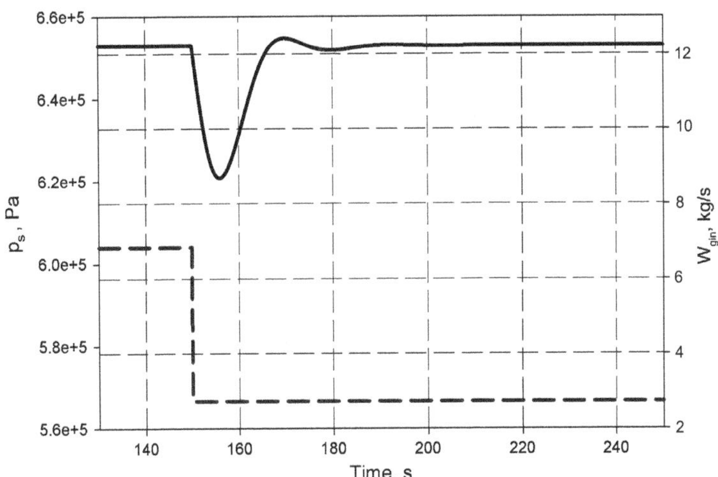


**Figure 24. Separator pressure (solid line) and step down inflow change (dashed line) with robust PI controller in the "minimum" mode.**

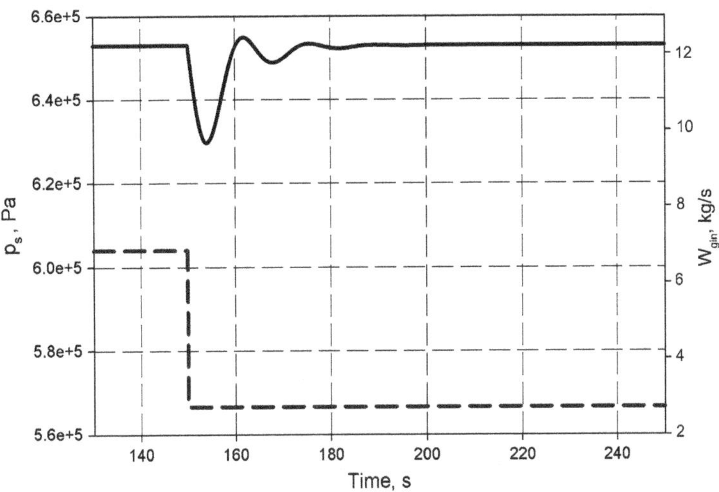


**Figure 25. Separator pressure (solid line) and step down inflow change (dashed line) with adaptive PI controller in the "minimum" mode.**

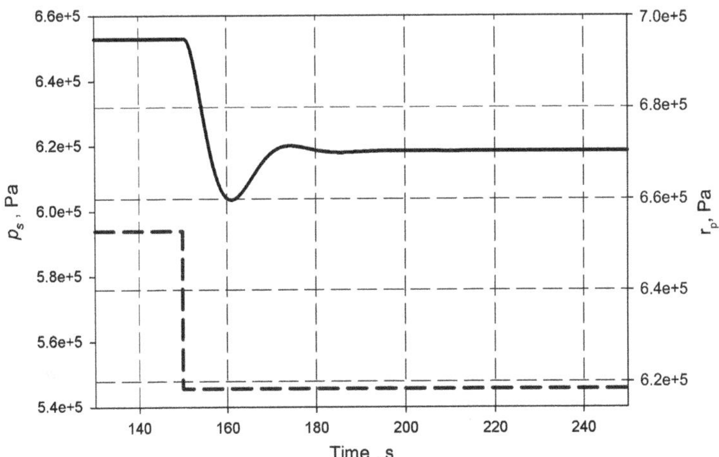


**Figure 26. Separator pressure (solid line) and step down change of set-point pressure (dashed line) with robust controller in the "minimum" mode.**

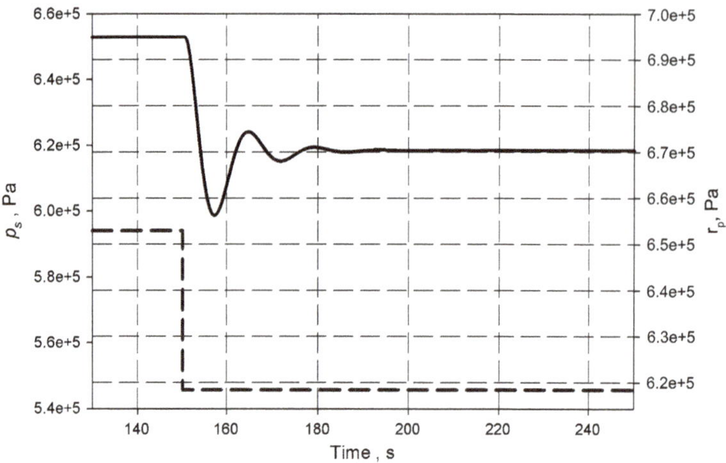


**Figure 27. Separator pressure (solid line) pressure and step down change of set-point pressure (dashed line) with adaptive controller in the "minimum" mode.**

**Table 3. IAE indices for step response**

| | | Controllers | |
|---|---|---|---|
| | | **Robust** | **Adaptive** |
| Figure reference | 12/13 | 0.00057 | 0.00056 |
| | 14/15 | 0.18675 | 0.18355 |
| | 16/17 | 0.30180 | 0.23395 |
| | 18/19 | 0.30165 | 0.21755 |
| | 20/21 | 0.19265 | 0.18850 |
| | 22/23 | 0.00022 | 0.00018 |
| | 24/25 | 0.30575 | 0.16035 |
| | 26/27 | 0.24520 | 0.20165 |

depressurization has to be done by sending the gas to flaring, which leads to a negative environmental impact. Therefore, a pressure controller with a good performance is highly recommended in this case. Thus, the choice of the adaptive PI controller over the robust controller is preferred.

## 6. Conclusions

Oil–gas separators are widely spread in the oil industry. As per authors' observations, gas pressure controllers often suffer poor performance due to operating mode change because once tuned in a certain operating mode, they may reveal either too-aggressive or too-sluggish closed-loop behavior in other operating modes. In the present research, a nonlinear model of the separator, which considers multiple operating modes, is presented. On the basis of MRFT, characterization of the process dynamics in various operating modes is proposed. This characterization allows one to use relatively simple methods of linear systems controller design and ensures overall stability through either conservative or gain scheduling strategies.

Respectively, two different approaches to the controller design/tuning are proposed: robust and adaptive. Methodology of realization of each of them is developed. Matlab/Simulink model of the system is developed. It is shown through simulations that both approaches are valid, with the robust approach being simple, whereas the adaptive approach providing higher performance.

**Funding**
The authors gratefully acknowledge the support of Project PIRC-14506 of the Petroleum Institute, Abu Dhabi, UAE.

**Author details**
Hamdati Al Shehhi[1]
E-mail: HaMAlShehhi1@pi.ac.ae
Igor Boiko[1]
E-mail: i.boiko@ieee.org
[1] The Petroleum Institute, Abu Dhabi, UAE.

**References**

Åström, K., & Hägglund, T. (1984). Automatic tuning of simple regulators with specifications on phase and amplitude margins. *Automatica, 20*, 645–651. http://dx.doi.org/10.1016/0005-1098(84)90014-1

Atherton, D. P. (1975). *Nonlinear control engineering-describing function analysis and design*. Workingham: Van Nostrand.

Beater, P. (2007). *Pneumatic drives*. Berlin: Springer-Verlag. http://dx.doi.org/10.1007/978-3-540-69471-7

Boiko, I. (2008). Autotune identification via the locus of a perturbed relay system approach. *IEEE Transactions on Control Systems Technology, 16*, 182–185. http://dx.doi.org/10.1109/TCST.2007.903108

Boiko, I. (2012). Loop tuning with specification on gain and phase margins via modified second-order sliding mode control algorithm. *International Journal of Systems Science, 43*, 97–104. http://dx.doi.org/10.1080/00207721003790344

Boiko, I. (2013). *Non-parametric tuning of PID controllers: A modified relay-feedback-test approach*. London: Springer-Verlag. http://dx.doi.org/10.1007/978-1-4471-4465-6

Hang, C., Astrom, K. J., & Wang, Q. G. (2002). Relay feedback auto-tuning of process controllers—A tutorial review. *Journal of Process Control, 12*, 143–162. http://dx.doi.org/10.1016/S0959-1524(01)00025-7

Kaya, I., & Atherton, D. P. (2001). Parameter estimation from relay autotuning with asymmetric limit cycle data. *Journal of Process Control, 11*, 429–439.

Liptak, B. G. (2005). *Instrument engineers' handbook* (Vol. 2). CRC Press.

Majhi, S., & Atherton, D. P. (1999). Autotuning and controller design for processes with small time delays. *IEE Proceedings D, 146*, 415–425.

Roussean, R. (1987). Phase segregation. In L. J. Jr., Jacobs, & W. Roy (Eds.), *Handbook of separation process technology* (pp. 129–190). New York, NY: Wiley.

Sayda, A. F., Taylor, J. H. (2007). Modeling and control of three-phase gravity separators in oil production facilities. In *Proceedings of 2007 American Control Conference*, Orlando, FL, USA.

Sayedain, S., Boiko, I. (2011). Optimal PI tuning rules for flow loop, based on modified relay feedback test. In *Proceedings of 2011 IEEE Conference on decision and control*, Orlando, FL, USA.

Stewart, M., & Arnold, K. (2009). *Gas-liquid and liquid-liquid separators*. Oxford: Elsevier.

Wang, Q., Lee, T. H., & Lin, C. (2003). *Relay feedback*. London: Springer. http://dx.doi.org/10.1007/978-1-4471-0041-6

Yu, C. (1998). Use of saturation relay feedback in PID control. *US Patent No. 5742503*.

Yu, C. (1999). *Automatic tuning of PID controllers: Relay feedback approach*. New York, NY: Springer.

Ziegler, J. G., & Nichols, N. B. (1942). Optimum settings for automatic controllers. *Transactions of the American Society of Mechanical Engineers, 64*, 759–768.

# 7

# Hybrid synchronization of hyperchaotic n-scroll Chua circuit using adaptive backstepping control

Suresh Rasappan[1]*

*Corresponding author: Suresh Rasappan, Department of Mathematics, Vel Tech University, No. 42 Avadi-Vel Tech Road, Avadi, Chennai 600062, Tamilnadu, India

E-mail: mrpsuresh83@gmail.com

Reviewing editor: James Lam, University of Hong Kong, Hong Kong

**Abstract:** In this paper, hybrid synchronization is investigated for n-scroll hyperchaotic Chua circuit using adaptive backstepping control. The theorem on hybrid synchronization for n-scroll hyperchaotic Chua circuit is established using Lyapunov stability theory. The backstepping scheme is recursive procedure that links the choice of Lyapunov function with the design of a controller and guarantees global stability performance of strict-feedback nonlinear systems. The backstepping control method is effective and convenient to hybrid synchronize the hyperchaotic systems which are mainly in this technique that gives the flexibility to construct a control law. Numerical simulations are also given to illustrate and validate the hybrid synchronization results derived in this paper.

**Subjects: Science; Technology; Systems & Control**

**Keywords: synchronization; Chaos; adaptive backstepping control; n-scroll hyperchaotic Chua circuit**

## ABOUT THE AUTHOR

Suresh Rasappan obtained his PhD degree in Mathematics from Vel Tech Rangarajan Dr Sakunthala R & D Institute of Science and Technology, Chennai, Tamil Nadu, India in 2013, and MPhil degree in Mathematics from Bharathiar University, Coimbatore, Tamil Nadu, India in 2008. He is currently working as an associate professor in the Department of Mathematics, Vel Tech Dr RR & Dr SR Technical University, Chennai, India. He has published over 28 papers in international journals and book chapters. His research interest is differential equations.

## PUBLIC INTEREST STATEMENT

Chaos synchronization can be applied in the areas of physics, engineering and biological science. Synchronization has been widely explored in a variety of fields including physical chemical, and ecological systems, secure communications etc. Synchronization of chaotic systems is a phenomenon that may occur when two or more chaotic oscillators are coupled, or when a chaotic oscillator drives another chaotic oscillator. Because the butterfly effect which causes the exponential divergence of the trajectories of two identical chaotic systems started with nearly the same initial conditions, synchronizing two chaotic systems is seemingly a challenging problem. In most synchronization approaches, the master-slave or drive-response formalism is used. If a particular chaotic system is called the master or drive system and another chaotic system is called the slave or response system, then the idea of synchronization is to use the output of the master system to control the slave system so that the output of the response system tracks the output of the master system asymptotically.

## 1. Introduction

Synchronization in chaos refers to the tendency of two or more systems which are coupled together to undergo closely related motion, even when the motions are chaotic.

The synchronization for chaotic systems has been widespread to the scope (Alligood, Sauer, & Yorke, 1997; Fujisaka & Yamada, 1983; Pecora & Carroll, 1990), such as generalized synchronization (Harmov, Koronovskii, & Moskalenko, 2005a; 2005b; Wang & Zhu, 2006), anti-synchronization, phase synchronization (Ge & Chen, 2006; Tokuda, Kurths, Kiss, & Hudson, 2008; Zhao, Lai, Wang, & Gao, 2004), lag synchronization, projective synchronization (Qiang, 2007), and generalized projective synchronization (Jian-Ping & Chang-Pin, 2006; Li, Xu & Li, 2007).

The property of anti-synchronization establishes a predominating phenomenon in symmetrical oscillators, in which the state vectors have the same absolute values but opposite signs.

When synchronization and anti-synchronization coexist, simultaneously, in chaotic systems, then that synchronization is called hybrid synchronization.

A variety of schemes to ensure the control and synchronization of such systems have been demonstrated based on their potential applications in various fields including chaos generator design, secure communication (Chen, 1996; Kanter, Kopelowitz, Kestler, & Kinzel, 2008; Yang & Chua, 1999), physical systems (Chern & Otsuka, 2012; Lakshmanan & Murali, 1996; Moreno & Pacheco, 2004), chemical reaction (Coffman, McCormick, Noszticzius, & Simoyi, 1987; Han, Kerrer, & Kuramoto, 1995), ecological systems (Blasius & Huppert, 1999), information science (Bauer, Atay, & Jost, 2010; Ghosh, Banerjee, & Chowdhury, 2007; Kocarev & Parlitz, 1995), energy resource systems, ghostburster neurons (Wang, Chen & Deng, 2009), biaxial magnet models (Moukam Kakmeni, Nguenang, & Kofane, 2006), neuronal models (Che, Wang, Tsang, & Chen, 2010; Hindmarsh & Rose, 1984; Qi, Huang, Chen, Wang, & Shen, 2008), IR epidemic models with impulsive vaccination (Zeng, Sun, Li, & Sun, 2005), and predicting the influence of solar wind to celestial bodies (Junxa, Dianchen, & Tian, 2006; Suresh & Sundarapandian, 2012a).

So far a variety of impressive approaches have been proposed for the synchronization of the chaotic systems such as OGY method (Ott, Grebogi, & Yorke, 1990), sampled feedback synchronization method (Murali & Lakshmaman, 2003), time delay feedback method (Park & Kwon, 2003), adaptive design method (Lu, Wu & Han, 2004; Park, 2008; Park, Lee, & Kwon, 2007), sliding mode control method (Ya, 2004), active control method (Sundarapandian & Suresh, 2010). and backstepping control design (Suresh & Sundarapandian, 2012b; Wu & Lu, 2003; Yu & Zhang, 2006).

Recently, backstepping method has been developed and designed to control the chaotic systems. A common concept in the method is to synchronize the chaotic system. The backstepping method is based on the mathematical model of the examined system, introducing new variables into a form depending on the state variables, controlling parameters, and stabilizing functions. The difficult work of synchronizing the chaotic system is to remove nonlinearities which were done in the system and influencing the stability of state operation. The use of backstepping method creates an additional nonlinearity and eliminates undesirable nonlinearities from the system (Suresh & Sundarapandian, 2012c; 2013; Wang, Zhang, & Guo, 2010; Wang 2011a, 2011b).

The uncertainties are commonly in chaos synchronization and other control system problems. The uncertainties are one of the main factors in leading the adaptive-based synchronization. Adaptive control design is a direct aggregation of control methodology with some form of recursive system which identifies the system to determine the control of linear or nonlinear systems.

Adaptive control design is studied and analyzed in theory of unknown, but fixed parameter systems. The controller feedback gain could be depending on the system parameter.

## 2. Problem statement

Consider the chaotic system described by the dynamics

$$\begin{aligned} \dot{x}_1 &= F_1(x_1, x_2, \dots, x_n, \alpha_i) \\ \dot{x}_2 &= F_2(x_1, x_2, \dots, x_n, \alpha_i) \\ \dot{x}_3 &= F_3(x_1, x_2, \dots, x_n, \alpha_i) \\ \vdots \; &\vdots \qquad \vdots \\ \dot{x}_n &= F_n(x_1, x_2, \dots, x_n, \alpha_i) \end{aligned} \tag{1}$$

where $x \in R^n$ is the state of the system, in which the system (1) is considered as the *master* system; and $\alpha_i$ is the unknown parameter, $\hat{\alpha}_i$ is the estimates as the parameter $\alpha_i$.

The *slave* system is a chaotic system with the controller $u = [u_1, u_2, u_3 ... u_n]^T$ described by the dynamics

$$\begin{aligned} \dot{y}_1 &= G_1(y_1, y_2, \dots, y_n, \alpha_i) + u_1(t) \\ \dot{y}_2 &= G_2(y_1, y_2, \dots, y_n, \alpha_i) + u_2(t) \\ \dot{y}_3 &= G_3(y_1, y_2, \dots, y_n, \alpha_i) + u_3(t) \\ \vdots \; &\vdots \qquad \vdots \\ \dot{y}_n &= G_n(y_1, y_2, \dots, y_n, \alpha_i) + u_n(t) \end{aligned} \tag{2}$$

where $u_i$ is the input to the system with parameter estimator $\hat{\alpha}_i$, $i = 1, 2, 3, \dots, n$, and $y \in R^n$ is the state of the slave system and $F_i$, $G_i (i = 1, 2, 3 ... n)$ linear or nonlinear functions with input from systems (1) and (2).

If $F_i = G_i$ for all i, then the system (1) and (2) are called *identical* and otherwise they are *nonidentical* chaotic systems.

The hybrid synchronization error is defined as

$$e_i = \begin{cases} y_i - x_i & \text{if} \quad \text{if i is odd} \\ y_i + x_i & \text{if} \quad \text{if i is even} \end{cases} \tag{3}$$

Then the synchronization error dynamics is obtained as

$$\begin{aligned} \dot{e}_1 &= G_1(y_1, y_2, \dots, y_n, \alpha_i) \\ &\quad -F_1(x_1, x_2, \dots, x_n, \alpha_i) + u_1 \\ \dot{e}_2 &= G_2(y_1, y_2, \dots, y_n, \alpha_i) \\ &\quad +F_2(x_1, x_2, \dots, x_n, \alpha_i) + u_2 \\ \vdots \; &\vdots \qquad \vdots \\ \dot{e}_n &= G_n(y_1, y_2, \dots, y_n, \alpha_i) \\ &\quad +(-1)^n F_n(x_1, x_2, \dots, x_n, \alpha_i) + u_n \end{aligned} \tag{4}$$

The parameter estimation error is defined as

$$e_{\alpha_i} = \alpha_i - \hat{\alpha}_i$$

The hybrid synchronization problem basically requires the global asymptotically stability of the error dynamics (4), i.e.

$$\lim_{t \to \infty} \| e(t) \| = 0 \tag{5}$$

for all initial conditions $e(0) \in R^n$.

Backstepping design procedure is recursive and guarantee global stability performance of strict-feedback chaotic systems. By using the backstepping design, at the $i$th step, the $i$th order subsystem is stabilized with respect to a Lyapunov function $V_i$, by the virtual control $\alpha_i$, and a control input function $u_i$.

Consider the global asymptotic stability of the system

$$\dot{e}_1 = G_1(y_1, y_2, \dots, y_n, \alpha_i) - F_1(x_1, x_2, \dots, x_n, \alpha_i) + u_1 \tag{6}$$

where $u_1$ is control input, which is the function of the error vector $e_i$, and the state variables $x(t) \in R^n$, $y(t) \in R^n$. As long as this feedback stabilizes, the system (6) will converge to zero as $t \to \infty$, where $e_2 = \alpha_1(e_1)$ is regarded as a virtual controller.

For the design of $\alpha_1(e_1)$ is to stabilize the subsystem (6), the Lyapunov function is defined by

$$V_1(e) = e_1^T P_1 e_1 + \sum_{i=1}^{k} e_{\alpha_i}^T R_1 e_{\alpha_i} \tag{7}$$

where $P_1$, *and* $R_1$ are positive definite matrices.

The derivative of $e_{\alpha_i}$ is

$$\dot{e}_{\alpha_i} = -\dot{\hat{\alpha}}_i \tag{8}$$

Suppose the derivative of $V_1$ is

$$\dot{V}_1 = -e_1^T Q_1 e_1 - \sum_{i=1}^{k} e_{\alpha_i}^T S_1 e_{\alpha_i} \tag{9}$$

where $Q_1$, *and* $S_1$ are positive definite matrices.

Then $\dot{V}_1$ is a negative definite function.

Thus by Lyapunov stability theory, the error dynamics(6) is globally asymptotically stable.

The function $\alpha_1(e_1)$ is an estimative function when $e_2$ is considered as a controller.

The error between $e_2$ and $\alpha_1(e_1)$ is

$$w_2 = e_2 - \alpha_1(e_1) \tag{10}$$

Consider the $(e_1, w_2)$ subsystem given by

$$\begin{aligned} \dot{e}_1 &= G_1(y_1, y_2, \dots, y_n, \alpha_i) \\ &\quad -F_1(x_1, x_2, \dots, x_n, \alpha_i) + u_1 \\ \dot{w}_2 &= G_2(y_1, y_2, \dots, y_n) \\ &\quad +F_2(x_1, x_2, \dots, x_n, \alpha_i) - \dot{\alpha}_1(e_1) + u_2 \end{aligned} \tag{11}$$

Let $e_3$ as a virtual controller in system (11).

Assume that when

$$e_3 = \alpha_2(e_1, w_2) \tag{12}$$

the system (11) is made globally asymptotically stable.

Consider the Lyapunov function defined by

$$V_2(e_2,\ w_2) = V_1(e_1) + w_2^T P_2 w_2 + \sum_{i=k+1}^{m} e_{\alpha_i}^T R_1 e_{\alpha_i} \tag{13}$$

where $P_2$, *and* $R_2$ are positive definite matrices.

Suppose the derivative of $V_2(e_1,\ w_2)$ is

$$\dot{V}_2 = -e_1^T Q_1 e_1 - w_2^T Q_2 w_2 - \sum_{i=k+1}^{m} e_{\alpha_i}^T S_2 e_{\alpha_i} \tag{14}$$

where $Q_1,\ Q_2$, *and* $S_2$ are positive definite matrices.

Then $\dot{V}_2(e_1,\ w_2)$ is a negative definite function.

Thus by Lyapunov stability theory, the error dynamics (11) is globally asymptotically stable. The virtual controller $e_3 = \alpha_2(e_1,\ w_2)$ and the state feedback input $u_2$ make the system (11) asymptotically stable.

For the $n$th state of the error dynamics, define the error variable $w_n$ as

$$w_n = e_n - \alpha_{n-1}(e_1,\ w_2, \ldots, w_{n-1}) \tag{15}$$

Considering the $(e_1,\ w_2, \ldots, w_n)$ subsystem given by

$$\begin{aligned}
\dot{e}_1 &= G_1(y_1,\ y_2, \ldots, y_n, \alpha_i) \\
&\quad -F_1(x_1,\ x_2, \ldots, x_n, \alpha_i) + u_1 \\
\dot{w}_2 &= G_2(y_1,\ y_2, \ldots, y_n, \alpha_i) \\
&\quad +F_2(x_1,\ x_2, \ldots, x_n) - \dot{\alpha}_1(e_1) + u_2 \\
\vdots\ &\ \vdots \qquad\qquad \vdots \\
\dot{w}_n &= G_n(y_1,\ y_2, \ldots, y_n, \alpha_i) \\
&\quad -F_n(x_1,\ x_2, \ldots, x_n, \alpha_i) \\
&\quad -\dot{\alpha}_{n-1}(e_1,\ w_2, \ldots, w_{n-1}) + u_n
\end{aligned} \tag{16}$$

Consider the Lyapunov function defined by

$$\begin{aligned}
V_n(e_2,\ w_2, \ldots, w_n) &= V_{n-1}(e_1,\ w_2, \ldots, w_{n-1}) \\
&\quad +w_n^T P_n w_n \sum_{i=m+1}^{n} e_{\alpha_i}^T R_n e_{\alpha_i}
\end{aligned} \tag{17}$$

where $P_n$, *and* $R_n$ are positive definite matrices.

Suppose the derivative of $V_n(e_1,\ w_2,\ w_3 .... w_n)$ is

$$\begin{aligned}
\dot{V}_n(e_1,\ w_2, \ldots, w_n) &= -e_1^T Q_1 e_1 - w_2^T Q_2 w_2 - \cdots \\
&\quad -w_n^T Q_n w_n - \sum_{i=k+1}^{m} e_{\alpha_i}^T S_n e_{\alpha_i}
\end{aligned} \tag{18}$$

where $Q_1,\ Q_2, \ldots, Q_n, S_n$ are positive definite matrices.

Then $\dot{V}_n(e_1,\ w_2, \ldots, w_n)$ is a negative definite function on $R^n$.

Thus by Lyapunov stability theory (Hahn, 1967), the error dynamics (16) is globally asymptotically stable.

The virtual controller is

$$e_n = \alpha_{n-1}(e_1,\ w_2, \dots, w_{n-1}) \tag{19}$$

and the state feedback input $u_n$ makes the system (16) globally asymptotically stable.

Hence, the state of master and slave systems are globally and asymptotically synchronized.

## 3. System description

Recently, theoretical design and hardware implementation of different kinds of chaotic oscillators have attracted increasing attention, aiming real-world applications of many chaos-based technologies and information systems.

The $n$-scroll hyperchaotic Chua circuit (Yu, Lu, & Chen, 2007) is given by the dynamics

$$\begin{aligned} \dot{x}_1 &= \alpha[g(x_2 - x_1) - x_3] \\ \dot{x}_2 &= \beta[-g(x_2 - x_1) - x_4] \\ \dot{x}_3 &= \gamma_0(x_1 + x_3) \\ \dot{x}_4 &= \gamma x_2 \end{aligned} \tag{20}$$

where $g(x_2 - x_1)$ is given by

$$\begin{aligned} g(x_2 - x_1) = m_{N-1}&(x_2 - x_1) \\ &+\tfrac{1}{2}\textstyle\sum_{i=1}^{N-1}(m_{i-1} - m_i) \\ &\times(|x_2 - x_1 + z_i| - |x_2 - x_1 + z_i|) \end{aligned} \tag{21}$$

The recursive positive switching points $z_i (i = 2,\ 3,\ 4, ..., N-1)$ can be deduced as

$$\begin{aligned} z_2 &= \tfrac{(1+k_1)\sum_{i=1}^{1}(m_{i-1}-m_i)x_i}{m_1-1} - k_1x_1 \\ z_3 &= \tfrac{(1+k_2)\sum_{i=1}^{2}(m_{i-1}-m_i)x_i}{m_2-1} - k_2x_2 \\ &\vdots \quad \vdots \qquad \vdots \\ z_{N-1} &= \tfrac{(1+k_{N-1})\sum_{i=1}^{N-1}(m_{i-1}-m_i)x_i}{m_{N-2}-1} - k_{N-2}x_{N-2} \end{aligned} \tag{22}$$

and the $k_i$ values are obtained as

$$k_i = \frac{x_{i+1} - x_i^E}{x_i^E - x_i}(1 \leq i \leq N-2) \tag{23}$$

in which $x_i^E$ are the positive equilibrium points of $g(x_2 - x_1)$.

### *3.1. Case 1 : 2-scroll hyperchaotic attractor*

The parameters of the systems (20) are taken in the case of hyperchaotic case as $\alpha = 2,\ \beta = 20$.

When $N = 2$, In Equation (20), the function $g(x_2 - x_1)$ is given by

$$\begin{aligned} g(x_2 - x_1) = m_1(x_2 - x_1) &+ \tfrac{1}{2}(m_0 - m_1) \\ &\times(|x_2 - x_1 + z_1| - |x_2 - x_1 + z_1|) \end{aligned} \tag{24}$$

When $m_0 = -0.2$, $m_1 = 3$ and $z_1 < 1$, the 2-scroll hyperchaotic attractor is generated. Figures 1–3 depict the 2-scroll hyperchaotic attractor.

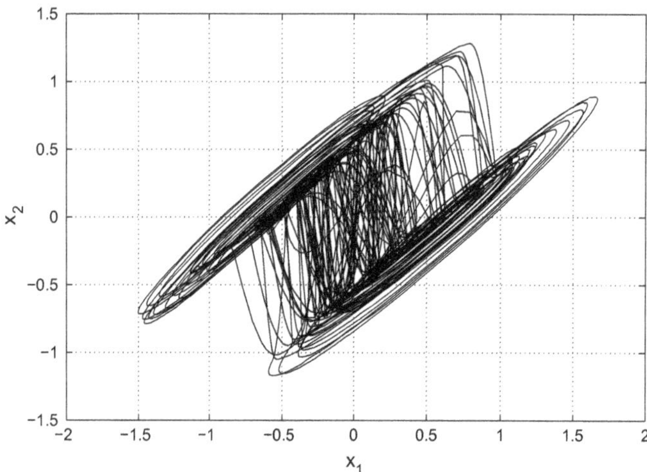


**Figure 1. 2-scroll hyperchaotic attractor.**

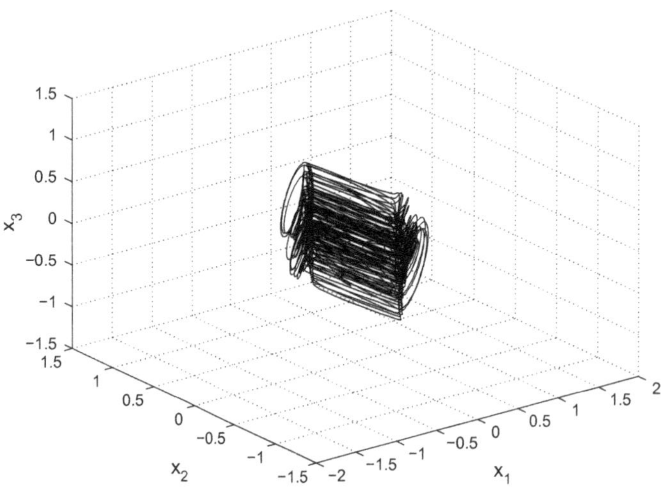


**Figure 2. 2-scroll hyperchaotic attractor.**

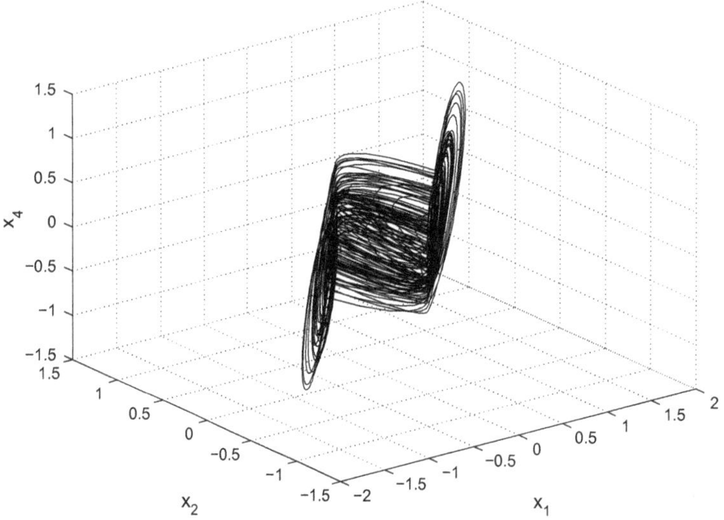


**Figure 3. 2-scroll hyperchaotic attractor.**

### 3.2. Case 2: 3-scroll hyperchaotic attractor

When $N = 3$, in Equation (20), the function $g(x_2 - x_1)$ is given by

$$\begin{aligned} g(x_2 - x_1) = m_2(x_2 - x_1) &+ \tfrac{1}{2}(m_0 - m_1) \\ &\times(|x_2 - x_1 + z_1| - |x_2 - x_1 + z_1|) \\ &+\tfrac{1}{2}(m_1 - m_2) \\ &\times(|x_2 - x_1 + z_2| - |x_2 - x_1 + z_2|) \end{aligned} \tag{25}$$

When $m_0 = 3,\ m_1 = -0.8,\ m_2 = 3,\ z_2 = 1.8333$ and $z_1 < 1$, the 3-scroll hyperchaotic attractor is generated. Figures 4–6 depict the 3-scroll hyperchaotic attractor.

### 3.3. Case 3: 4-scroll hyperchaotic attractor

When $N = 4$, in Equation (20), the function $g(x_2 - x_1)$ is given by

$$\begin{aligned} g(x_2 - x_1) = m_3(x_2 - x_1) &+ \tfrac{1}{2}(m_0 - m_1) \\ &\times(|x_2 - x_1 + z_1| - |x_2 - x_1 + z_1|) \\ &+\tfrac{1}{2}(m_1 - m_2) \\ &\times(|x_2 - x_1 + z_2| - |x_2 - x_1 + z_2|) \\ &+\tfrac{1}{2}(m_2 - m_3) \\ &\times(|x_2 - x_1 + z_3| - |x_2 - x_1 + z_3|) \end{aligned} \tag{26}$$

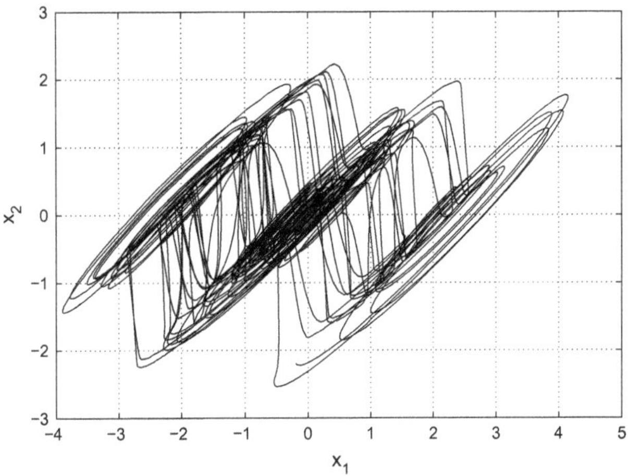


**Figure 4. 3-scroll hyperchaotic attractor.**

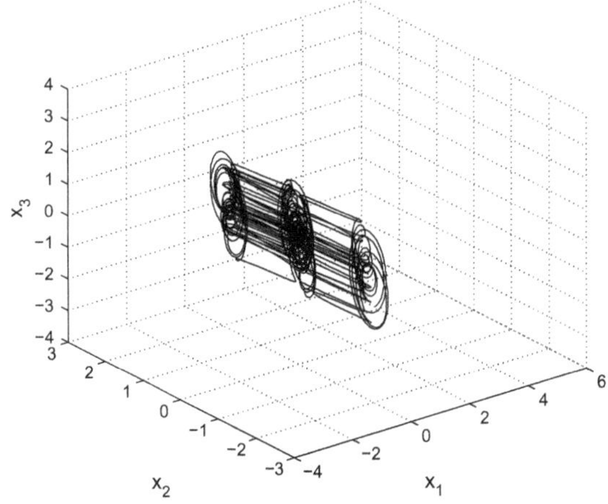


**Figure 5. 3-scroll hyperchaotic attractor.**

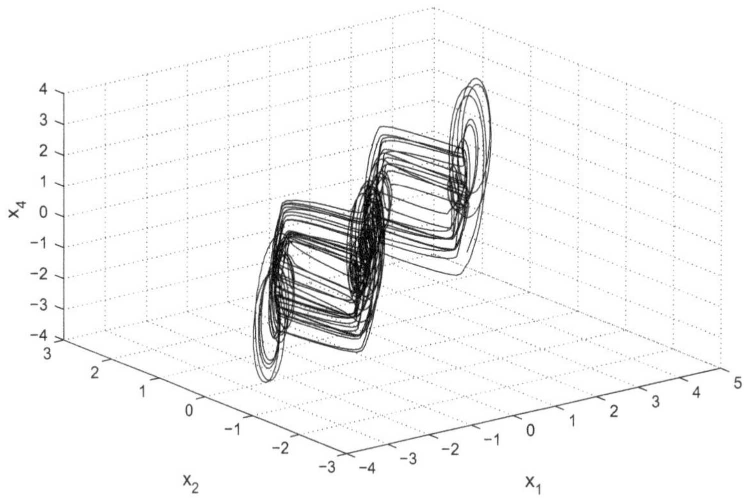


**Figure 6. 3-scroll hyperchaotic attractor.**

When $m_0 = m_2 = -0.7$, $m_1 = m_3 = 2.9$, $m_2 = 3$, $z_2 = 1.5289$, $z_3 = 3.0239$ and $z_1 < 1$, the 4-scroll hyperchaotic attractor is generated.

Figures 7–9 depict the 4-scroll hyperchaotic attractor.

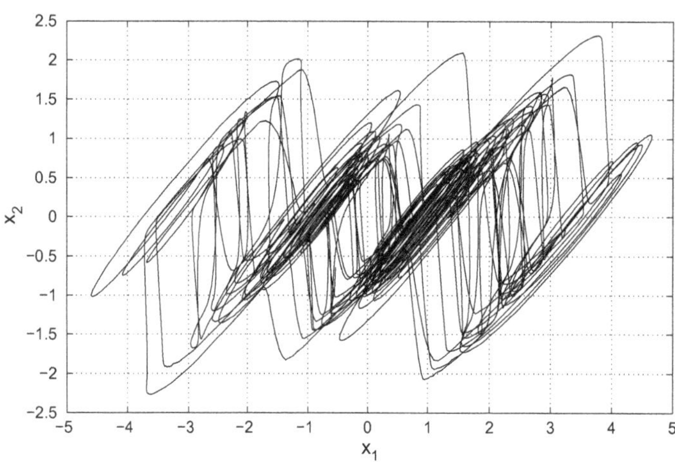


**Figure 7. 4-scroll hyperchaotic attractor**

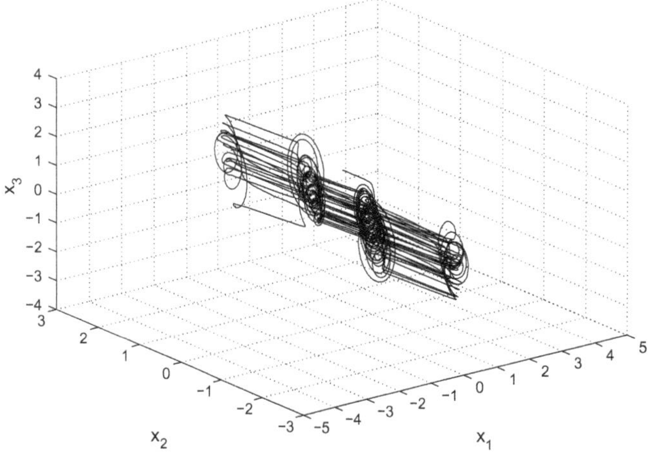


**Figure 8. 4-scroll hyperchaotic attractor.**

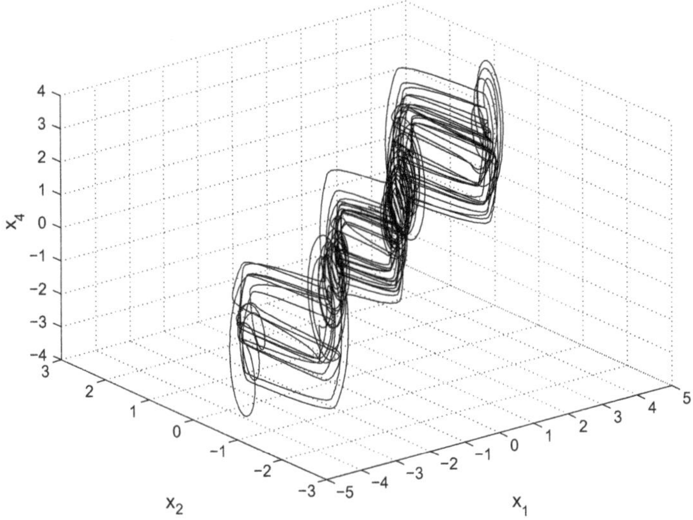


**Figure 9. 4-scroll hyperchaotic attractor.**

## 4. Hybrid Synchronization of n-scroll hyperchaotic Chua circuits via backstepping control with recursive feedback

In this section, the backstepping method with recursive feedback function is applied for the hybrid synchronization of identical hyperchaotic $n$-scroll Chua circuits (Yu et al., 2007).

The $n$-scroll hyperchaotic Chua circuit is taken as the master system, which is described by

$$\begin{aligned} \dot{x}_1 &= \alpha[g(x_2 - x_1) - x_3] \\ \dot{x}_2 &= \beta[-g(x_2 - x_1) - x_4] \\ \dot{x}_3 &= \gamma_0(x_1 + x_3) \\ \dot{x}_4 &= \gamma x_2 \end{aligned} \tag{27}$$

where $g(x_2 - x_1)$ is given by

$$\begin{aligned} g(x_2 - x_1) = m_{N-1}(x_2 - x_1) + \\ \tfrac{1}{2}\textstyle\sum_{i=1}^{N-1}(m_{i-1} - m_i) \\ \times(|x_2 - x_1 + z_i| - |x_2 - x_1 + z_i|) \end{aligned} \tag{28}$$

where $x(t)(i = 1, 2, 3, 4) \in R^4$ are state variables.

The $n$-scroll hyperchaotic Chua circuit is also taken as the slave system, which is described by

$$\begin{aligned} \dot{y}_1 &= \alpha[g(y_2 - y_1) - y_3] + u_1 \\ \dot{y}_2 &= \beta[-g(y_2 - y_1) - y_4] + u_2 \\ \dot{y}_3 &= \gamma_0(y_1 + y_3) + u_3 \\ \dot{y}_4 &= \gamma y_2 + u_4 \end{aligned} \tag{29}$$

where $g(y_2 - y_1)$ is given by

$$\begin{aligned} g(y_2 - y_1) = m_{N-1}(y_2 - y_1) + \tfrac{1}{2}\textstyle\sum_{i=1}^{N-1}(m_{i-1} - m_i) \\ \times(|y_2 - y_1 + z_i| - |y_2 - y_1 + z_i|) \end{aligned} \tag{30}$$

where $y(t)(i = 1, 2, 3, 4) \in R^4$ are state variables.

The hybrid synchronization error is defined by

$$e_1 = y_1 - x_1,\ e_2 = y_1 + x_2,\ e_3 = y_3 - x_3,\ e_4 = y_4 + x_4 \tag{31}$$

The error dynamics is obtained as

$$\begin{aligned}
\dot{e}_1 &= \alpha[g(y_2 - y_1) - g(x_2 - x_1)] - \alpha e_3 + u_1 \\
\dot{e}_2 &= -\beta[g(x_2 - x_1) + g(y_2 - y_1)] - \beta e_4 + u_2 \\
\dot{e}_3 &= \gamma_0(e_1 + e_3) + u_3 \\
\dot{e}_4 &= \gamma e_2 + u_4
\end{aligned} \tag{32}$$

The modified error dynamics is defined by

$$\begin{aligned}
\dot{e}_1 &= \alpha[g(y_2 - y_1) - g(x_2 - x_1)] - \alpha e_3 + u_1 \\
\dot{e}_2 &= -\beta[g(x_2 - x_1) + g(y_2 - y_1)] \\
&\quad -\beta e_4 + e_1 - y_1 + x_1 + u_2 \\
\dot{e}_3 &= \gamma_0(e_1 + e_3) + e_2 - y_2 - x_2 + u_3 \\
\dot{e}_4 &= \gamma e_2 + e_3 - y_3 + x_3 + u_4
\end{aligned} \tag{33}$$

Now the objective is to find control law $u_i,\ i = 1,\ 2,\ 3,\ 4$ and for the parameter update law $\hat{\alpha},\ \hat{\beta},\ \hat{\gamma},\ \hat{\gamma}_0$ for stabilizing the system (32) at the origin.

First consider the stability of the system

$$\dot{e}_4 = \gamma e_2 + e_3 - y_3 + x_3 + u_4 \tag{34}$$

where $e_3$ is regarded as virtual controller.

Consider the Lyapunov function defined by

$$V_1(e_4) = \frac{1}{2}e_4^2 + \frac{1}{2}e_\gamma^2 \tag{35}$$

Let define the parameter estimation error as

$$e_\gamma = \gamma - \hat{\gamma} \tag{36}$$

Differentiating the Equation (36)

$$\dot{e}_\gamma = -\dot{\hat{\gamma}} \tag{37}$$

Differentiate $V_1$ along with the Equation (37)

$$\dot{V}_1 = e_4(\gamma e_2 + e_3 - y_3 + x_3 + u_4) + e_\gamma(-\dot{\hat{\gamma}}) \tag{38}$$

Assume the controller $e_3 = \alpha_1(e_4)$.

If

$$\alpha_1(e_4) = -k_1 e_4, \quad \text{and } u_4 = y_3 - x_3 - \hat{\gamma} e_2 \tag{39}$$

and the parameter update law $\dot{\hat{\gamma}}$ is taken as

$$\dot{\hat{\gamma}} = e_2 e_4 + k_2 e_\gamma \tag{40}$$

then

$$\dot{V}_1 = -k_1 e_4^2 - k_2 e_\gamma^2 \tag{41}$$

which is a negative definite function. Hence, the system (34) is globally asymptotically stable.

The function $\alpha_1(e_4)$ is an estimative function when $e_3$ is considered as a controller.

The error between $e_3$ and $\alpha_1(e_4)$ is

$$w_2 = e_3 - \alpha_1(e_4) = e_3 + k_1 e_4 \tag{42}$$

Consider the $(e_1,\ w_2)$ subsystem given by

$$\begin{aligned} \dot{e}_1 &= e_\gamma e_2 + e_3 \\ \dot{w}_2 &= \gamma_0(e_1 + w_2 - k_1 e_4) + (k_1 e_\gamma + 1)e_2 \\ &\quad + k_1(w_2 - k_1 e_4) - y_2 - x_2 + u_3 \end{aligned} \tag{43}$$

Let $e_2$ be a virtual controller in system (43).

Assume that when $e_2 = \alpha_2(e_4, w_2)$ and the system (43) is made globally asymptotically stable.

Consider the Lyapunov function defined by

$$V_2(e_4,\ w_2) = V_1(e_4) + \frac{1}{2}w_2^2 + \frac{1}{2}e_{\gamma_0}^2 \tag{44}$$

Let us define the parameter estimation error as

$$e_{\gamma_0} = \gamma_0 - \hat{\gamma}_0 \tag{45}$$

Differentiating the Equation (45), we get

$$\dot{e}_{\gamma_0} = -\dot{\hat{\gamma}}_0 \tag{46}$$

The derivative of $V_2(e_4,\ w_2)$ is

$$\begin{aligned} \dot{V}_2 &= \dot{V}_1 + w_2 \dot{w}_2 + e_{\gamma_0} \dot{e}_{\gamma_0} \\ &= e_4(e_\gamma e_2 + w_2 - k_1 e_4) + e_\gamma(\dot{\hat{\gamma}}) \\ &\quad + w_2(\gamma_0(e_1 + w_2 - k_1 e_4) + (k_1 e_\gamma + 1)e_2 \\ &\quad + k_1(w_2 - k_1 e_4) - y_2 - x_2 + u_3) + e_{\gamma_0}(-\dot{\hat{\gamma}}_0) \end{aligned} \tag{47}$$

Substituting for $e_3$ from (42) into (47) and simplifying, we get

$$\begin{aligned} \dot{V}_2 &= -k_1 e_4^2 - k_2 e_\gamma^2 \\ &\quad + w_2(e_4 + \gamma_0(e_1 + w_2 - k_1 e_4) + (k_1 e_\gamma + 1)e_2 \\ &\quad + k_1(w_2 - k_1 e_4) - y_2 - x_2 + u_3) + e_{\gamma_0}(-\dot{\hat{\gamma}}_0) \end{aligned} \tag{48}$$

Assume the virtual controller $e_2 = \alpha_2(e_4, w_2)$

$$\begin{aligned} &\alpha_2(e_1, w_2) = 0 \\ &u_3 = y_2 + x_2 - e_4 - k_1(w_2 \\ &\quad -k_1 e_4) - k_3 w_2 - \hat{\gamma}_0(e_1 + w_2 - k_1 e_4) \end{aligned} \tag{49}$$

The parameter update law $\dot{\hat{\gamma}}_0$ is

$$\dot{\hat{\gamma}}_0 = w_2(e_1 + w_2 - k_1 e_4) + k_4 e_{\gamma_0} \tag{50}$$

Then it follows that

$$\dot{V}_2 = -k_1 e_4^2 - k_2 e_\gamma^2 - k_3 w_2^2 - k_4 e_{\gamma_0}^2 \tag{51}$$

Thus, $\dot{V}_2$ is a negative definite function and hence the system (43) is globally asymptotically stable.

Define the error variable $e_2$ and $\alpha_2(e_4, w_2)$ as

$$w_3 = e_2 - \alpha_2(e_4, w_2) \tag{52}$$

Consider the $(e_4, w_2, w_3)$ subsystem given by

$$\begin{aligned} \dot{e}_1 &= e_\gamma e_2 + e_3 \\ \dot{w}_2 &= (k_1 e_\gamma + 1)e_2 - e_4 - k_3 w_2 \\ &\quad + e_{\gamma_0}(e_1 + w_2 - k_1 e_4) \\ \dot{w}_3 &= -\beta[g(x_2 - x_1) + g(y_2 - y_1)] \\ &\quad -\beta e_4 + e_1 - y_1 + x_1 + u_2 \end{aligned} \tag{53}$$

Let $e_1$ be a virtual controller in system (53).

Assume when it is equal to $e_1 = \alpha_3(e_4, w_2, w_3)$, the system (53) is made globally asymptotically stable.

Consider the Lyapunov function defined by

$$V_3(e_1, w_2, w_3) = V_2(e_1, w_2) + \frac{1}{2}w_3^2 + \frac{1}{2}e_\beta^2 \tag{54}$$

Let us define the parameter estimation error as

$$e_\beta = \beta - \hat{\beta} \tag{55}$$

The derivative of (55) is

$$\dot{e}_\beta = -\dot{\hat{\beta}} \tag{56}$$

The derivative of $V_3(e_3, w_2, w_3)$ is

$$\dot{V}_3 = \dot{V}_2(e_1, w_2) + w_3 \dot{w}_3 + e_\beta \dot{e}_\beta \tag{57}$$

i.e.

$$\begin{aligned} \dot{V}_3 &= e_4(e_\gamma e_2 + w_2 - k_1 e_4) + e_\gamma(-e_2 e_4 - k_2 e_\gamma) \\ &\quad + w_2[(k_1 e_\gamma + 1)w_3 - e_4 - k_3 w_2 \\ &\quad + e_{\gamma_0}(e_1 + w_2 - k_1 e_4)] \\ &\quad + e_{\gamma_0}[-w_2(e_1 + w_2 - k_1 e_4) - k_4 e_{\gamma_0}] \\ &\quad + w_3[-\beta[g(x_2 - x_1) + g(y_2 - y_1)] \\ &\quad -\beta e_4 + e_1 - y_1 + x_1 + u_2] + e_\beta(-\dot{\hat{\beta}}) \end{aligned} \tag{58}$$

Substituting for $e_2$ from (52) into (58) and simplifying, we get

$$\begin{aligned}\dot{V}_3 = &-k_1e_1^2 - k_2e_\gamma^2 - k_3w_2^2 - k_4e_{\gamma_0}^2\\ &+w_3[w_2(k_1e_\gamma + 1)\\ &-\beta[g(x_2 - x_1) + g(y_2 - y_1)]\\ &-\beta e_4 + e_1 - y_1 + x_1 + u_2] + e_\beta(-\dot{\hat{\beta}})]\end{aligned} \tag{59}$$

Assume the virtual controller $e_1 = \alpha_3(e_4, w_2, w_3)$.

choose

$$\begin{aligned}&\alpha_3(e_4, w_2, w_3) = 0\\ &u_2 = y_1 - x_1 + \hat{\beta}e_4\\ &\quad +\beta[g(x_2 - x_1) + g(y_2 - y_1)]\\ &\quad -k_5w_3 - w_2(k_1e_\gamma + 1)\end{aligned} \tag{60}$$

The parameter update law $\dot{\hat{\beta}}$ is

$$\dot{\hat{\beta}} = -w_3w_4 + k_6e_\beta \tag{61}$$

Then it follows that

$$\dot{V}_3 = -k_1e_4^2 - k_2e_\gamma^2 - k_3w_2^2 - k_4e_{\gamma_0}^2 - k_5w_3^2 - k_6e_\beta^2 \tag{62}$$

Thus, $\dot{V}_3$ is a negative definite function and hence the system (53) is globally asymptotically stable.

The error between $e_1$ and $\alpha_3(e_4, w_2, w_3)$ is

$$w_4 = e_1 - \alpha_3(e_1, w_2, w_3) \tag{63}$$

Consider $(e_4, w_2, w_3, w_4)$ subsystem given by

$$\begin{aligned}\dot{e}_1 &= e_\gamma e_2 + w_2 - k_1e_4\\ \dot{w}_2 &= (k_1e_\gamma + 1)e_2 - e_4 - k_3w_2\\ &\quad +e_{\gamma_0}(e_1 + w_2 - k_1e_4)\\ \dot{w}_3 &= -e_\beta e_4 + e_1 - k_5w_3 - w_2(k_1e_\gamma + 1)\\ \dot{w}_4 &= \alpha[g(y_2 - y_1) - g(x_2 - x_1)] - \alpha e_3 + u_1\end{aligned} \tag{64}$$

Consider the Lyapunov function defined by

$$V_4(e_1, w_2, w_3, w_4) = V_3(e_1, w_2, w_3) + \frac{1}{2}w_4^2 + \frac{1}{2}e_\alpha^2 \tag{65}$$

Let define the parameter error as

$$e_\alpha = \alpha - \hat{\alpha} \tag{66}$$

The derivative of $e_\alpha$ is

$$\dot{e}_\alpha = -\dot{\hat{\alpha}} \tag{67}$$

The derivative of $V_4(e_3, w_2\ w_3, w_4)$ is

$$\dot{V}_4 = \dot{V}_3(e_1, w_2, w_3) + \dot{w}_4w_4 + e_\alpha\dot{e}_\alpha \tag{68}$$

i.e.

$$\begin{aligned}\dot{V}_4 = & -k_1 e_4^2 - k_2 e_\gamma^2 - k_3 w_2^2 - k_4 e_{\gamma_0}^2 - k_5 w_3^2 - k_6 e_\beta^2 \\ & + w_4(w_3 + \alpha[g(y_2 - y_1) - g(x_2 - x_1)] \\ & - \alpha e_3 + u_1) + e_\alpha(-\dot{\hat{\alpha}})\end{aligned} \tag{69}$$

Choose the controller

$$u_1 = -w_3 - \alpha[g(y_2 - y_1) - g(x_2 - x_1)] + \hat{\alpha} e_3 - k_7 w_4 \tag{70}$$

and the parameter update law $\dot{\hat{\alpha}}$ is

$$\dot{\hat{\alpha}} = -e_3 w_4 + k_8 e_\alpha \tag{71}$$

Then

$$\begin{aligned}\dot{V}_4 = & -k_1 e_4^2 - k_2 e_\gamma^2 - k_3 w_2^2 - k_4 e_{\gamma_0}^2 - k_5 w_3^2 - k_6 e_\beta^2 \\ & - k_7 w_4^2 - k_8 e_\alpha^2\end{aligned} \tag{72}$$

Thus, $\dot{V}_4$ is a negative definite function.

Thus, by Lyapunov stability theory (Hahn, 1967), the error dynamics (64) is globally asymptotically stable for all initial condition.

Thus, the states of master and slave systems are globally and asymptotically hybrid synchronized.

## 5. Theorem

The identical n-scroll hyperchaotic Chua's circuit (27) and (29) are globally and asymptotically hybrid synchronized with the adaptive backstepping controls

$$\begin{aligned} u_1 = & -w_3 - \alpha[g(y_2 - y_1) - g(x_2 - x_1)] \\ & + \hat{\alpha} e_3 - k_7 w_4 \\ u_2 = & \, y_1 - x_1 + \hat{\beta} e_4 + \beta[g(x_2 - x_1) + g(y_2 - y_1)] \\ & - k_5 w_3 - w_2(k_1 e_\gamma + 1) \\ u_3 = & -w_2 - \gamma_0 e_1 - 2\gamma_0 w_3 + e_4 \\ u_4 = & \, y_2 + x_2 - e_4 - k_1(w_2 - k_1 e_4) - k_3 w_2 \\ & - \hat{\gamma}_0(e_1 + w_2 - k_1 e_4)\end{aligned} \tag{73}$$

and with the parameter update laws

$$\begin{aligned}\dot{\hat{\alpha}} &= -e_3 w_4 + k_8 e_\alpha \\ \dot{\hat{\beta}} &= -w_3 w_4 + k_6 e_\beta \\ \dot{\hat{\gamma}}_0 &= w_2(e_1 + w_2 - k_1 e_4) + k_4 e_{\gamma_0} \\ \dot{\hat{\gamma}} &= e_2 e_4 + k_2 e_\gamma\end{aligned} \tag{74}$$

## 6. Numerical simulation

For the numerical simulations, the fourth order Runge–Kutta method is used to solve the differential Equations (27) and (29) with the backstepping controls $u_1$, $u_2$, $u_3$, and $u_4$ given by (29).

### *6.1. Case 1: 2-scroll hyperchaotic attractor*

The parameters of the systems (27) are taken in the case of hyperchaotic case as $\alpha = 2,\ and \beta = 20.$

When $N = 2,\ m_0 = -0.2,\ m_1 = 3$ and $z_1 = 0.5$, the double scroll hyperchaotic attractor is generated.

The initial values of the master system (27) are chosen as $x_1(0) = 0.947,\ x_2(0) = 0.234,\ x_3(0) = 0.472,\ x(4) = 0.198$ and the initial values of the slave system (29) are chosen as $y_1(0) = 0.157,\ y_2(0) = 0.648,\ y_3(0) = 0.810,\ y(4) = 0.108.$

The initial values of the estimated parameters are $\hat{\alpha} = 2.3,\ \hat{\beta} = 4.9,\ \hat{\gamma} = 10,\ \hat{\gamma_0} = 5.5.$

Figure 10 depicts the hybrid synchronization of 2-scroll hyperchaotic Chua's circuits (27) and (29).

Figure 11 depicts the hybrid synchronization error between 2-scroll hyperchaotic Chua's circuits (27) and (29).

Figure 12 depicts the parameter estimation of 2-scroll hyperchaotic Chua's circuits (27) and (29).

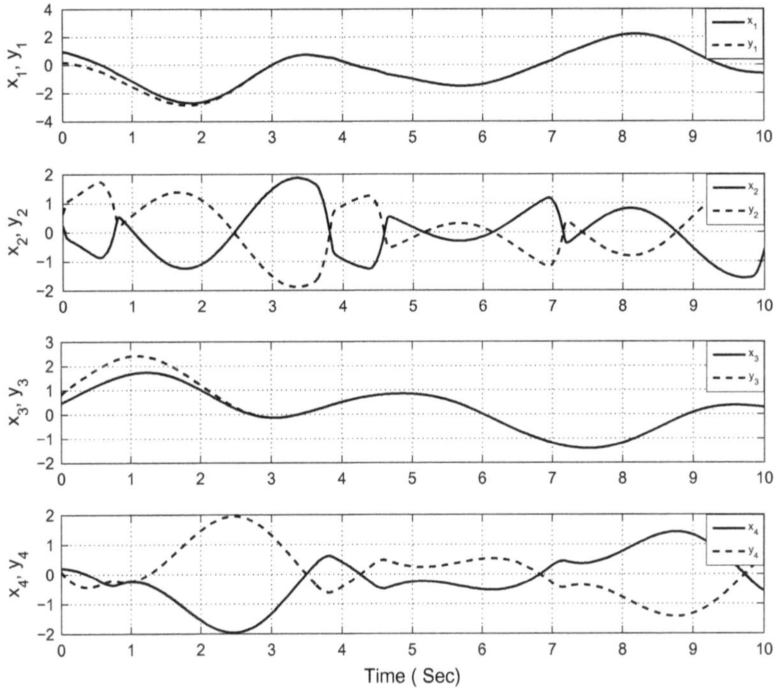


**Figure 10. Hybrid synchronization of 2-scroll hyperchaotic Chua's circuits (27) and (29).**

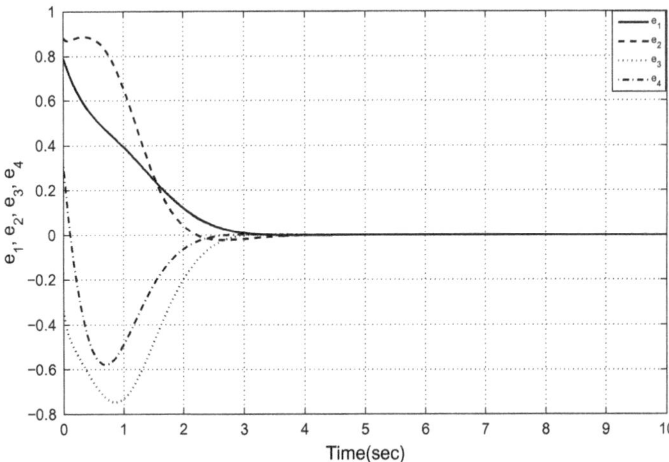


**Figure 11. Hybrid synchronization error between 2-scroll hyperchaotic Chua's circuits (27) and (29).**

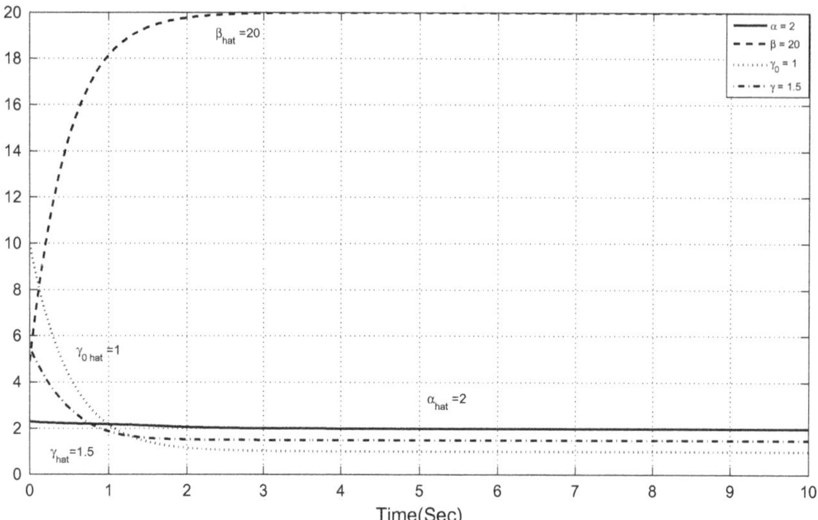


**Figure 12. Parameter estimation of 2-scroll hyperchaotic Chua's circuits (27) and (29). The estimated values of the parameters** $\hat{\alpha}\hat{\beta}, \hat{\gamma}$and$\hat{\gamma}$**converge to system parameters** $\alpha = 2, \beta = 20, \gamma = 1.5,$ **and** $\gamma_0 = 1$.

The estimated values of the parameters $\hat{\alpha}, \hat{\beta}, \hat{\gamma},$ and $\hat{\gamma_0}$ converge to system parameters $\alpha = 2, \beta = 20, \gamma = 1.5,$ and $\gamma_0 = 1$.

### *6.2. Case 2: 3-scroll hyperchaotic attractor*

When $N = 3, m_0 = 3, m_1 = -0.8, m_2 = 3, z_2 = 1.8333,$ and$z_1 = 0.5$ the 3-scroll hyperchaotic attractor is generated.

The initial values of the master system (27) are chosen as $x_1(0) = 0.431, x_2(0) = 0.281, x_3(0) = 0.983, x(4) = 0.731$ and the initial values of the slave system (29) are chosen as $y_1(0) = 1.012, y_2(0) = 3.012, y_3(0) = 2.018, y(4) = 0.112$.

The initial values of the estimated parameters are $\hat{\alpha} = 10.318, \hat{\beta} = 3.121, \hat{\gamma} = 5.000, \hat{\gamma_0} = 3$.

Figure 13 depicts the hybrid synchronization of 3-scroll hyperchaotic Chua's circuits (27) and (29).

Figure 14 depicts the hybrid synchronization error between 3-scroll hyperchaotic Chua's circuits (27) and (29).

Figure 15 depicts the parameter estimation of 3-scroll hyper chaotic Chua's circuits (27) and (29).

The estimated values of the parameters $\hat{\alpha}, \hat{\beta}, \hat{\gamma},$ and $\hat{\gamma_0}$ converge to system parameters $\alpha = 2, \beta = 20, \gamma = 1.5,$ and $\gamma_0 = 1$.

### *6.3. Case 3: 4-scroll hyperchaotic attractor*

When $N = 4, m_0 = m_2 = -0.7, m_1 = m_3 = 2.9, m_2 = 3, z_2 = 1.5289, z_3 = 3.0239$ and $z_1 = 0.5$ the 4-scroll hyperchaotic attractor is generated.

The initial values of the master system (27) are chosen as $x_1(0) = 1.938, x_2(0) = 2.138, x_3(0) = 1.708, x(4) = 3.325$ and the initial values of the slave system (29) are chosen as $y_1(0) = 0.125, y_2(0) = 0.986, y_3(0) = 0.065, y(4) = 1.363$.

The initial values of the estimated parameters are $\hat{\alpha} = 20.363, \hat{\beta} = 0.563, \hat{\gamma} = 10.613, \hat{\gamma_0} = 9$.

Figure 16 depicts the hybrid synchronization of 4-scroll hyperchaotic Chua's circuits (27) and (29).

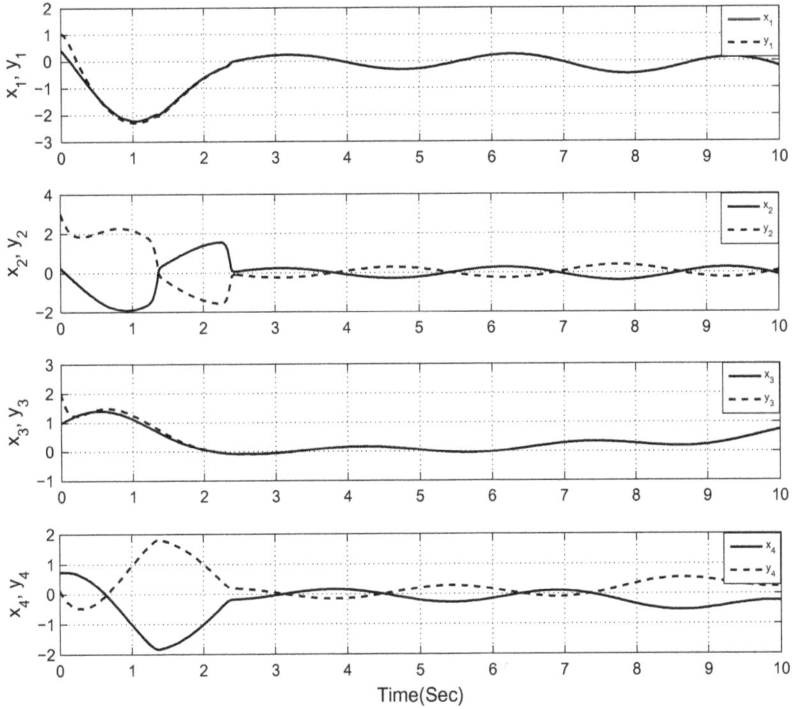


**Figure 13. Hybrid synchronization of 3-scroll hyperchaotic Chua's circuits (27) and (29).**

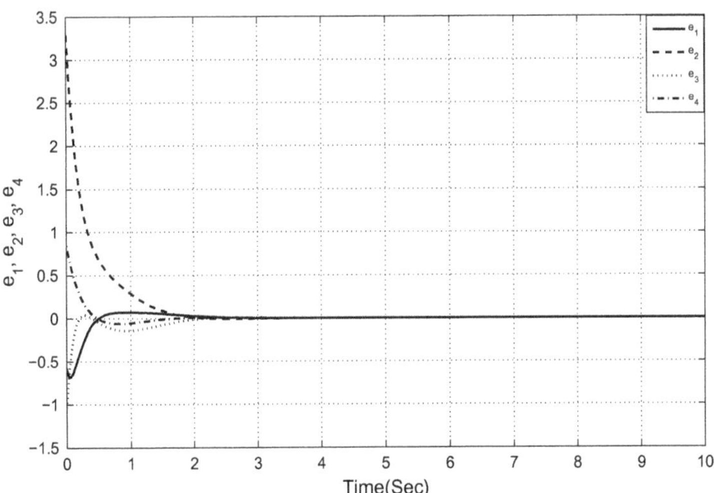


**Figure 14. Hybrid synchronization error between 3-scroll hyperchaotic Chua's circuits (27) and (29).**

Figure 17 depicts the hybrid synchronization error between 4-scroll hyperchaotic Chua's circuits (27) and (29).

Figure 18 depicts the parameter estimation of 4-scroll hyperchaotic Chua's circuits (27) and (29).

The estimated values of the parameters $\hat{\alpha}$, $\hat{\beta}$, $\hat{\gamma}$, and $\hat{\gamma}_0$ converge to system parameters $\alpha = 2$, $\beta = 20, \gamma = 1.5$, and $\gamma_0 = 1$.

## 7. Conclusion

In this paper, the adaptive backstepping control method has been applied to achieve global chaos hybrid synchronization for a family of n-scroll hyperchaotic Chua circuits. The backstepping control is a systematic procedure for hybrid synchronizing hyperchaotic systems and there is no derivative

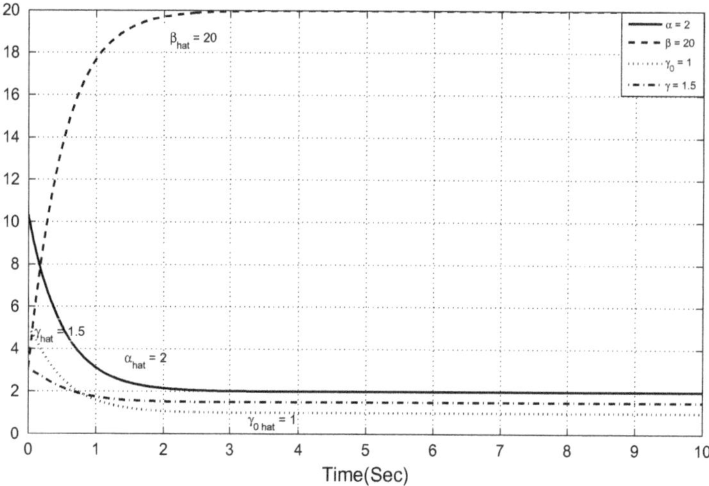


**Figure 15. Parameter estimation of 3-scroll hyperchaotic Chua's circuits (27) and (29). The estimated values of the parameters $\hat{\alpha}\hat{\beta}$, $\hat{\gamma}$and$\hat{\gamma}$converge to system parameters $\alpha = 2, \beta = 20, \gamma = 1.5,$ and $\gamma_0 = 1$.**

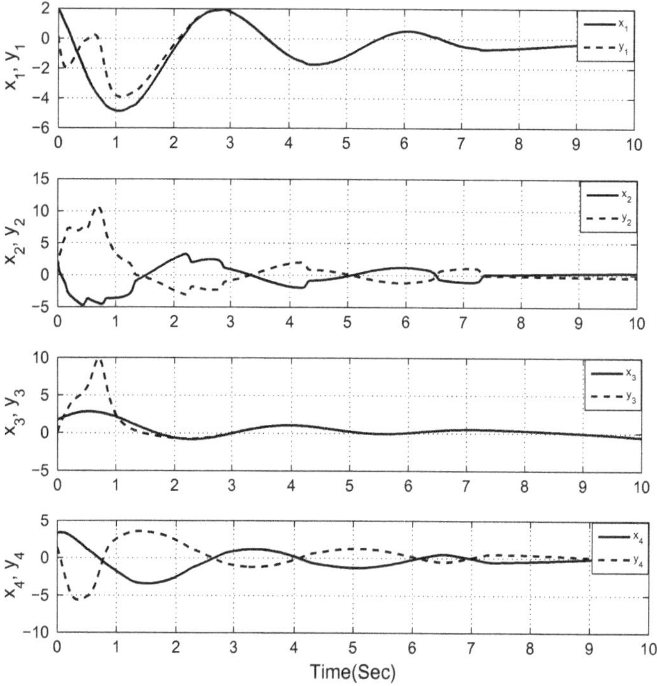


**Figure 16. Hybrid synchronization of 4-scroll hyperchaotic Chua's circuits (27) and (29).**

in controller. The adaptive backstepping control design has been demonstrated to class of n-scroll hyperchaotic Chua circuits. Numerical simulations have been given to illustrate and validate the effectiveness of the proposed hybrid synchronization schemes of the chaotic circuit. The adaptive backstepping method is very effective and convenient to achieve global chaos hybrid synchronization.

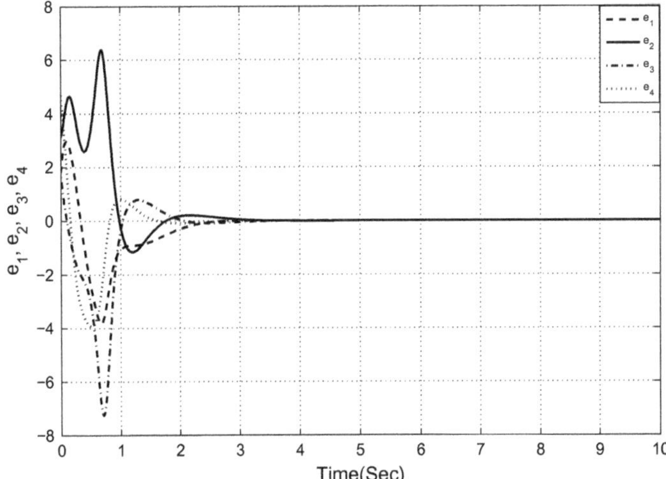


**Figure 17. Hybrid synchronization error between 4-scroll hyperchaotic Chua's circuits (27) and (29).**

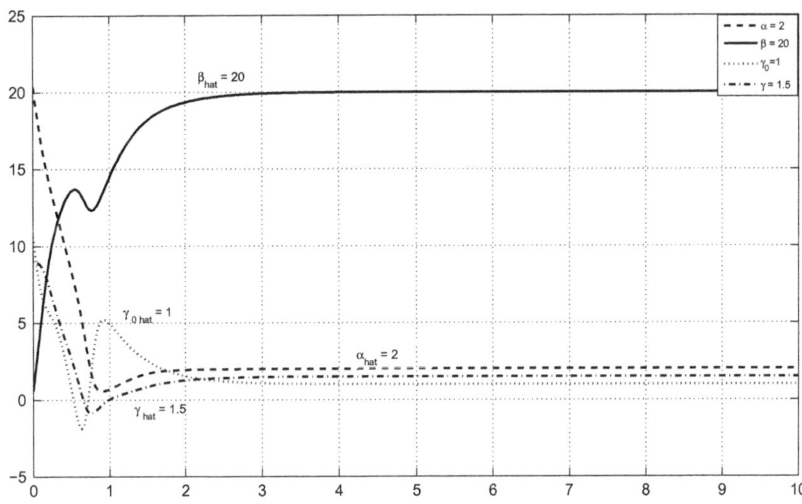


**Figure 18. Parameter estimation of 4-scroll hyperchaotic Chua's circuits (27) and (29). The estimated values of the parameters $\hat{\alpha}\hat{\beta}$, $\hat{\gamma}$and$\hat{\gamma}_0$converge to system parameters $\alpha = 2, \beta = 20, \gamma = 1.5$, and $\gamma_0 = 1$.**

**Funding**

The authors received no direct funding for this research.

**Author details**

Suresh Rasappan[1]

E-mail: mrpsuresh83@gmail.com

1 Department of Mathematics, Vel Tech University, No. 42 Avadi-Vel Tech Road, Avadi, Chennai 600062 Tamilnadu, India.

**References**

Alligood, K. T., Sauer, T., & Yorke, J. A. (1997). *Chaos: An introduction to dynamical systems*. Berlin: Springer-Verlag.

Bauer, F., Atay, F. M., & Jost, J. (2010). Synchronized chaos in network of simple unit. *Europhysics Letters, 89*, 20002. doi:10.1209/0295-5075/89/20002

Blasius, B., Huppert, A., & Stone, L. (1999). Complex dynamics and phase synchronization in spatially extended ecological system. *Nature, 399*, 354–359.

Che, Y. Q., Wang., J., Tsang, K. M., & Chen, W. L. (2010). Unidirectional synchronization for Hindmarsh-Rose neurons via robust adaptive sliding mode control. *Nonlinear Analysis: Real world Application, 11*, 1096–1104.

Chen, Y.-Y. (1996). Masking messages in chaos. *Europhysics Letters, 34*, 245–250.

Coffman, K. G., McCormick, W. D., Noszticzius, Z., Simoyi, R. H., & Swinney, H. L. (1987). Universality, multiplicity and the effect of iron impurities in the Belousov–Zhabotinskii reaction. *Journal of Chemical Physics, 86*, 119–129.

Fujisaka, H., & Yamada, T. (1983). Stability theory of synchronized motion in coupled-oscillator systems. *Progress of Theoretical Physics, 69*, 32–37.

Ge, Z. M., & Chen, C. C. (2006). Projective synchronization of a new hyper chaotic Lorenz systems. *Chaos, Solitons Fractals, 20*, 639–647.

Ghosh, D., Banerjee, S., & Chowdhury, A. R. (2007). Synchronization between variable time-delayed systems and cryptography. *Europhysics Letters, 80*, 30006. doi:10.1209/0295-5075/30006

Hahn, W. (1967). *The stability of motion.* Berlin: Springer-Verlag.

Han, S. K., Kerrer, C., & Kuramoto, Y. (1995). D-phasing and bursting in coupled neural oscillators. *Physical Review Letters, 75*, 3190–3193.

Harmov, A. E., Koronovskii, A. A., & Moskalenko, O. I. (2005a). Generalized synchronization onset. *Europhysics Letters, 72*, 901–907.

Harmov, A. E., Koronovskii, A. A., & Moskalenko, O. I. (2005b). Intermittent generalized synchronization in unidirectionally coupled chaotic oscillators. *Europhysics Letters, 70*, 169–174.

Hindmarsh, J. L., & Rose, R. M. (1984). A model of neuronal bursting using 3-coupled First order differential equations. *Proceedings of the Royal Society London B: Biological Sciences, 221*, 81–102.

Jian-Ping, Y., & Chang-Pin, L. (2006). Generalized projective synchronization for the chaotic Lorenz systems and the chaotic Chen system. *Journal of Shanghai University, 10*, 299–304.

Kanter, I., Kopelowitz, E., Kestler, J., & Kinzel, W. (2008). Chaos synchronization with dynamics filter: Two-way is better than one way. *Europhysics Letters, 83,* 50005. doi:10.1009/0295-5075/50005

Kocarev, L., & Parlitz, U. (1995). General approach for chaotic synchronization with application to communication. *Physical Review Letters, 74*, 5028–5030.

Lakshmanan, M., & Murali, K. (1996). *Chaos in nonlinear oscillators: Controlling and synchronization.* Singapore: World Scientific.

Li, R. H., Xu, W., & Li, S. (2007). Adaptive generalized projective synchronization in different chaotic systems based on parameter identifiction. *Physics Letters A, 367*, 119–206.

Lih, J. S., Chern, J. L., & Otsuka, K. (2012). Joint time-frequency analysis of synchronized chaos. *Europhysics Letters, 57*, 810–816.

Lu, J., Wu, X., & Han, X. (2004). Adaptive feedback synchronization of a unified chaotic system. *Physics Letters A, 329*, 327–333.

Moreno, Y., & Pacheco, A. F. (2004). Synchronization of Kuramato oscillators in scale-free networks. *Europhysics Letters, 68*, 603–609.

Moukam Kakmeni, F. M., Nguenang, J. P., & Kofane, T. C. (2006). Chaos synchronization in bi-axial magnets modeled by bloch equation. *Chaos, Solitons Fractals, 30*, 690–699.

Murali, K., & Lakshmaman, M. (2003). Secure communication using a compound signalmusing sampled-data feedback. *Applied Mathematics and Mechanics, 11*, 1309–1315.

Ott, E., Grebogi, C., & Yorke, J. A. (1990). Controlling chaos. *Physical Review Letters, 64*, 1196–1199.

Park, J. H. (2008). Adaptive control for modified projective synchronization of a four dimentional chaotic system with uncertain parameters. *Journal of Computational and Applied Maths, 213*, 288–293.

Park, J. H., & Kwon, O. M. (2003). A novel criterion for delayed feedback control of time-delay chaotic systems. *Chaos, Soliton Fractals, 17*, 709–716.

Park, J. H., Lee, S. M., & Kwon, O. M. (2007). Adaptive synchronization of Genesio-Tesi chaotic system. *Physics Letters A, 371*, 263–270.

Pecora, L. M., & Carroll, T. L. (1990). Synchronization in chaotic systems. *Physical Review Letters, 64*, 821–824.

Qi, G. X., Huang, H. B., Chen, L., Wang, H. J., & Shen, C. K. (2008). Fast synchronization in neuronal networks. *Europhysics Letters, 83*, 38003. doi:10.1209/0295-5075/82/38003

Qiang, J. (2007). Phase synchronization of coupled chaotic multiple time scales systems. *Physics Letters A, 370*, 40–45.

Sundarapandian, V., & Suresh, R. (2010). Global chaos synchronization for Rossler and Arneodo chaotic systems by nonlinear control. *Far East Journal of Applied Mathematics, 44*, 137–148.

Suresh, R., & Sundarapandian, V. (2012a). Global chaos synchronization of WINDMI and Coullet chaotic systems by backstepping control. *Far East Journal of Mathematical Sciences, 67*, 265–287.

Suresh, R. & Sundarapandian, V. (2012b). Hybrid synchronization of *n*-scroll Chua and Lur'e chaotic systems via backstepping control with novel feedback. *Archives of Control Sciences, 22*, 73–95.

Suresh, R., & Sundarapandian, V. (2012c). Global chaos synchronization of hyperchaotic ChenLee system using backstepping contol via recursive feedback. In J. Mathew, P. Patra, D. K. Pradhan, & A. J. Kuttyamma (Eds), *Lecture notes in computer science* (Vol. 305, CCIS pp. 212–221.) Berlin Heidelbreg: Springer Verlag.

Suresh, R., & Sundarapandian, V. (2013). Global chaos synchronization of a family of *n*-scroll hyperchaotic Chua circuits using backstepping control with recursive feedback. *Far East Journal of Mathematical Sciences, 73*, 73–95.

Tokuda, I. T., Kurths, J., Kiss, I. Z., & Hudson, J. L. (2008). Predicting phase synchronization of non-phase-coherent chaos. *Europhysics Letters, 83*, 50003. doi:1209/0295-5075/83/50003

Wang, J., Chen, L., & Deng, B. (2009). Synchronization of ghostburster neurons in external electrical stimulation via $H_\infty$ variable universe fuzzy adaptive control. *Chaos, Solitons Fractals, 39*, 2076–2085.

Wang, J., Dianchen, L., & Tian, L. (2006). Global synchronization for time delay WINDMI system. *Chaos, Solitons and Fractals, 30*, 629–635.

Wang, Z., Zhang, W., & Guo, Y. (2010). *Robust consensus for uncertain multi-agent systems on directed communication topologies.* In *IEEE International Conference on Decision and Control* (pp. 6317–6322). Atlanta, GA.

Wang, Z., Zhang, W., & Guo, Y. (2011a, June 29–July 1). *Adaptive output consensus tracking of a class of uncertain multi-agent systems. American Control Conference.* (pp. 3387–3392). San Francisco, CA.

Wang, Z., Zhang, W., & Guo, Y. (2011b, June 29–July 1). Adaptive backstepping-based synchronization of networked uncertain Lagrangian systems. In *American Control Conference.* (pp. 1057–1062). San Francisco, CA.

Wang, Y. M., & Zhu, H. (2006). Generalized synchronization of continuous chaotic systems. *Chaos Solitons Fractals, 27*, 97–101.

Wu, X., & Lu, J. (2003). Parameter identification and backstepping control of uncertain chaotic system. *Chaos Soliton Fractals, 18*, 721–729.

Ya, H. T. (2004). Design of adaptive sliding mode controller for chaos synchronizationwith uncertainities. *Chaos Soliton and Fractals, 22*, 341–347.

Yang, T., & Chua, L. O. (1999). Generalized synchronization of chaos via linear transformations. *International Journal of Bifurcation Chaos, 9*, 215–219.

Yu, S., Lu, J., & Chen, G. (2007). A family of n-scroll hyperchaotic attractors and their realization. *Physics Letters A, 364*, 244–251.

Yu, Y., & Zhang, S. C. (2006). Adaptive backstepping synchronization of uncertain chaotic systems. *Chaos Soliton and Fractals, 27*, 1369–1375.

Zhao, L., Lai, Y.-C., Wang, R., & Gao, J.-Y. (2004). Limits to chaotic phase synchronization. *Europhysics Letters, 66*, 324. doi:10.1209/epl/i2003-10220-2

Zeng, G. Z., Chen, L. S., & Sun, L. H. (2005). Complexity of an SIR epidemic dynamics model with impulsive vaccination control. *Chaos, Solitons and Fractals, 26*, 495–505.

# 8

# Group consensus in generic linear multi-agent systems with inter-group non-identical inputs

Yilun Shang[1*]

*Corresponding author: Yilun Shang, Einstein Institute of Mathematics, Hebrew University of Jerusalem, Jerusalem 91904, Israel; SUTD-MIT International Design Center, Singapore University of Technology and Design, Singapore 138682, Singapore

E-mail: shylmath@hotmail.com

Reviewing editor: James Lam, The University of Hong Kong, Hong Kong

**Abstract:** This paper studies the group consensus problem for generic linear multi-agent systems under directed information flow. External adapted inputs are introduced to realize the intra-group synchronization as well as the inter-group separation. Without imposing complicated algebraic criteria or restrictive graphic conditions on the interaction topology, we show that the group consensus can be achieved by designing appropriate gains given any magnitude of the coupling strengths among the agents. Numerical examples are presented to illustrate the availability of our results.

**Subjects: Automation Control, Linear & Multilinear Algebra, Robotics & Cybernetics**

**Keywords: network of multiple agents, linear time-invariant system, group consensus, cooperative control**

### 1. Introduction

Consensus problems for multi-agent systems have been attracting many researchers in recent years, due to their broad applications in various areas such as swarming/flocking (Gazi & Passino, 2003; Shang & Bouffanais, 2014; Tanner, Jadbabaie, & Pappas, 2007), distributed sensor networks (Kar & Moura, 2009), and multi-vehicle formation control (Lian & Deshmukh, 2006; Ren & Sorensen, 2008). As a fundamental issue in cooperative control of multiple agents, the main goal of consensus problem is to design appropriate distributed algorithms (referred to as consensus protocols), such

## ABOUT THE AUTHOR

Yilun Shang is a researcher at Hebrew University of Jerusalem. He completed his doctoral studies at Shanghai Jiao Tong University in 2010, and held appointments at University of Texas at San Antonio and Singapore University of Technology and Design. His research work involves the development of network science and complex system control. He is particularly interested in random graph, self-organization, and agent-based modeling.

## PUBLIC INTEREST STATEMENT

The field of consensus problem has evolved rapidly in the last decades. More complicated consensus patterns are explored including group (or cluster) consensus, where certain components inside the network show isochronous synchronization phenomenon. This paper addresses the group consensus problem of a network of continuous-time agents with generic linear dynamics. Such systems include single-integrator, double-integrator, and higher order integrator dynamics as special cases. By introducing external adapted inputs, group consensus can be realized exponentially fast. Here, the external control inputs contribute to the intra-group synchronization and inter-group separation, without imposing complicated algebraic criteria or restrictive topological constraints. Moreover, only non-negative weights are assigned to the communication links, which have practical applicability.

that the states of all agents converge to a common value with information exchanges between each other. In Olfati-Saber and Murray (2004), a theoretical framework for consensus problems was provided for continuous-time networked dynamical systems under both fixed and switching topologies. In Jadbabaie, Lin, and Morse (2003), asymptotic consensus protocols based on nearest neighbor rules were designed for first-order discrete-time systems. Up to now, numerous papers concerning consensus protocols, including average consensus (Fagnani & Zampieri, 2009; Lin & Jia, 2008), finite-time consensus (Shang, 2012), stochastic consensus (Hatano & Mesbahi, 2005; Huang, Dey, Nair, & Manton, 2010), leader-following consensus (Wen, Duan, Chen, & Yu, 2014; Wen, Li, Duan, & Chen, 2013), and quantized consensus (Kashyap, Başar, & Srikant, 2007), have been published. For details, we refer the readers to survey papers (Cao, Yu, Ren, & Chen, 2013; Olfati-Saber, Fax, & Murray, 2007) and the references therein.

The aforementioned results focus attention on the complete consensus of all agents in a network. However, when carrying out a cooperative task, a group of agents should be capable of coping with unanticipated situations, and may evolve into several subgroups with the changes of environments, situations, or even time. This phenomenon widely exists in engineering and biological systems, from military reconnaissance to heterogeneous robots sorting (Kumar, Garg, & Kumar, 2010), from predator-evasion behaviors of a herd of animals (Schellinck & White, 2011) to opinion formation in social networks (Shang, 2013a, 2014). Suitable protocols have been designed recently to ensure *group consensus*, i.e. the states of all agents within the same subgroup asymptotically converge to a consistent value, while there is no agreement between different subgroups. It is clear that the complete consensus is a special case of group consensus. Group consensus problems for continuous-time single-integrator agents under switching topologies were explored in Yu and Wang (2010) using the Lyapunov direct method and double-tree-form transformations. In Xia and Cao (2011), sufficient and necessary conditions for group consensus were provided for continuous-time multi-agent systems under a couple of different mechanisms concerning whether or not the agents' self-dynamics are identical. Second-order group consensus was addressed in Ma, Wang, and Miao (2014) and Feng, Xu, and Zhang (2014). Algebraic criteria for group consensus were reported in Han, Lu, and Chen (2013) and Shang (2013b) for discrete-time single-integrator dynamics. In addition, a group of continuous-time agents with non-linear self-dynamics (Sun, Bai, Jia, Xiong, & Chen, 2011), time delay (Shang, 2013c), linear time-invariant dynamics (Qin & Yu, 2013; Tan, Liu, & Duan, 2011), and choice-based protocols (Liu & Wong, 2013) can also reach group consensus under some conditions.

It is widely known that the weak coupling strength among agents may lead to instability and inhibits the convergence of the state trajectories. In most realistic systems, however, it is literally impossible to make the coupling strengths arbitrarily large. Thus, it is inevitable to study the sufficient/necessary conditions for group consensus concerning the coupling strengths. In the previously mentioned work, some algebraic criteria were proposed to ensure group consensus. For example, linear matrix inequality conditions were introduced in Yu and Wang (2010) and Xia and Cao (2011), and conditions involving eigenvalues of interaction topologies were proposed in Sun et al. (2011, Shang (2013c), and Tan et al. (2011). The feasibility of these algebraic conditions turns out to be very difficult to check. Alternatively, the authors in Qin and Yu (2013) imposed a graphic constraint on the interaction topology. It is shown that the group consensus can be achieved via pinning control irrespective of how weak or strong the couplings among agents are, if the underlying network has an acyclic partition.

In this paper, continuing with previous works, we study the group consensus of a network of agents with continuous-time generic linear dynamics. By introducing inter-group non-identical inputs, we show that the system can achieve group consensus by designing suitable consensus gains given any magnitude of the coupling strengths. Our result differs from the existing literature in that we impose neither complicated algebraic criteria (Shang, 2013c; Sun et al., 2011; Tan et al., 2011; Xia & Cao, 2011; Yu & Wang, 2010) nor graphic constraints (Liu & Wong, 2013; Ma et al., 2014; Qin & Yu, 2013) on the coupling. Moreover, in the aforementioned works (Qin & Yu, 2013; Shang, 2013c; Sun et al., 2011; Tan et al., 2011; Xia & Cao, 2011; Yu & Wang, 2010), the couplings among agents in different groups may be negatively weighted to desynchronize the states of agents in different groups. Nevertheless, negative weights are difficult to find practical applications. In our framework, the external adapted inputs help realize inter-group separation. Thanks to that, we only require non-negative weights. Convergence rate and ultimate consensus state can be specified as well. It is worthwhile to mention that similar external inputs mechanisms were dealt with in Han et al. (2013) and Shang (2013b) for discrete-time single-integrator agents, where the coupling strengths need to be sufficiently strong. The approaches used are totally different.

The rest of the paper is organized as follows. The problem to be investigated is formulated in Section 2. Main results are given in Section 3. Simulations are performed in Section 4 to illustrate theoretical results. Conclusions are drawn in Section 5.

The following notations will be used throughout the paper. $1_n \in \mathbb{R}^n$ (resp. $0_n \in \mathbb{R}^n$) is the $n$-dimensional column vector with all entries equal to one (resp. zero). $I_n \in \mathbb{R}^{n \times n}$ is the $n$-dimensional identity matrix. $A^T$ is the transpose of matrix $A$. $\mathrm{diag}(A_1, \ldots, A_p)$ is the "block diagonal" matrix with the $k$-th main diagonal block being matrix $A_k$. $||x||$ stands for the Euclidean norm of a vector $x$. $A \otimes B$ refers to the Kronecker product of two matrices $A$ and $B$ (Horn & Johnson, 1985).

## 2. Problem formulation

Let $\mathcal{G} = (\mathcal{V}, \mathcal{E}, \mathcal{A})$ be a weighted directed graph of order $N$, where $\mathcal{V} = \{v_1, v_2, \cdots, v_N\}$ is the set of nodes (or agents), $\mathcal{E} \subseteq \mathcal{V} \times \mathcal{V}$ is the set of directed edges, and $\mathcal{A} = (a_{ij}) \in \mathbb{R}^{N \times N}$ is the weighted adjacency matrix. A directed edge of $\mathcal{G}$ is denoted by $(v_i, v_j) \in \mathcal{E}$, indicating that agent $v_j$ can obtain information from agent $v_i$. The entry $a_{ij} > 0$ if $(v_j, v_i) \in \mathcal{E}$; $a_{ij} = 0$ otherwise. Moreover, we assume $a_{ii} = 0$ for all $i$. The set of neighbors of agent $v_i$ is denoted by $\mathcal{N}_i = \{v_j \in \mathcal{V} : (v_j, v_i) \in \mathcal{E}\}$. $\mathcal{L} = (l_{ij}) \in \mathbb{R}^{N \times N}$ is the Laplacian matrix of $\mathcal{G}$, where $l_{ij} = -a_{ij}$, $i \neq j$, and $l_{ii} = \sum_{j=1, j \neq i}^{N} a_{ij}$. $\sum_{j=1}^{N} a_{ij}$ is called the in-degree of agent $v_i$. A directed path is a sequence of distinct nodes $v_{i_1}, v_{i_2}, \ldots, v_{i_k}$ in $\mathcal{G}$ such that $(v_{i_{j-1}}, v_{i_j}) \in \mathcal{E}$ for $j = 2, \ldots, k$. $\mathcal{G}$ is said to contain a spanning tree if there exists an agent (referred to as root) such that every other agent can be connected via a directed path starting from the root. The Laplacian matrix has the following property (see e.g. [Ren and Beard (2005), Lemma 3.3] and [Horn and Johnson (1985), Theorem 8.3.1]).

Lemma 1 *If $\mathcal{G}$ has a spanning tree, then the Laplacian matrix $\mathcal{L}$ has exactly one zero eigenvalue with corresponding eigenvector $1_N$, and all of the non-zero eigenvalues are with positive real parts. Moreover, $\mathcal{L}$ has a non-negative left eigenvalue $\eta \in \mathbb{R}^N$ associated with the zero eigenvalue, satisfying $\eta^T \mathcal{L} = 0_N^T$ and $\eta^T 1_N = 1$.*

Consider a multi-agent system consisting of $N$ agents with interaction graph delineated by a directed graph $\mathcal{G}$, where the group of agents is partitioned into $p$ subgroups for some integer $p$. Without loss of generality, we assume $\mathcal{V}_1 = \{v_1, \ldots, v_{N_1}\}$, $\mathcal{V}_2 = \{v_{N_1+1}, \ldots, v_{N_1+N_2}\}, \cdots$, $\mathcal{V}_p = \left\{ v_{\sum_{k=1}^{p-1} N_k + 1}, \ldots, v_N \right\}$ and $\sum_{k=1}^{p} N_k = N$. Set $N_0 = 0$. The dynamics of agent $v_i$ takes the following form

$$\dot{x}_i(t) = Ax_i(t) + Bu_i(t), \quad t \geq 0, \quad i = 1, 2, \ldots, N \tag{1}$$

where $x_i(t) \in \mathbb{R}^n$ and $u_i(t) \in \mathbb{R}^m$ represent the state and control input of agent $v_i$; $A \in \mathbb{R}^{n \times n}$ and $B \in \mathbb{R}^{n \times m}$ are constant system matrices. Motivated by Han et al. (2013) and Shang (2013b), we introduce the following consensus protocol with inter-group non-identical inputs:

$$u_i(t) = K \left( \sum_{v_j \in \mathcal{N}_i} a_{ij}(x_j(t) - x_i(t)) + w_i(t) \right) \tag{2}$$

where $K \in \mathbb{R}^{m \times n}$ is the constant consensus gain matrix to be designed, and $w_i(t) \in \mathbb{R}^n$ are external inputs satisfying $w_i(t) = w_j(t)$ if and only if $v_i, v_j \in \mathcal{V}_k$, $k = 1, \ldots, p$. In addition, we assume that $w_i(t)$, $i = 1, \ldots, N$ are bounded, i.e. $||w_i(t)|| \leq c$ for all $t \geq 0$, where $c$ is a constant. It is worth noting that the linear time-invariant dynamics (1) with protocol (2) addresses not only the self-dynamics of the agent but also the interactions between the neighboring agents. Hence, it is more general than the integrator cases; see, e.g. Yu and Wang (2010), Xia and Cao (2011), and Sun et al. (2011).

*Definition 1* The multi-agent system (1) under the control law (2) is said to achieve group consensus if there exists a consensus gain $K$, such that for any $x_i(0) \in \mathbb{R}^n$,

$$\lim_{t\to\infty} ||x_i(t) - x_j(t)|| = 0, \quad \text{for all } v_i, v_j \in \mathcal{V}_k, \quad k = 1, \dots, p$$

and

$$\lim_{t\to\infty} ||x_i(t) - x_j(t)|| > 0, \quad \text{for each pair of } v_i \in \mathcal{V}_k \text{ and } v_j \in \mathcal{V}_l \text{ with } k \neq l$$

In addition, if there exist positive numbers $\kappa$, $C$, and $t_0$ such that $||x_i(t) - x_j(t)|| \le Ce^{-\kappa t}$ for all $v_i, v_j \in \mathcal{V}_k$ ($k = 1, \dots, p$) and $t > t_0$, we say the consensus is achieved exponentially fast (with rate at least $\kappa$).

The Laplacian matrix of $\mathcal{G}$ takes the following block matrix form:

$$\mathcal{L} = \begin{bmatrix} \mathcal{L}_{11} & \mathcal{L}_{12} & \cdots & \mathcal{L}_{1p} \\ \mathcal{L}_{21} & \mathcal{L}_{22} & \cdots & \mathcal{L}_{2p} \\ \vdots & \vdots & \ddots & \vdots \\ \mathcal{L}_{p1} & \mathcal{L}_{p2} & \cdots & \mathcal{L}_{pp} \end{bmatrix} \tag{3}$$

where $\mathcal{L}_{kk} \in \mathbb{R}^{N_k \times N_k}$ specifies the information exchange within subgroup $\mathcal{V}_k$, and $\mathcal{L}_{kl} \in \mathbb{R}^{N_k \times N_l}$ specifies the information exchange from subgroup $\mathcal{V}_l$ to $\mathcal{V}_k$.

Assumption 1 For all $k, l = 1, \dots, p$, $\mathcal{L}_{kl}$ has a constant row sum.

Note that Assumption 1 means $\mathcal{L}_{kl}\mathbf{1}_{N_l} = c_{kl}\mathbf{1}_{N_l}$ for some constant $c_{kl} \in \mathbb{R}$. Such an assumption is widely made in most of the literature pertaining to group consensus problems; among them, many further assume that $c_{kl} = 0$ for all $k$, $l$—called the in-degree balanced condition [see e.g. (Feng et al., 2014; Qin & Yu, 2013; Sun et al., 2011; Xia & Cao, 2011; Yu & Wang, 2010)].

## 3. Main results

In this section, we tackle multi-agent system (1) under protocol (2). The objective is to derive simple sufficient conditions for achieving group consensus for any coupling strength among agents.

Below, we first present a lemma, which will be needed in the convergence analysis in Theorem 1. We believe that the result is also interesting in its own right.

Lemma 2 *Consider an $nN \times nN$ matrix $\tilde{\Omega}$ given by*

$$\tilde{\Omega} = \begin{bmatrix} \tilde{\Omega}_{11} & \tilde{\Omega}_{12} & \cdots & \tilde{\Omega}_{1p} \\ \tilde{\Omega}_{21} & \tilde{\Omega}_{22} & \cdots & \tilde{\Omega}_{2p} \\ \vdots & \vdots & \ddots & \vdots \\ \tilde{\Omega}_{p1} & \tilde{\Omega}_{p2} & \cdots & \tilde{\Omega}_{pp} \end{bmatrix}$$

*where* $\tilde{\Omega}_{kl} = \begin{bmatrix} \Omega_{11} & \Omega_{12} & \cdots & \Omega_{1N_l} \\ \Omega_{21} & \Omega_{22} & \cdots & \Omega_{2N_l} \\ \vdots & \vdots & \ddots & \vdots \\ \Omega_{N_k1} & \Omega_{N_k2} & \cdots & \Omega_{N_kN_l} \end{bmatrix}$ *and* $\Omega_{ij} \in \mathbb{R}^{n\times n}$ *for* $1 \le i \le N_k$, $1 \le j \le N_l$, *and* $1 \le k, l \le p$. *Define an nN-dimensional vector* $\tilde{\Gamma} = ((\mathbf{1}_{N_1} \otimes \Gamma_1)^T, \dots, (\mathbf{1}_{N_p} \otimes \Gamma_p)^T)^T$ *with* $\Gamma_k \in \mathbb{R}^n$ *for* $1 \le k \le p$. *Assume that for all k, l, there exist some* $\alpha_{kl} \in \mathbb{R}^{n\times n}$ *satisfying* $\tilde{\Omega}_{kl}(1_{N_l} \otimes I_n) = 1_{N_k} \otimes \alpha_{kl}$, *i.e.* $\tilde{\Omega}_{kl}$ *"as a block matrix" has a constant row sum. Let* $\alpha = (\alpha_{kl}) \in \mathbb{R}^{np\times np}$. *Then*

$$e^{\tilde{\Omega}} \cdot \tilde{\Gamma} = \left( (1_{N_1} \otimes b_1)^T, \dots, (1_{N_p} \otimes b_p)^T \right)^T$$

*where* $b_k = \sum_{l=1}^{p} (e^{\alpha})_{kl} \Gamma_l$ *for* $1 \le k \le p$. *Here,* $(e^{\alpha})_{kl}$ *means the (k, l)-block if we partition* $e^{\alpha}$ *in conformity with that of* $\alpha$.

*Proof* From the straightforward calculation, we know that

$$\tilde{\Omega} \cdot \left( \mathrm{diag}(1_{N_1}, \dots, 1_{N_p}) \otimes I_n \right) = \left( \mathrm{diag}(1_{N_1}, \dots, 1_{N_p}) \otimes I_n \right) \cdot \alpha$$

Since the expansion $e^M = \sum_{i=0}^{\infty} \frac{1}{i!} M^i$ holds for any square matrix $M$, we further obtain

$$e^{\tilde{\Omega}} \cdot \left( \mathrm{diag}(1_{N_1}, \dots, 1_{N_p}) \otimes I_n \right) = \left( \mathrm{diag}(1_{N_1}, \dots, 1_{N_p}) \otimes I_n \right) \cdot e^{\alpha}$$

If we partition $e^{\tilde{\Omega}}$ as $e^{\tilde{\Omega}} = \begin{bmatrix} \Theta_{11} & \Theta_{12} & \cdots & \Theta_{1p} \\ \Theta_{21} & \Theta_{22} & \cdots & \Theta_{2p} \\ \vdots & \vdots & \ddots & \vdots \\ \Theta_{p1} & \Theta_{p2} & \cdots & \Theta_{pp} \end{bmatrix}$ in conformity with that of $\tilde{\Omega}$, then it is easy to check

$$\Theta_{kl}(1_{N_l} \otimes I_n) = 1_{N_k} \otimes (e^{\alpha})_{kl} \tag{4}$$

where $(e^{\alpha})_{kl}$ is defined as above.

Using (4) and the properties of Kronecker product, we derive that

$$e^{\tilde{\Omega}} \cdot \tilde{\Gamma} = \begin{bmatrix} \sum_{l=1}^{p} \Theta_{1l}(1_{N_l} \otimes \Gamma_l) \\ \vdots \\ \sum_{l=1}^{p} \Theta_{pl}(1_{N_l} \otimes \Gamma_l) \end{bmatrix} = \begin{bmatrix} \sum_{l=1}^{p} \Theta_{1l}(1_{N_l} \otimes I_n)(1 \otimes \Gamma_l) \\ \vdots \\ \sum_{l=1}^{p} \Theta_{pl}(1_{N_l} \otimes I_n)(1 \otimes \Gamma_l) \end{bmatrix}$$

$$= \begin{bmatrix} \sum_{l=1}^{p} (1_{N_1} \otimes (e^{\alpha})_{1l})(1 \otimes \Gamma_l) \\ \vdots \\ \sum_{l=1}^{p} (1_{N_p} \otimes (e^{\alpha})_{pl})(1 \otimes \Gamma_l) \end{bmatrix} = \begin{bmatrix} 1_{N_1} \otimes b_1 \\ \vdots \\ 1_{N_p} \otimes b_p \end{bmatrix}$$

as desired.

Set $x(t) = (x_1^T(t), \dots, x_N^T(t))^T$. The system (1) under protocol (2) can be recast in the following matrix form

$$\dot{x}(t) = (I_N \otimes A - \mathcal{L} \otimes BK)x(t) + (I_N \otimes BK)w(t) \tag{5}$$

where $w(t) = \left( (1_{N_1} \otimes w_1(t))^T, \dots, (1_{N_p} \otimes w_p(t))^T \right)^T$ and $\mathcal{L}$ is given by (3). If $(A, B)$ is stabilizable, there exists a non-negative definite matrix $P$, such that the algebraic Riccati equation

$$A^T P + PA - PBB^T P + I_n = 0 \tag{6}$$

holds, and all eigenvalues of $A - BB^T P$ are in the open left half-plane (Ogata, 2010). Let $\lambda_1 = 0, \lambda_2, \dots, \lambda_N$ be the eigenvalues of Laplacian $\mathcal{L}$.

Based on the above preparations, we are now in a position to get the main result concerning group consensus of system (5).

THEOREM 1 *Under Assumption 1, if (A, B) is stabilizable and $\mathcal{G}$ contains a spanning tree, then the gain matrix can be designed as $K=\sigma\cdot B^TP$, where $\sigma\geq\left(\min_{2\leq i\leq N}\{Re(\lambda_i)\}\right)^{-1}$ and P is given in (6), such that the multi-agent system (1) under protocol (2) can achieve group consensus exponentially fast.*

*Proof* The solution of (5) can be written as

$$x(t)=e^{(I_N\otimes A-\mathcal{L}\otimes BK)t}x(0)+\int_0^t e^{(I_N\otimes A-\mathcal{L}\otimes BK)(t-s)}(I_N\otimes BK)w(s)ds \tag{7}$$

Since $\mathcal{G}$ contains a spanning tree, by Lemma 1, we know that $\lambda_2,\dots,\lambda_N$ are in the open right half-plane. Furthermore, there exists an invertible matrix $Q$ taking the form $Q=[\,1_N\ \ Y\,]$ and $Q^{-1}=[\,\eta\ \ Z\,]^T$ with $\eta=(\eta_1,\dots,\eta_N)^T$ defined in Lemma 1, $Y,Z\in\mathbb{R}^{N\times(N-1)}$ such that $\mathcal{L}$ is similar to a Jordan canonical form, i.e.

$$Q^{-1}\mathcal{L}Q=\operatorname{diag}(0,J)$$

where $J$ is a $(N-1)\times(N-1)$-dimensional upper triangular matrix, whose principal diagonal elements consist of $\lambda_i$, $i=2,\dots,N$.

Since (A, B) is stabilizable, from the comments above Theorem 1, $\sigma\geq\left(\min_{2\leq i\leq N}\{\mathrm{Re}(\lambda_i)\}\right)^{-1}$, and the property that all the eigenvalues of $A-(a+\imath b)BB^TP$, $(\imath^2=-1)$ are in the open left half-plane for any $a\geq1$ and $b\in\mathbb{R}$ (Ma & Zhang, 2010), we derive that all the eigenvalues of $A-\lambda_iBK$, $i=2,\dots,N$ are in the open left half-plane. Thus, we have

$$\begin{aligned}e^{(I_N\otimes A-\mathcal{L}\otimes BK)t}&=(Q\otimes I_n)\begin{bmatrix}e^{At} & 0\\ 0 & e^{(I_{N-1}\otimes A-J\otimes BK)t}\end{bmatrix}(Q^{-1}\otimes I_n)\\ &\rightarrow(1_N\eta^T)\otimes e^At\end{aligned}$$

as $t\rightarrow\infty$. Hence,

$$e^{(I_N\otimes A-\mathcal{L}\otimes BK)t}x(0)\rightarrow\left((1_N\eta^T)\otimes e^{At}\right)x(0) \tag{8}$$

exponentially fast as $t\rightarrow\infty$.

Note that

$$\begin{aligned}&\int_0^t e^{(I_N\otimes A-\mathcal{L}\otimes BK)(t-s)}(I_N\otimes BK)w(s)ds\\ &=\int_0^t(Q\otimes I_n)\begin{bmatrix}e^{A(t-s)} & 0\\ 0 & e^{(I_{N-1}\otimes A-J\otimes BK)(t-s)}\end{bmatrix}(Q^{-1}\otimes I_n)(I_N\otimes BK)w(s)ds\\ &=\int_0^t(Q\otimes I_n)\begin{bmatrix}e^{A(t-s)} & 0\\ 0 & 0\end{bmatrix}(Q^{-1}\otimes I_n)(I_N\otimes BK)w(s)ds\\ &+\int_0^t(Q\otimes I_n)\begin{bmatrix}0 & 0\\ 0 & e^{(I_{N-1}\otimes A-J\otimes BK)(t-s)}\end{bmatrix}(Q^{-1}\otimes I_n)(I_N\otimes BK)w(s)ds\\ &:=T_1(t)+T_2(t)\end{aligned}$$

Recall that $N_0=0$, and by straightforward calculation, we obtain

$$T_1(t)=1_N\otimes\left(\sum_{k=1}^{p}\left(\sum_{i=N_0+\dots+N_{k-1}+1}^{N_0+\dots+N_k}\eta_i\right)\int_0^t e^{A(t-s)}BKw_k(s)ds\right) \tag{9}$$

Based on Assumption 1 and the fact that $w_i(t)$, $i=1, \dots, N$ are intra-group identical and inter-group non-identical inputs, we can think of $I_N \otimes A - \mathcal{L} \otimes BK$ as $\tilde{\Omega}$ and think of $(I_N \otimes BK)w(s)$ as $\tilde{\Gamma}(s) = \left((1_{N_1} \otimes \Gamma_1(s))^T, \dots, (1_{N_p} \otimes \Gamma_p(s))^T\right)^T$ with $\Gamma_k(s) = BKw_k(s)$ for $k=1, \dots, p$ in Lemma 2. In the light of Lemma 2, the second term on the right-hand side of (7) can be written as $\Psi(t) := \left((1_{N_1} \otimes \Psi_1(t))^T, \dots, (1_{N_p} \otimes \Psi_p(t))^T\right)^T$, where $\Psi_k(t) \in \mathbb{R}^n$, $1 \le k \le p$. In other words,

$$T_1(t) + T_2(t) = \Psi(t) \tag{10}$$

Since all the eigenvalues of $A - \lambda_i BK$, $i=2, \dots, N$ are in the open left half-plane and $w(t)$ is bounded, we know that the limit $\lim_{t\to\infty} T_2(t)$ exists and is a constant matrix. It is clear that the convergence is exponentially fast, and we write $\Phi := \lim_{t\to\infty} T_2(t)$.

Therefore, by virtue of (7–10), we obtain that

$$\begin{aligned} x(t) &\to \left((1_N \eta^T) \otimes e^{At}\right) x(0) \\ &+ 1_N \otimes \left( \sum_{k=1}^{p} \left( \sum_{i=N_0+\dots+N_{k-1}+1}^{N_0+\dots+N_k} \eta_i \right) \int_0^t e^{A(t-s)} BKw_k(s) ds \right) + \Phi \\ &= \left((1_N \eta^T) \otimes e^{At}\right) x(0) + \left((1_{N_1} \otimes \Psi_1(t))^T, \dots, (1_{N_p} \otimes \Psi_p(t))^T\right)^T \end{aligned}$$

exponentially fast as $t \to \infty$. Thus, for $v_i, v_j \in \mathcal{V}_k (k=1, \dots, p)$, $||x_i(t) - x_j(t)|| \to 0$ as $t$ tends to infinity. By the basic inequality $||x_i - x_j|| \le ||x_i - c|| + ||x_j - c||$, it is easy to see that the convergence is exponentially fast. For $v_i \in \mathcal{V}_k$ and $v_j \in \mathcal{V}_l$, $k \ne l$, with the inter-group non-identical inputs $w_i(t)$, one can have $\lim_{t\to\infty} ||x_i(t) - x_j(t)|| > 0$. This completes the proof.

*Remark 1* The design of feedback matrix $K$ in Theorem 1 has a highly desirable feature—it uncouples the effects of the agent dynamics and the network topology. Specifically, each agent constructs a gain $B^T P$ by only using (6), and scales it by a multiplicative factor $\sigma$ taking into account of the interaction topology.

*Remark 2* The control signals $\{w_i(t)\}_{i=1}^N$ play a key role in achieving group consensus. The intra-group identical and inter-group non-identical property facilitates the transformation to an explicitly solvable system via Lemma 2. Moreover, if each subgroup $\mathcal{V}_k$ $(k=1, \dots, p)$ admits a control node (or *leader*), which generates the desired target trajectory $s_k(t)$ autonomously, then our consensus protocol (2) becomes similar to the general track control protocol studied in (Zhang, Lewis, & Das, 2011) by taking $w_i(t) = c_k(s_k(t) - x_i(t))$ for any $v_i \in \mathcal{V}_k$ $(i=1, \dots, N; k=1, \dots, p)$, where $c_k>0$. It is reasonable to conjecture that $s_k(t)$ would be the ultimate consensus trajectory for subgroup $\mathcal{V}_k$ under some connectivity conditions.

If we take $n=m$, $A=0$, and $B=I_n$, the system (5) reduces to the case of integrator agents. We then have the following corollary.

COROLLARY 1 *Under Assumption 1, if $\mathcal{G}$ contains a spanning tree, then the gain matrix can be designed as $K=\sigma I_n$, where $\sigma \ge \left(\min_{2\le i\le N}\{Re(\lambda_i)\}\right)^{-1}$, such that the multi-agent system*

$$\dot{x}_i(t) = u_i(t), \quad t \ge 0, \quad i = 1, 2, \dots, N$$

*under protocol (2) can achieve group consensus exponentially fast.*

## 4. Simulation

In this section, we present numerical simulations to illustrate the validity of the proposed theoretical results.

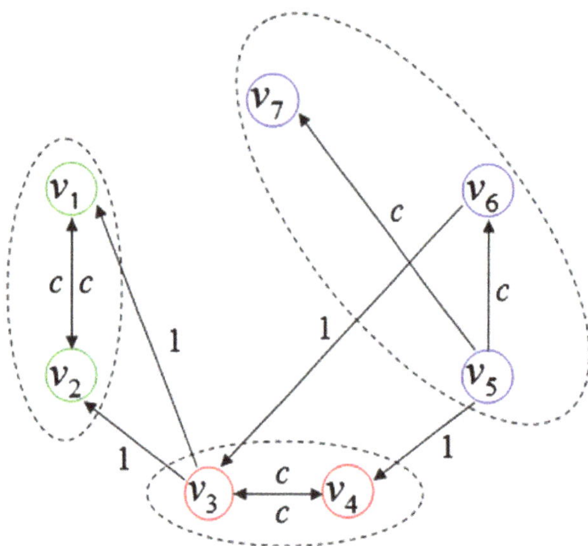


**Figure 1. Network topology $\mathcal{G}$.**

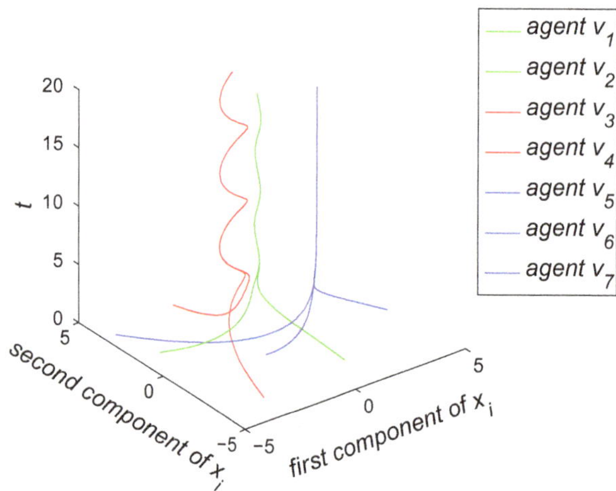


**Figure 2. The state trajectories for the case of c = 1.**

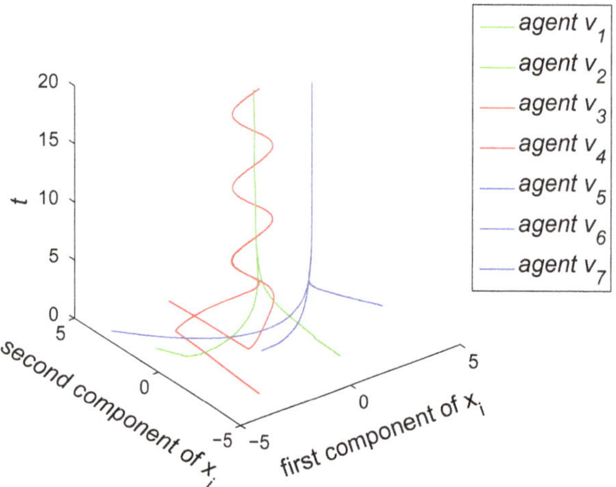


**Figure 3. The state trajectories for the case of c = .01.**

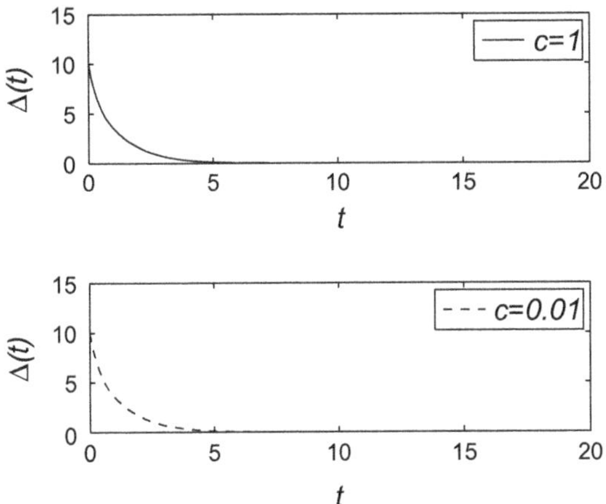


**Figure 4. The dynamical behavior of Δ(t).**

Consider the multi-agent system with $N=7$ agents divided into three subgroups $\mathcal{V}_1=\{v_1, v_2\}$, $\mathcal{V}_2=\{v_3, v_4\}$, and $\mathcal{V}_3=\{v_5, v_6, v_7\}$. The communication topology among agents is shown in Figure 1. The inter-group weights and intra-group weights are set to be 1 and $c$, respectively, for some positive $c$. The Laplacian matrix is given by

$$\mathcal{L}=\left[\begin{array}{cc:cc:ccc} 1+c & -c & -1 & 0 & 0 & 0 & 0 \\ -c & 1+c & -1 & 0 & 0 & 0 & 0 \\ \hdashline 0 & 0 & 1+c & -c & 0 & -1 & 0 \\ 0 & 0 & -c & 1+c & -1 & 0 & 0 \\ \hdashline 0 & 0 & 0 & 0 & 0 & 0 & 0 \\ 0 & 0 & 0 & 0 & -c & c & 0 \\ 0 & 0 & 0 & 0 & -c & 0 & c \end{array}\right]$$

Take $n=2$, $m=1$, and let the agent dynamics (1) be specified as

$$A=\begin{bmatrix} -1 & 0 \\ 1 & -1 \end{bmatrix} \text{ and } B=\begin{bmatrix} 0 \\ 1 \end{bmatrix}$$

The pair $(A, B)$ is stabilizable. The inputs are chosen as $w_1(t)=w_2(t)=((t+1)^{-1}, 2(t+1)^{-1})^T$, $w_3(t)=w_4(t)=(\sin(t), 3+\cos(t))^T$, and $w_5(t)=w_6(t)=w_7(t)=(2, -2)^T$. Therefore, all conditions in Theorems 1 hold. From (6), we solve $P=\begin{bmatrix} 0.6569 & 0.1716 \\ 0.1716 & 0.4142 \end{bmatrix}$. The initial states $x_i(0)$, $i=1, \dots, 7$ are taken randomly in $[-5, 5]^2$.

We consider two cases of coupling strength: $c=1$ and $c=.01$. In view of Theorem 1, the gain matrices are solved as $K=(.1716, .4142)$ and $K=(17.16, 41.42)$, respectively. Define $\Delta(t)=\max_{1\le k\le 3} \max_{v_i, v_j\in\mathcal{V}_k} ||x_i(t)-x_j(t)||$ as a measure of the discrepancy of states within the same groups. The dynamical behaviors of the states $x_i(t)$, $i=1, \dots, 7$ are shown in Figures 2 and 3, respectively. Figure 4 shows the corresponding error trajectories, indicating that the group consensus is achieved for both cases.

## 5. Conclusion

This paper studies the group consensus problem of a network of continuous-time agents with linear dynamics, whose interaction topology is directed and fixed. Based on algebraic graph theory and matrix theory, external adapted inputs are introduced to realize group consensus exponentially fast. Comparing the existing works on group consensus, we highlight the following features of our result:

- We study the case where each agent has dynamics of a continuous linear time-invariant system. Such systems include the single-integrator, double-integrator, and higher order integrator dynamics as special cases.
- The external control inputs contribute to the intra-group synchronization and inter-group separation, without imposing complicated algebraic criteria (Shang, 2013c; Sun et al., 2011; Tan et al., 2011; Xia & Cao, 2011; Yu & Wang, 2010) or restrictive topological constraints (Liu & Wong, 2013; Ma et al., 2014; Qin & Yu, 2013).
- Only non-negative weights are assigned to the communication links, which have practical applicability. In Yu and Wang (2010), Xia and Cao (2011), Sun et al. (2011), Shang (2013c), Tan et al. (2011), and Qin and Yu (2013), negative weights are essentially needed for group consensus.

Moreover, the group consensus can be achieved by designing appropriate control gains for any given magnitude of the coupling strengths among the agents. Numerical simulations are presented to illustrate the effectiveness of our theoretical results.

**Funding**
The authors received no direct funding for this research.

**Author details**
Yilun Shang[1]
E-mail: shylmath@hotmail.com
[1] Einstein Institute of Mathematics, Hebrew University of Jerusalem, Jerusalem 91904, Israel; SUTD-MIT International Design Center, Singapore University of Technology and Design, Singapore 138682, Singapore.

**References**

Cao, Y., Yu, W., Ren, W., & Chen, G. (2013). An overview of recent progress in the study of distributed multi-agent coordination. *IEEE Transactions on Industrial Informatics, 9*, 427–438. http://dx.doi.org/10.1109/TII.2012.2219061

Fagnani, F., & Zampieri, S. (2009). Average consensus with packet drop communication. *SIAM Journal on Control and Optimization, 48*, 102–133. http://dx.doi.org/10.1137/060676866

Feng, Y., Xu, S., & Zhang, B. (2014). Group consensus control for double-integrator dynamic multiagent systems with fixed communication topology. *International Journal of Robust and Nonlinear Control, 24*, 532–547. http://dx.doi.org/10.1002/rnc.v24.3

Gazi, V., & Passino, K. M. (2003). Stability analysis of swarms. *IEEE Transactions on Automatic Control, 48*, 692–697. http://dx.doi.org/10.1109/TAC.2003.809765

Han, Y., Lu, W., & Chen, T. (2013). Cluster consensus in discrete-time networks of multiagents with inter-cluster nonidentical inputs. *IEEE Transactions on Neural Networks and Learning Systems, 24*, 566–578. http://dx.doi.org/10.1109/TNNLS.2013.2237786

Hatano, Y., & Mesbahi, M. (2005). Agreement over random networks. *IEEE Transactions on Automatic Control, 50*, 1867–1872. http://dx.doi.org/10.1109/TAC.2005.858670

Horn, R. A., & Johnson, C. R. (1985). *Matrix analysis*. Cambridge: Cambridge University Press. http://dx.doi.org/10.1017/CBO9780511810817

Huang, M., Dey, S., Nair, G. N., & Manton, J. H. (2010). Stochastic consensus over noisy netowrks with Markovian and arbitrary switches. *Automatica, 46*, 1571–1583. http://dx.doi.org/10.1016/j.automatica.2010.06.016

Jadbabaie, A., Lin, J., & Morse, A. S. (2003). Coordination of groups of mobile autonomous agents using nearest neighbor rules. *IEEE Transactions on Automatic Control, 48*, 988–1001. http://dx.doi.org/10.1109/TAC.2003.812781

Kar, S., & Moura, J. M. F. (2009). Distributed consensus algorithms in sensor networks with imperfect communication: Link failures and channel noise. *IEEE Transactions on Signal Processing, 57*, 355–369. http://dx.doi.org/10.1109/TSP.2008.2007111

Kashyap, A., Başar, T., & Srikant, R. (2007). Quantized consensus. *Automatica, 43*, 1192–1203. http://dx.doi.org/10.1016/j.automatica.2007.01.002

Kumar, M., Garg, D. P., & Kumar, V. (2010). Segregation of heterogeneous units in a swarm of robotic agents. *IEEE Transactions on Automatic Control, 55*, 743–748. http://dx.doi.org/10.1109/TAC.2010.2040494

Lian, Z., & Deshmukh, A. (2006). Performance prediction of an unmanned airborne vehicle multi-agent system. *European Journal of Operational Research, 172*, 680–695. http://dx.doi.org/10.1016/j.ejor.2004.10.015

Lin, P., & Jia, Y. (2008). Average consensus in networks of multi-agents with both switching topology and coupling time-delay. *Physica A: Statistical Mechanics and its Applications, 387*, 303–313. http://dx.doi.org/10.1016/j.physa.2007.08.040

Liu, Z., Wong, W. S. (2013). Choice-based cluster consensus in multi-agent systems. In *Proceeding of 32nd Chinese Control Conference* (pp. 7285–7290). Xi'an.

Ma, C. Q., & Zhang, J. F. (2010). Necessary and sufficient conditions for consensusability of linear multi-agent systems. *IEEE Transactions on Automatic Control, 55*, 1263–1268.

Ma, Q., Wang, Z., & Miao, G. (2014). Second-order group consensus for multi-agent systems via pinning leader-following approach. *Journal of the Franklin Institute, 351*, 1288–1300. http://dx.doi.org/10.1016/j.jfranklin.2013.11.002

Ogata, K. (2010). *Modern control engineering*. Upper Saddle River, NJ: Prentice Hall.

Olfati-Saber, R., Fax, J. A., & Murray, R. M. (2007). Consensus and cooperation in networked multi-agent systems. *Proceedings of the IEEE, 95*, 215–233. http://dx.doi.org/10.1109/JPROC.2006.887293

Olfati-Saber, R., & Murray, R. M. (2004). Consensus problem in networks of agents with switching topology and time-delays. *IEEE Transactions on Automatic Control, 49*, 1520–1533. http://dx.doi.org/10.1109/TAC.2004.834113

Qin, J., & Yu, C. (2013). Cluster consensus control of generic linear multi-agent systems under directed topology with acyclic partition. *Automatica, 49*, 2898–2905. http://dx.doi.org/10.1016/j.automatica.2013.06.017

Ren, W., & Beard, R. W. (2005). Consensus seeking in multiagent systems under dynamically changing interaction topologies. *IEEE Transactions on Automatic Control, 50*, 655–661. http://dx.doi.org/10.1109/TAC.2005.846556

Ren, W., & Sorensen, N. (2008). Distributed coordination architecture for multi-robot formation control. *Robotics and Autonomous Systems, 56*, 324–333. http://dx.doi.org/10.1016/j.robot.2007.08.005

Schellinck, J., & White, T. (2011). A review of attraction and repulsion models of aggregation: Methods, findings and a discussion of model validation. *Ecological Modelling, 222*, 1897–1911. http://dx.doi.org/10.1016/j.ecolmodel.2011.03.013

Shang, Y. (2012). Finite-time consensus for multi-agent systems with fixed topologies. *International Journal of Systems Science, 43*, 499–506. http://dx.doi.org/10.1080/00207721.2010.517857

Shang, Y. (2013a). Deffuant model with general opinion distributions: First impression and critical confidence bound. *Complexity, 19*, 38–49. http://dx.doi.org/10.1002/cplx.v19.2

Shang, Y. (2013b). $L^1$ group consensus of multi-agent systems with switching topologies and stochastic inputs. *Physics Letters A, 377*, 1582–1586. http://dx.doi.org/10.1016/j.physleta.2013.04.054

Shang, Y. (2013c). Group consensus of multi-agent systems in directed networks with noises and time delays. *International Journal of Systems Science* Retrieved from http://dx.doi.org/10.1080/00207721.2013.862582

Shang, Y. (2014). Consensus formation of two-level opinion dynamics. *Acta Mathematica Scientia, 34*, 1029–1040. http://dx.doi.org/10.1016/S0252-9602(14)60067-9

Shang, Y., & Bouffanais, R. (2014). Influence of the number of topologically interacting neighbors on swarm dynamics. *Scientific Reports, 4*, Article number: 4184.

Sun, W., Bai, Y. Q., Jia, R., Xiong, R., & Chen, J. (2011). Multi-group consensus via pinning control with non-linear heterogeneous agents. In *Proceeding of the 8th Asian Control Conference* (pp. 323–328). Taiwan.

Tan, C., Liu, G. P., & Duan, G. R. (2011). Couple-group consensus of multi-agent systems with directed and fixed topology. In *Proceeding of the 30th Chinese Control Conference* (pp. 6515–6520). Yantai.

Tanner, H. G., Jadbabaie, A., & Pappas, G. J. (2007). Flocking in fixed and switching networks. *IEEE Transactions on Automatic Control, 52*, 863–868. http://dx.doi.org/10.1109/TAC.2007.895948

Wen, G., Duan, Z., Chen, G., & Yu, W. (2014). Consensus tracking of multi-agent systems with Lipschitz-type node dynamics and switching topologies. *IEEE Transactions on Circuits and Systems Part I, 61*, 499–511.

Wen, G., Li, Z., Duan, Z., & Chen, G. (2013). Distributed consensus control for linear multi-agent systems with discontinuous observations. *International Journal of Control, 86*, 95–106. http://dx.doi.org/10.1080/00207179.2012.719637

Xia, W., & Cao, M. (2011). Clustering in diffusively coupled networks. *Automatica, 47*, 2395–2405. http://dx.doi.org/10.1016/j.automatica.2011.08.043

Yu, J., & Wang, L. (2010). Group consensus in multi-agent systems with switching topologies and communication delays. *Systems & Control Letters, 59*, 340–348. http://dx.doi.org/10.1016/j.sysconle.2010.03.009

Zhang, H., Lewis, F. L., & Das, A. (2011). Optimal design for synchronization of cooperative systems: State feedback, observer and output feedback. *IEEE Transactions on Automatic Control, 56*, 1948–1952. http://dx.doi.org/10.1109/TAC.2011.2139510

# 9

# Power quality analysis of hybrid renewable energy system

Rinchin W. Mosobi[1]*, Toko Chichi[1] and Sarsing Gao[1]

*Corresponding author: Rinchin W. Mosobi, Department of Electrical Engineering, North Eastern Regional Institute of Science and Technology, Nirjuli, Arunachal Pradesh 791 109, India

E-mail: wangzom123@gmail.com

Reviewing editor: Duc Pham, University of Birmingham, UK

**Abstract:** An hybrid renewable energy sources consisting of solar photovoltaic, wind energy system, and a microhydro system is proposed in this paper. This system is suitable for supplying electricity to isolated locations or remote villages far from the grid supply. The solar photovoltaic system is modeled with two power converters, the first one being a DC-DC converter along with an maximum power point tracking to achieve a regulated DC output voltage and the second one being a DC-AC converter to obtain AC output. The wind energy system is modeled with a wind-turbine prime mover with varying wind speed and fixed pitch angle to drive an self excited induction generator (SEIG). Owing to inherent drooping characteristics of the SEIG, a closed loop turbine input system is incorporated. The microhydro system is modeled with a constant input power to drive an SEIG. The three different sources are integrated through an AC bus and the proposed hybrid system is supplied to R, R-L, and induction motor loads. A static compensator is proposed to improve the load voltage and current profiles; it also mitigates the harmonic contents of the voltage and current. The static synchronous compensator is realized by means of a three-phase IGBT-based current-controlled voltage source inverter with a self-supporting DC bus. The complete system is modeled and simulated using Matlab/Simulink. The simulation results obtained illustrate the feasibility of the proposed system and are found to be satisfactory.

**Subjects: Technology; Engineering & Technology; Electrical & Electronic Engineering; Electrical Engineering Communications**

## ABOUT THE AUTHOR

Sarsing Gao is working as associate professor in the Department of Electrical Engineering at North Eastern Regional Institute of Science and Technology, Nirjuli, Arunachal Pradesh, India. He is a senior member of IEEE. His areas of interest are electrical machines, energy, distributed generation and power quality.

## PUBLIC INTEREST STATEMENT

This paper details the possible ways of integrating renewable energy sources which may be called a hybrid renewable energy system (HRES). The recent research in the field of HRES shows the possibility of integrating either photovoltaic (PV)-wind or PV-microhydro or wind-microhydro as a hybrid system. In this paper, the possibility of integrating three renewable energy sources namely, PV-wind-microhydro power generation system is presented. It aims at supplying electricity to remote villages which are far from the grid supply. Since power quality is a major issue in such a system, the same is analyzed. MATLAB/Simulink is used for modeling the system and studying various performance characteristics of the system under different electric loads. A static synchronous compensator (STATCOM) is incorporated to improve the voltage profile at the load end thereby achieving an improved power quality of the supply.

**Keywords: Renewable energy system; maximum power point tracking; boost converter; solar photovoltaic; wind energy; self-excited induction generator; STATCOM; total harmonic distortion**

## 1. Introduction

Unlike the conventional energy sources, the non-conventional energy sources are clean, reliable, and abundant in nature. The environmental degradation such as pollution, global warming, and greenhouse gas emissions which are caused by conventional sources of energy and accelerated by ever-growing industrial activities throughout the world is a concern for all. The current researches, therefore, lay emphasis on harnessing renewable energy sources (RES) for generating electricity to supply power especially, to rural consumers where grid connection is not available. For such locations, decentralized power generation using available dispersed RES is a better and workable solution. It is shown that power generation from combined RES such as wind and hydro could make a power system more cost-effective and environmental friendly (Nejad, Radzi, Kadir, & Hizam, 2012). A technological innovation, however, is needed to utilize these energy sources to an optimum level and to obtain greater efficiency. The combined system is likely to improve the generating capacity as well the reliability of the power supply. A feasibility study of solar-wind-diesel hybrid system is carried out in (Lipu, Uddin, & Miah, 2013) with promising results but for the increased cost of diesel oil and associated environmental pollution caused due to operation of diesel engine. Thus, the system considered may not be of much help. As reported, a wind-microhydro hybrid system when operated in grid connected mode using back-to-back connected power converters may improve the power system reliability (Goel, Singh, Murthy, & Kishore, 2011). The reliability and cost-effectiveness of integrated renewable energy system at remote and distant places is evaluated using optimization technique (Bansal, Khatod, & Saini, 2014) with fairly good results. However, it is found that not many works have been done on the power quality analysis of integrated renewable energy system using solar, wind, and microhydro energy sources. This system could be very successful especially in the sub-tropical region where there is sufficient rainfall.

A solar PV arrays with maximum power point tracking (MPPT) can generate electricity on large scale (Gera, Rai, Parvej, & Soni, 2013). With the development of large wind turbine generators and advancement in solid-state devices, wind energy system (WES) has become one of the most viable options of generating energy. However, due to the intermittent nature of wind speed, it calls for tracking of maximum power point of operation (Villalva, Gazoli, & Ruppert, 2009; Villalva, Gazoli, & Filho, 2009). Power generation from mini- or microhydro is a well-proven technology (International Renewable Energy Agency (IRENA), 2012).

In this paper, a HRES comprising photovoltaic (PV) system, wind-turbine generator system, and a microhydro system is proposed. The proposed system is a combination of 15 × 2 PV arrays with variable temperature and irradiance, a variable-speed wind-turbine coupled to self-excited induction generator (SEIG) and a constant power microhydro system coupled to SEIG. The performances of the hybrid system under various loads namely, induction motor (IM) load, resistive (R), and R-L load are presented. The reactive power compensation using static synchronous compensator (STATCOM) is presented. The STATCOM is observed to be working well thereby regulating the terminal voltage and reducing the harmonic contents of the system voltage and current.

## 2. Solar PV system

### *2.1. System description*

A PV cell is basically a semiconductor diode whose p-n junction is exposed to light. When the irradiance falls on the surface of the solar cell, it absorbs photons and excites electron-hole pair separation to produce electromotive force.

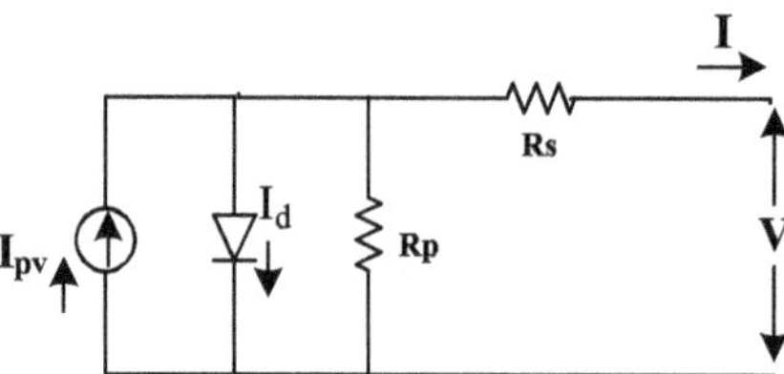


**Figure 1. Equivalent circuit diagram of a PV cell.**

Figure 1 shows the equivalent circuit of the ideal PV cell including series ($R_s$) and parallel ($R_p$) resistances which actually represents the contact and junction resistances. From the basic equation of semiconductors, the I–V characteristics of the ideal PV cell may be written as

$$I = I_{pv,cell} - \underbrace{I_{o,cell}\left[\exp\left(\frac{qV}{aKT}\right) - 1\right]}_{I_d} \tag{1}$$

where $I_{pv,cell}$ is the current generated by the incident light, $I_d$ is the diode current, $I_{o,cell}$ is the reverse saturation current or leakage current of the diode, $q$ is the electron charge, $k$ is Boltzmann constant, $T$ (in Kelvin) is the temperature of the p-n junction, and $a$ is the diode ideality factor. Figure 2 shows the ideal I–V characteristics of a solar PV module.

### 2.2. Modeling PV array

The actual PV array is designed using series and parallel combination of various solar cells to meet the desired performance. The modeling of a PV module and an array, therefore, depends on number of cells connected in series/parallel in a module and the number of modules connected in series/parallel in an array. Series connection of solar array is employed to increase the system voltage, while parallel connection gives higher total current (Villalva et al., 2009).

Actual PV arrays, therefore, requires inclusion of additional parameters to Equation 1. Thus,

$$I = I_{pv} - I_o\left[\exp\left(\frac{V + R_S I}{V_t a}\right) - 1\right] - \frac{V + R_S I}{R_p} \tag{2}$$

where, $I_{pv}$ and $I_o$ are the *PV* and saturation currents of the array and $V_t = \frac{N_s KT}{q}$ is the thermal voltage of the array with $N_s$ connected in series. If the array is composed of $N_{pp}$ parallel connection of modules, the *PV* and saturation currents may be expressed as $I_{pv} = I_{pv,cell}\, N_{pp}$, $I_o = I_{o,cell}\, N_{pp}$. In Equation 2, $R_s$ is the equivalent series resistance of the array and $R_p$ is the equivalent parallel resistance. $R_s$ and $R_p$ are expressed as $\left(\frac{N_{ss}}{N_{pp}}\right) R_S$ and $\left(\frac{N_{ss}}{N_{pp}}\right) R_p$, respectively. Equation 2 gives rise to I–V curve shown in Figure 3. This curve depends on internal characteristics of the device such as $R_s$ and $R_p$ and on external influences such as irradiance level and temperature. The solar PV array in this paper is modeled for $N_{pp} = 15$ and $N_{ss} = 2$. The light-generated current of the *PV* cell depends linearly on solar irradiation and is influenced by temperature as,

$$I_{pv} = \left(I_{pv,n} + K_I \Delta T\right)\frac{G}{G_n} \tag{3}$$

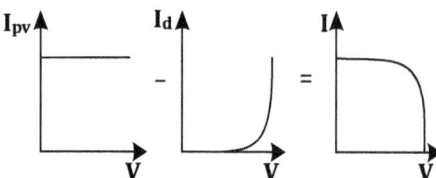

**Figure 2. I–V characteristics of the PV cell.**

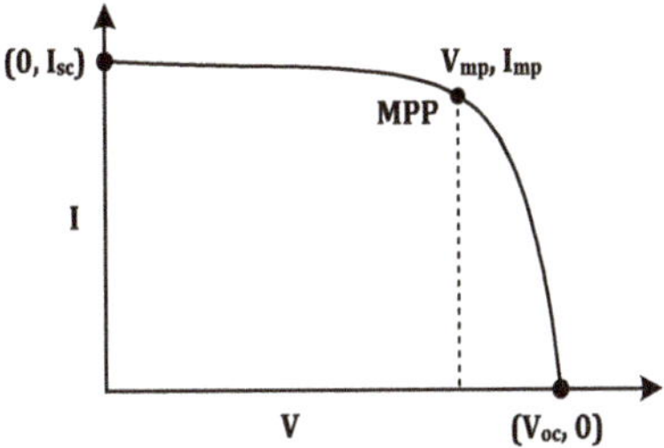


**Figure 3. I-V characteristics of the PV array.**

where, $I_{pv,n}$ is the light-generated current at the nominal condition, $\Delta T = T - T_n$, where $T$ and $T_n$ are the actual and nominal temperatures, $G$ (in W/m²) is the irradiance on the device surface and $G_n$ is the nominal irradiance.

The diode saturation current $I_o$ and its dependence on the temperature may be expressed as

$$I_o = \frac{I_{sc,n} + K_I \Delta T}{\exp\left(\frac{V_{oc,n} + K_V \Delta T}{a V_t}\right) - 1} \tag{4}$$

where, $K_I$ is current coefficient, $K_V$ is voltage coefficient, $I_{sc,n}$ is short circuit current at nominal condition, $V_{oc,n}$ is open circuit voltage at nominal condition, $a$ is ideality factor, and $V_t$ is thermal voltage.

Figures 4 and 5 show the simulated values of I-V and P-V characteristics of the PV array considered in this work. From Figure 4, it is observed that with decreasing irradiance, there is a negligible change in the maximum output voltage and there is a marginal decrement in output current and a noticeable drop in the maximum power point (MPP).

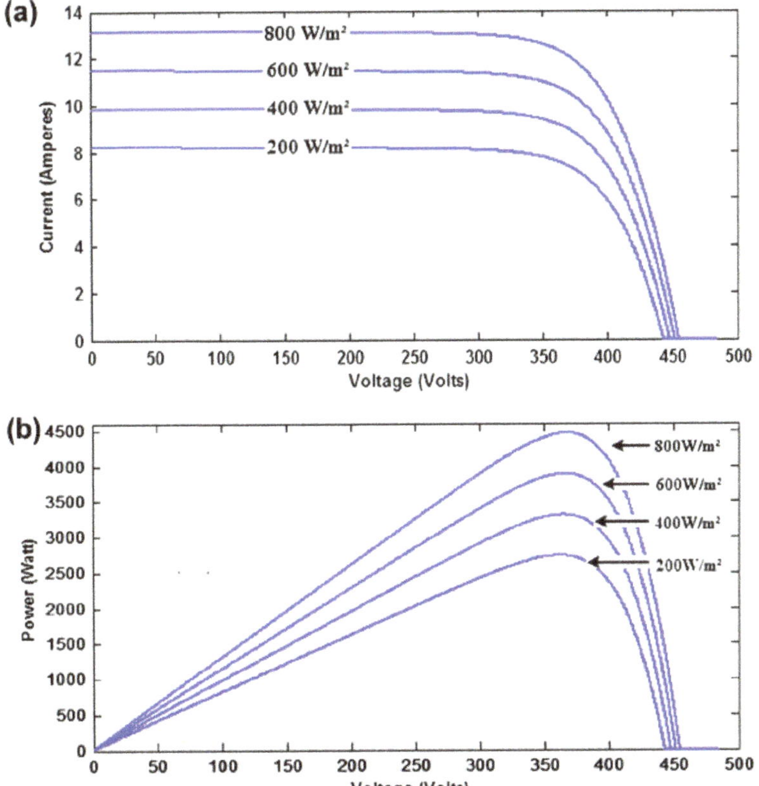


**Figure 4. (a) I-V and (b) P-V curve of a PV array with variable irradiance.**

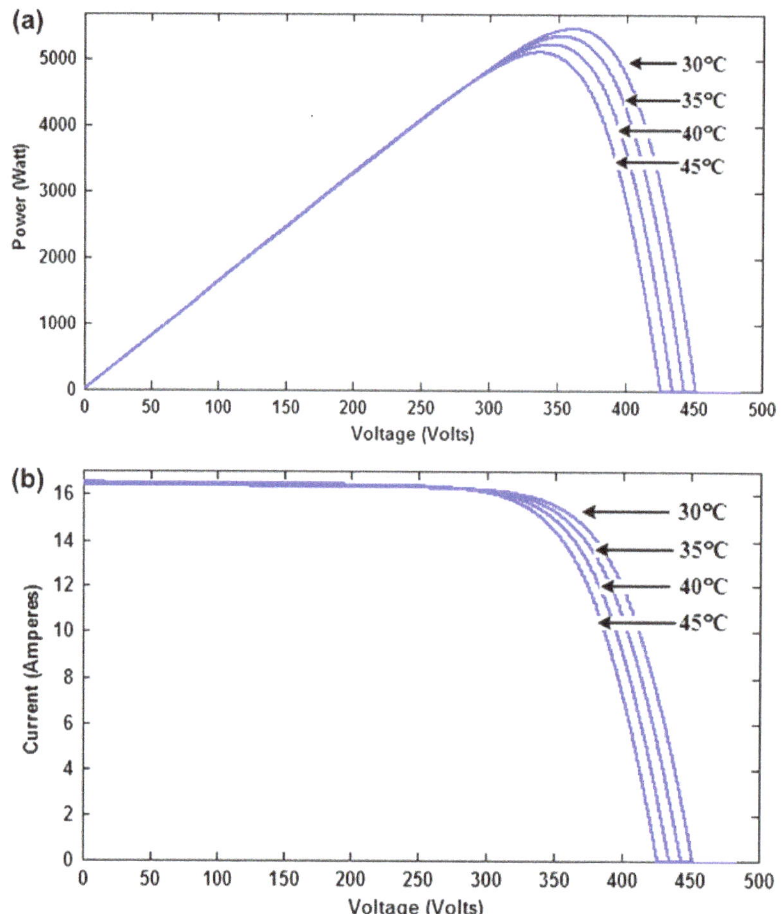


**Figure 5. (a) P–V and (b) I–V curve of a PV array with variable temperature.**

Figure 5 shows that for a fixed irradiance value as the cell temperature decreases, a small decrement in the output current occurs with an increment in output voltage and corresponding increase in MPP.

Table 1 shows the parameters of the PV array obtained from datasheet of Kyocera KC200GT at standard test condition (STC) of 25 °C and 1,000 W/m$^2$. The KC Series solar modules of Kyocera have multicrystalline silicon solar cell with an efficiency of over 16%. The same PV modules have been considered in this study (Kyocera KC200GT, xxxx).

**Table 1. Parameters of KC200GT array at 25 °C, 1000 W/m$^2$ (Kyocera KC200GT, xxxx)**

| Parameters | KC200GT |
|---|---|
| $V_{oc}$ | 32.9 V |
| $I_{sc}$ | 8.21 A |
| $K_V$ | −0.1230 V/K |
| $K_I$ | 0.0032 A/K |
| $P_{max}$ | 200.143 W |
| $R_p$ | 415.405 Ω |
| $R_s$ | 0.221 Ω |
| $a$ | 1.3 |
| $N_s$ | 54 |

where, $V_{oc}$ = open circuit voltage, $I_{sc}$ = short circuit current, $K_V$ = temperature coefficient of $V_{oc}$, $K_I$ = temperature coefficient of $I_{sc}$, $P_{max}$ = maximum power, $R_s$ = series resistance, $R_p$ = parallel resistance, $a$ = ideality factor and $N_s$ = number of cells in series. Combining Equations 3 and 4, a new expression is obtained as

$$I = I_{pv}N_{pp} - I_o N_{pp}\left[\exp\left(\frac{V + R_s\left(\frac{N_{ss}}{N_{pp}}\right)}{V_t a N_{ss}} I\right) - 1\right] - \frac{V + R_s\left(\frac{N_{ss}}{N_{pp}}\right)}{R_p\left(\frac{N_{ss}}{N_{pp}}\right)} I \tag{5}$$

where, $I_{pv}$ is PV current of the array, $I_o$ is diode saturation current, $a$ is ideality factor, $V_t$ is thermal voltage, $I$ is output current and $V$ is output voltage.

### 2.3. Power conditioning system

The power-conditioning system includes a boost converter with MPPT, a three-phase CC-VSI, and an LC filter. The input to boost converter is an unregulated DC voltage obtained directly from the PV array and therefore, it is likely to fluctuate due to variations in the incident solar radiation, also called irradiance and temperature. Thus, the average DC output voltage must be regulated to get the desired constant value throughout its operation. The MPPT used is based on perturbation and observation (P & O) technique (Kondawar & Vaidya, 2012). For a PV array with operating voltage as $V$ and current as $I$, the power is $P = VI$. At the maximum power point, $\frac{dP}{dV} = 0$ with its sign defined by

$$\frac{1}{V}\frac{dP}{dV} = \frac{dI}{dV} + \frac{I}{V} \tag{6}$$

Figure 6 shows the flowchart of P & O algorithm. The MPPT feeds the desired PV array voltage to the boost converter through change in duty cycle as a gating signal (Villalva et al., 2009). A change in applied voltage and current results in changed value of power. If the power increases, voltage and current increases in the same direction and vice versa.

The boost converter shown in Figure 7 works in two modes:

Mode 1 ($0 < t \leq t_{on}$):

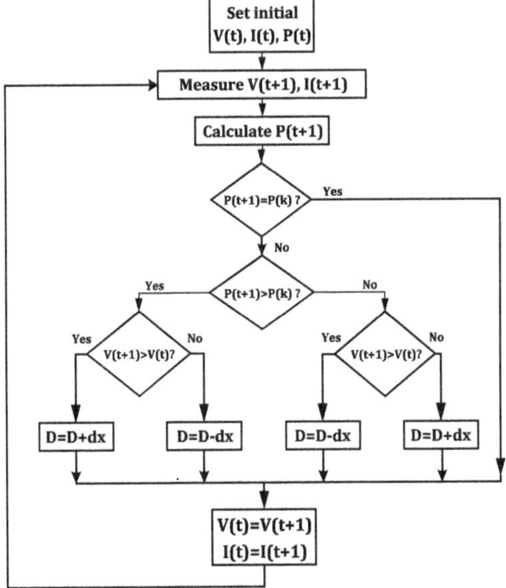


**Figure 6. Flowchart of P & O method.**

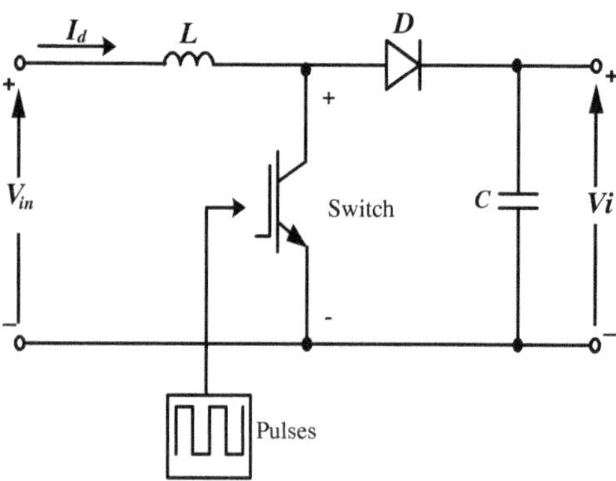


**Figure 7. Circuit diagram of step-up DC-DC converter.**

Mode 1 begins when IGBT's is switched on at $t = 0$ and terminates at $t = t_{on}$. The inductor current IL($t$) greater than zero and ramp up linearly. The inductor voltage is $V_i$.

Mode 2 ($t_{on} < t < T_s$):

Mode 2 begins when IGBT's is switched off at $t = t_{on}$ and terminates at $t = t_s$. The inductor current decrease until the IGBT is turned on again during the next cycle. The voltage across the inductor in this period is $V_i - V_o$ (Hasaneen & Mohammed, 2008; Rashid, 1993).

In steady-state condition, time integral of the inductor voltage over one time period must be zero.

$$V_i t_{on} + (V_i - V_o)\, t_{off} = 0 \tag{7}$$

where, $V_i$ is the input voltage, $V_o$ is the average output voltage, $t_{on}$ is the switching-on time of the IGBT, and $t_{off}$ is the switching-off time of the IGBT.

Dividing both sides by $T_s$ and rearranging the terms yield

$$\frac{V_o}{V_i} = \frac{T_s}{t_{off}} = \frac{1}{1 - D} \tag{8}$$

where, $T_s$ is the switching period (s) and $D$ is the duty cycle.

The output of three-phase CC-VSI is fed to the AC bus through LC filter. The configuration of three-phase CC-VSI is shown in Figure 8 (Rashid, 1993).

## 3. Wind and micro hydro system

A SEIG is employed for both wind and micro hydro systems. The reason for using SEIG is due to its lower unit cost compared to the conventional synchronous generator, maintenance free operation, and its inherent protective characteristics in case of short circuit on lines (Murthy, Jose, & Singh, 1998). For modeling SEIG the main flux path saturation is accounted for while the saturation in the leakage flux path of magnetic core of the machine, the iron loss, and the rotational losses are neglected (Molina, dos Santos, & Pacas, 2010). The equivalent circuit of SEIG is shown in Figure 9.

From the equivalent circuit, it can be deduced as,

$$\left(\frac{R_s}{F} + jX_{ls} + Z_L + Z_r\right)\bar{I}_s = 0 \tag{9}$$

where, $Z_L = \frac{-jX_C R_L}{(F^2 R_L - jFX_C)}$ and $Z_r = \frac{jX_m R_r - X_m X_{lr}(F-v)}{j(X_m + X_{lr})(F-v) + R_r}$

Under self excitation conditions,

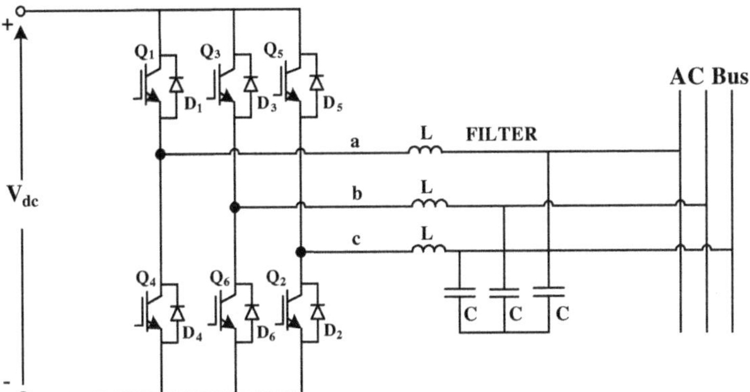


**Figure 8. Configuration of three-phase VSI with LC filter.**

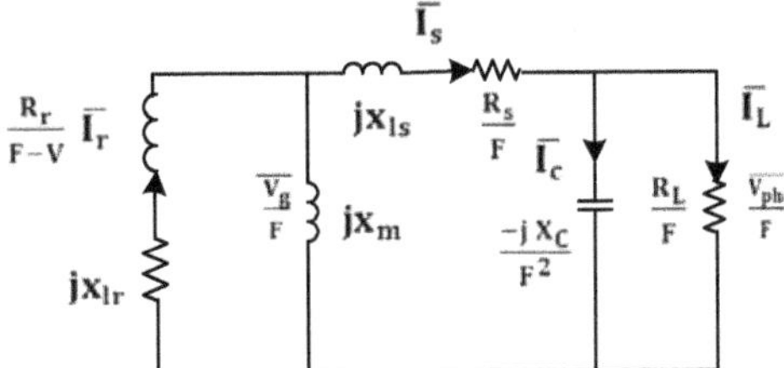

**Figure 9. Equivalent circuit of CAG.**

$$\frac{R_s}{F} + jX_{ls} + Z_L + Z_r = 0 \tag{10}$$

where, $V_g$ is the air-gap voltage, $V_{ph}$ is the phase voltage, $v$ is the speed, and $F$ is the frequency.

For any speed $v$, Equation 10 may be solved for $X_m$ and $F$ using Newton–Raphson iteration method which is ideal for finding solution of non-linear equations or any other known methods of solving non-linear equations with the starting values being taken as the unsaturated value of $X_m$, and $F$ equal to $v$. Using these values of $X_m$ and $F$, the total performance of the machine can be evaluated in conjunction with the measured variation of $X_m$ with $V_g/F$ for the generator being considered.

In wind energy conversion system, a variable speed variable frequency scheme is proposed due to its ability of efficient energy conversion. The system is operated over a wide range of speed without employing pitch control mechanism for the wind turbine. Voltage build-up is achieved by connecting excitation capacitors across the stator terminals. According to wind speed cube law, the power, $P$ developed by the wind turbine of blade diameter, $d$ at a wind speed, $s$ is given by (Raina & Malik, 1983).

$$P = \frac{1}{8} \rho \pi d^2 C_p s^3 \tag{11}$$

where, $\rho$ is the density of air and $C_p$ is the power coefficient which is the ratio of shaft power to wind power.

The microhydro system is modeled with a hydro turbine driven SEIG. The input to the turbine is from flowing stream with proper civil works in order to maintain constant flow so that the input power to the hydro turbine remains constant. Finally, the AC output from wind and microhydro systems is integrated with the PV system and the combination is brought out to a common AC bus.

## 4. Static synchronous compensator (STATCOM)

The proposed STATCOM is a three-legged IGBT based CC-VSI with a DC link capacitor and an AC filtering inductances ($Lf$, $R_f$). The DC bus capacitor is used to offer self-support to the DC bus of the compensator. The fluctuation in capacitance voltage is due to power consumed by the devices in the VSI and filter resistance. It is essential to find out the real and reactive power requirement of the hybrid system to maintain balanced voltages and current at the load side. First, the fluctuation in the DC link capacitor voltage is monitored. This is an index of the imbalance in real power or indirectly it is an indication of direct axis component of current, $I_d$. Next, the peak value of the line-to-line voltage from the integrated system is computed. This is compared with the reference or expected value. The difference between these two quantities is an indication of the reactive power required by the system or indirectly this will be the amount of quadrature current, $I_q$ to be supplied by the system.

These two axes reference currents namely $I_d$ and $I_q$ are converted into three-phase form by Inverse Park's transformation. The cos($\omega t$) and sin($\omega t$) terms needed for Park's transformation are derived with the help of a phase locked loop which is fed with unit templates of line voltages from the source. The three-phase reference currents thus obtained are compared with the actual currents from the integrated system in a hysteresis current controller to yield the firing signals for the six devices in the STATCOM. Figure 10 shows the schematic diagram of the CC-VSI unit (Gao, Bhuvaneswari, Murthy, & Kalla, 2014; Goel et al., 2011).

Inverse Park's transformation (dq0-abc transformation) is represented as

$$I_{dq0} = TI_{abc} = \sqrt{\frac{2}{3}}\begin{bmatrix} \cos(\omega t) & \cos\left(\omega t - \frac{2\pi}{3}\right) & \cos\left(\omega t + \frac{2\pi}{3}\right) \\ \sin(\omega t) & \sin\left(\omega t - \frac{2\pi}{3}\right) & \sin\left(\omega t + \frac{2\pi}{3}\right) \\ \frac{\sqrt{2}}{2} & \frac{\sqrt{2}}{2} & \frac{\sqrt{2}}{2} \end{bmatrix}\begin{bmatrix} I_a \\ I_b \\ I_c \end{bmatrix} \quad (12)$$

## 5. Proposed integrated system and simulation results

The proposed scheme of integration of solar, wind, and microhydro is modeled and simulated in Matlab/Simulink environment. The three different RES are connected to a common AC bus on to which varying loads are connected (Mohammed, 2006; Sharaf & El-Sayed, 2009). The power quality issues (Anees, 2012) like the system voltage and current regulation and harmonic reductions have been analyzed with the help of a STATCOM.

Figure 11 shows the hybrid system proposed in this study. The simulink model of the system is shown in Figure 12. Simulation is carried out initially with the three different sources independently under no load conditions and eventually with the hybrid system through three-phase AC bus connected to three-phase IM, R, and RL loads. The results are presented through Figures 13–16.

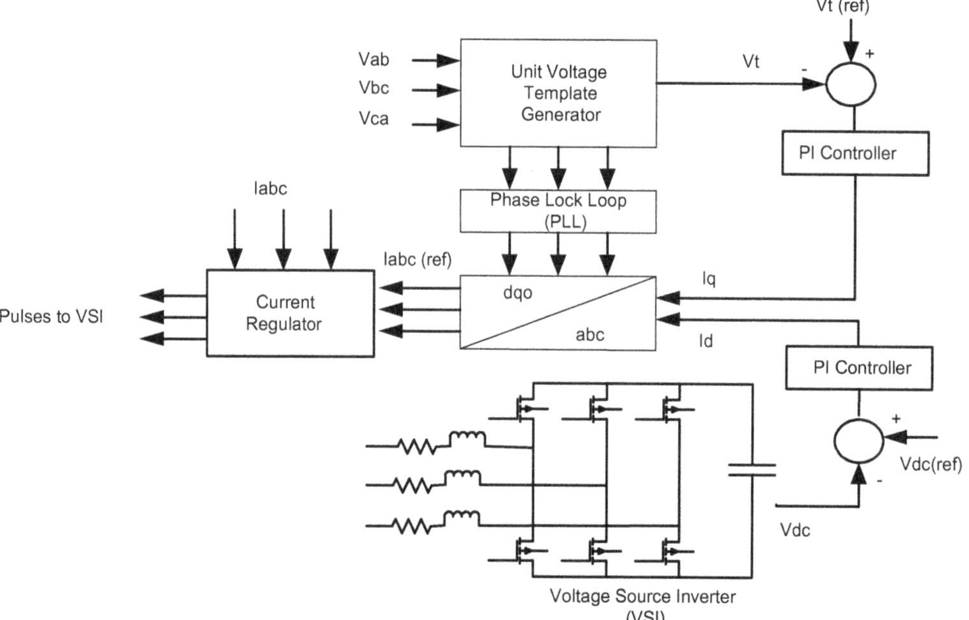


**Figure 10. Schematic diagram of current-controlled voltage source inverter (CC-VSI).**

It is clearly observed that the three different RES generates little over rated voltage of 459 V (rms) or 650 V (peak) under no load conditions. The frequency is maintained at 50 Hz. As seen from Figure 16, on connecting loads namely, IM, R, and R-L loads at 1.0, 1.2 and 1.4 s, respectively, there is a voltage drop in the system. The loads are switched out at 1.6 s and the system rated voltage is regained after sometime. A STATCOM is, therefore, connected to the system at 0.6 s and the loads are connected at the same instants. It is observed that the system voltage and frequency is maintained at its rated value; in this case the voltage is 415 V (rms) or 586 V (peak). Figure 17 shows the voltage and current waveforms of the integrated system with STATCOM. It is observed that the AC bus voltage is regulated with the use of STATCOM though a small amount of transient current occurs at the time of switching in the STATCOM. The overall performance of the system is observed to have been improved.

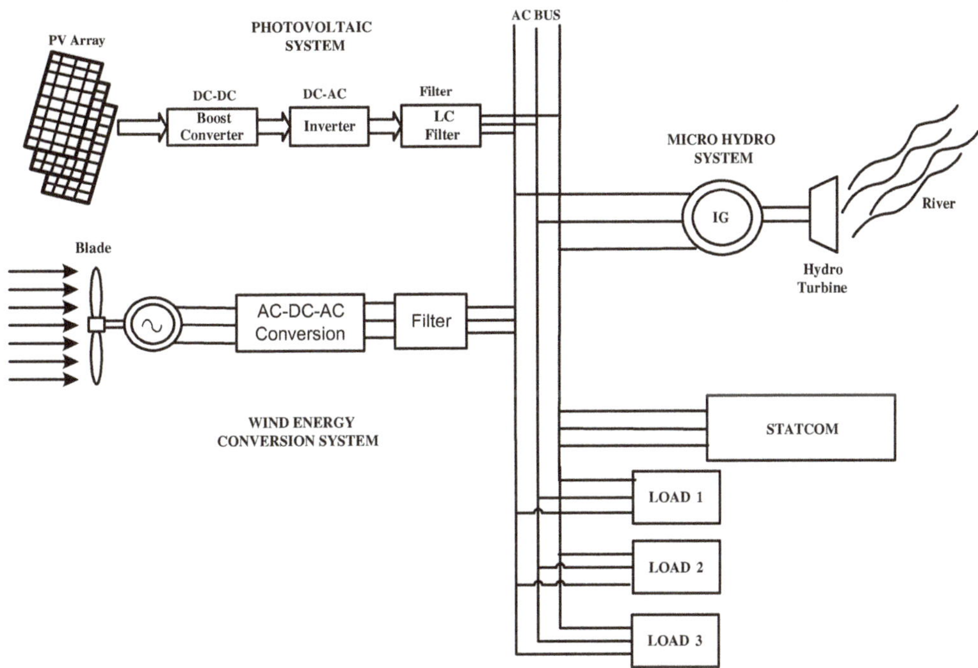


**Figure 11. Model of the proposed standalone hybrid solar, wind, and microhydro system feeding three-phase loads.**

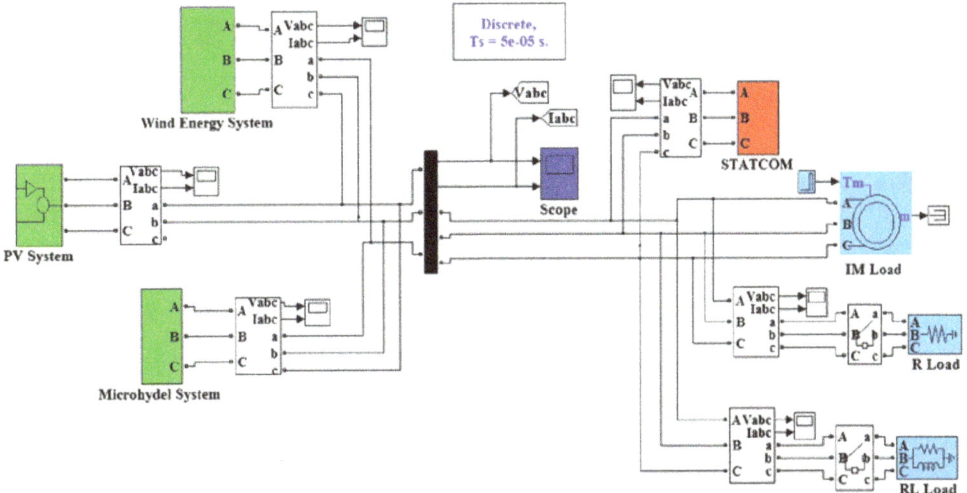


**Figure 12. Simulink model of hybrid PV, wind, and microhydro system.**

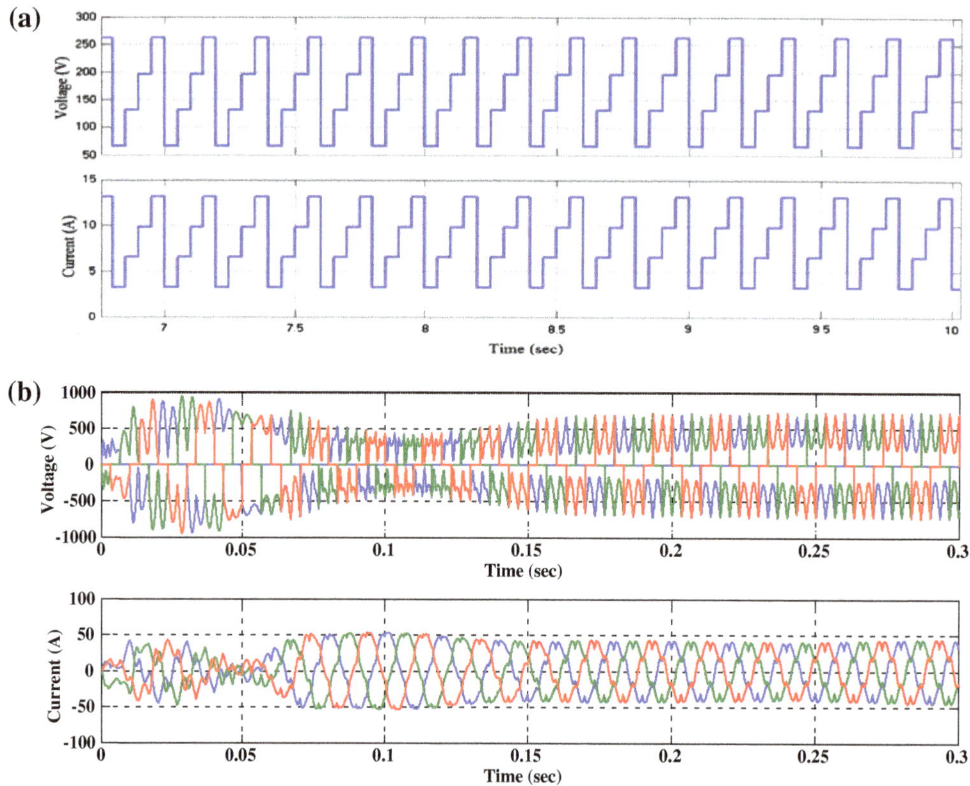


**Figure 13. Voltage and current waveforms of PV array (a) before DC-AC converter and (b) after DC-AC converter under no load conditions.**

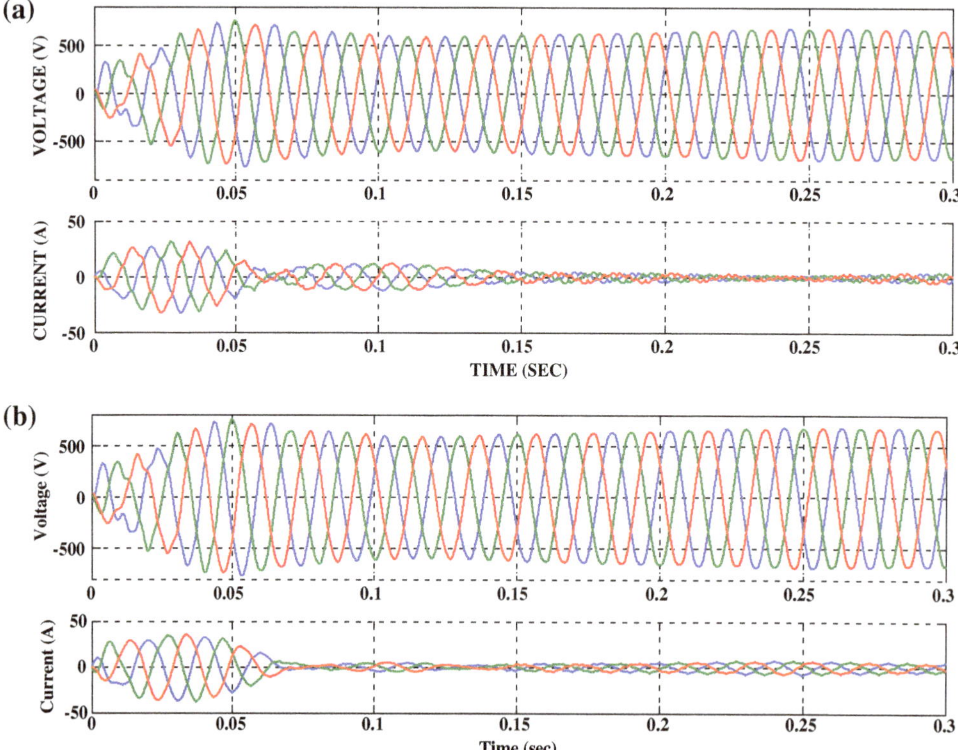


**Figure 14. Voltage and current waveforms of (a) WES and (b) MHS under no load conditions.**

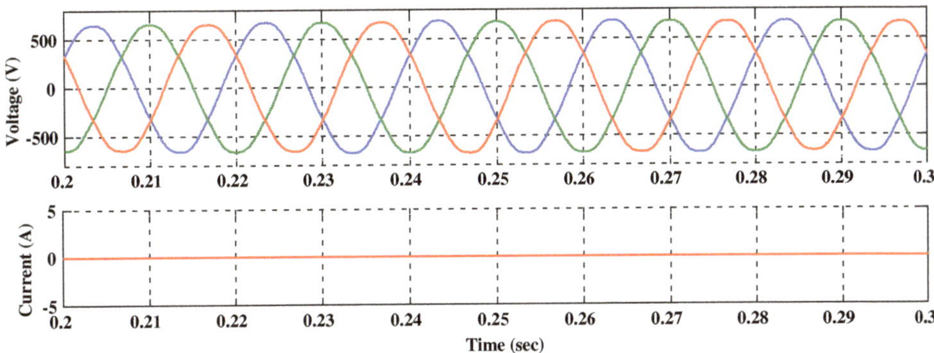


**Figure 15. Voltage and current waveforms of hybrid system under no-load condition.**

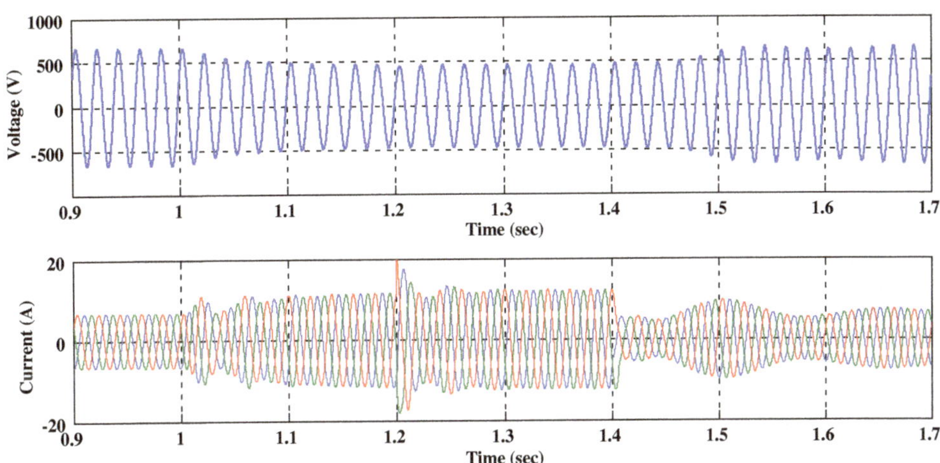


**Figure 16. Voltage and current waveforms of hybrid system under IM, R, and RL loads.**

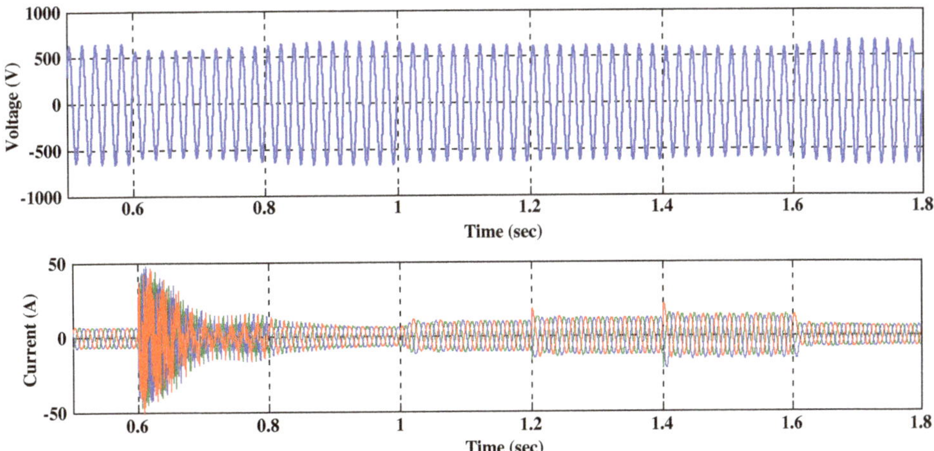


**Figure 17. Voltage and current waveforms of hybrid system with STATCOM under IM, R, and RL loads.**

The total harmonic distortion (THD) is an important figure of merit used to quantify the level of harmonics in voltage and current waveform. There exist two different definitions of THD, in the first one the harmonic content is compared to the fundamental and in the second one to the rms value of the waveform (Similovitz, 2005).

The THD (compared to fundamental) of the system voltages and currents under load conditions are analyzed here and they are observed to be well within IEEE Standard 519-1992. Figures 18 and 19 show the THD of system voltages and currents under IM, R, and R-L loads.

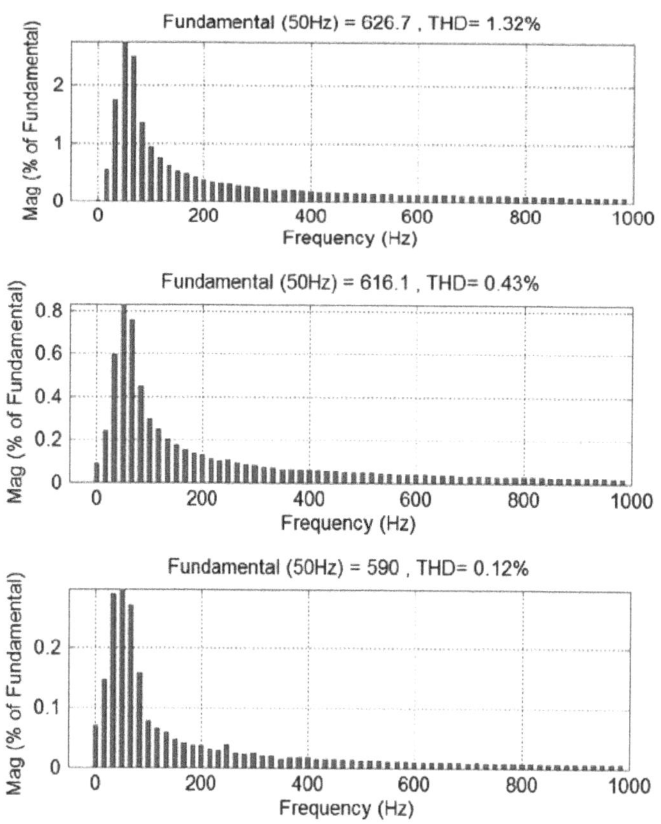


**Figure 18. THD of system voltages under IM, R, and RL loads.**

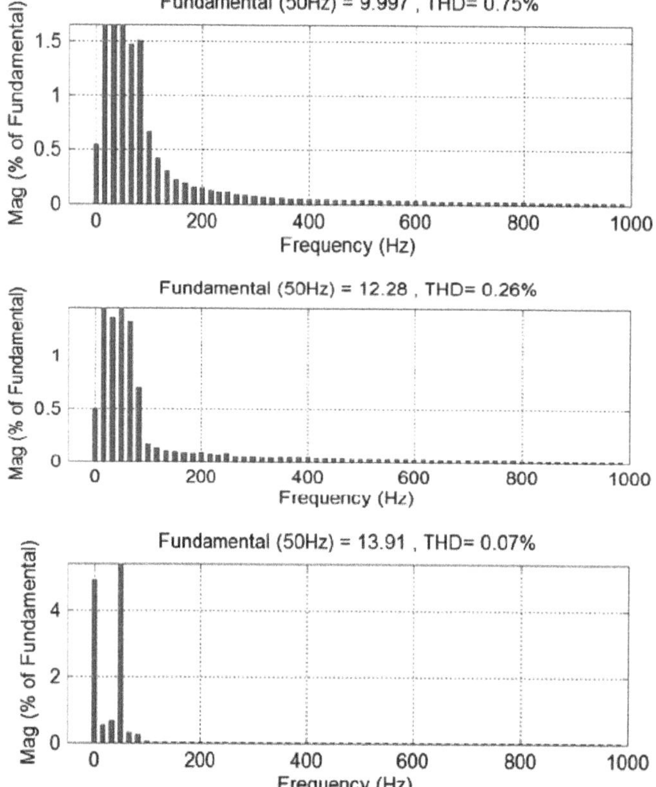


**Figure 19. THD of system currents under IM, R, and RL loads.**

## 6. Conclusions

In this work, an attempt has been made to investigate in detail the possibility of HRES as one of the most effective ways of decentralized power generation. Besides increasing the generating capacity of the energy sources, the scheme is envisaged to greatly liberate the rural populace of the perennial energy demands. With the energy sources connected together, the reliability of the power system increases. The power quality aspects are studied and with the use of STATCOM the voltage and current regulations improves. The system voltage is well maintained at 415 V (rms) under different load conditions. The harmonic content of the system voltages and currents is greatly reduced when loads are connected. The THD of the voltages and currents are of the order of 0.1–1.3% and 0.07–0.7% for IM, R, and RL loads, respectively. The voltage and frequency has been maintained constant under different dynamic conditions. The integrated system along with the addition of STATCOM can be seen as a viable option for supplying electricity to far flung areas.

**Funding**
The authors received no direct funding for this research.

**Author details**
Rinchin W. Mosobi[1]
E-mail: wangzom123@gmail.com
Toko Chichi[1]
E-mail: toko.chichi7@gmail.com
Sarsing Gao[1]
E-mail: sarsing@nerist.ac.in
[1] Department of Electrical Engineering, North Eastern Regional Institute of Science and Technology, Nirjuli, Arunachal Pradesh 791 109, India.

**References**

Anees, A. S. (2012). Grid integration of renewable energy sources: Challenges, issues and possible solutions. In *5th India International Conference on Power Electronics (IICPE)* (pp. 1–6). Delhi: IEEE.

Bansal, M., Khatod, D. K., & Saini, R. P. (2014, February). Modeling and optimization of integrated renewable energy system for a rural site. In *International Conference on Reliability, Optimization and Information Technology (ICROIT)* (pp. 25–28). Faridabad. http://dx.doi.org/10.1109/ICROIT.2014.6798289

Gao, S., Bhuvaneswari, G., Murthy, S. S., & Kalla, U. K. (2014, March). Efficient voltage regulation scheme for three-phase self-excited induction generator feeding single-phase load in remote locations. *IET Renewable Power Generation, 8*, 100–108. http://dx.doi.org/10.1049/iet-rpg.2012.0204

Gera, R. K., Rai, H. M., Parvej, Y., & Soni, H. (2013, January–June). Renewable energy scenario in India: Opportunities and challenges. *Indian Journal of Electrical And Biomedical Engineering, 1*, 10–16.

Goel, P. K., Singh, B., Murthy, S. S., & Kishore, N. (2011, April). Isolated wind-hydro hybrid system using cage generator and battery storage. *IEEE Transactions on Industrial Electronics, 58*, 1141–1153.

Hasaneen, B. M., & Mohammed, A. A. E. (2008, March). Design and simulation of DC/DC boost converter. In *IEEE Power System Conference, MEPCON* (pp. 335–340). Aswan.

International Renewable Energy Agency (IRENA). (2012, June). *Renewable energy technologies: Cost analysis of hydropower, 1*(3/5). Retrieved from www.irena.org/

Kondawar, S. S., & Vaidya, U. B. (2012, August). A comparison of two MPPT techniques for pv system in Matlab simulink. *International Journal of Engineering Research and Development, 2*, 73–79.

Kyocera KC200GT. (xxxx). Solar photovoltaic power modules datasheet. *High efficiency multi crystal photo voltaic module.*

Lipu, M. S. H., Uddin, M. S., & Miah, M. A. R. (2013). A feasibility study of solar-wind-diesel hybrid system in rural and remote areas of Bangladesh. *International Journal on Renewable Energy Resource, 3*, 895–900.

Mohammed, A. A. E. (2006). *Study of interconnecting issues of photovoltaic/wind hybrid system with electric utility using artificial intelligence* (PhD Thesis). Faculty of Engineering, Electrical Engineering Department, Minia University, Minya.

Molina, M. G., dos Santos, E. C., & Pacas, M. (2010). Improved power conditioning system for grid integration of photovoltaic solar energy conversion systems, In *Transmission and Distribution Conference and Exposition in Latin America* (pp. 163–170). Sao Paulo. http://dx.doi.org/10.1109/TDC-LA.2010.5762877

Murthy, S. S., Jose, R., & Singh, B. (1998). Experience in the development of microhydel grid independent power generation scheme using induction generators for Indian conditions. In *IEEE Conference on Global Connectivity in Energy, Computer, Communication and Control*, (Vol. 2, pp. 461–465).

Nejad, M. F., Radzi, M. A. M., Kadir, M. Z. A. A., & Hizam, H. (2012, December). Hybrid renewable energy systems in remote areas of equatorial countries. In *IEEE Student Conference on Research & Development (SCOReD)* (pp. 11–16). Pulau Pinang.

Raina, G., & Malik, O. P. (1983, December). Wind energy conversion using a self-excited induction generator. *IEEE Transactions on Power Apparatus and System, PAS-102*, 3933–3936.

Rashid, M. H. (1993). *Power electronics: Circuits, devices and applications* (2nd ed.). Englewood Cliffs, NJ: Prentice-Hall.

Sharaf, A. M., & El-Sayed, M. A. H. (2009). A novel hybrid integrated wind-pv micro co-generation energy scheme for village electricity. In *Electric Machine and Drives Conference IEMDC* (pp. 1244–1249).

Similovitz, Doron (2005, January). On the defination of total harmonic distortion and its effect on measurement interpretation. *IEEE Transaction on Power Delivery, 20*, 526–528.

Villalva, M. G., Gazoli, J. R., & Filho, E. (2009). Comprehensive approach to modeling and simulation of photovoltaic arrays. *IEEE Transactions on Power Electronics, 24*, 1198–1208. http://dx.doi.org/10.1109/TPEL.2009.2013862

Villalva, M. G., Gazoli, J. R., & Ruppert, E. (2009). *Modelling and circuit -based simulation of photovoltaic arrays* (pp. 1244–1245). Brazil: Power Electronics Conference.

## Appendix

| | |
|---|---|
| PV arrayTemperature (K)Solar irradiance (W/m$^2$) | 25 + 273.15, 30 + 273.15, 35 + 273.15, 40 + 273.15200, 400, 600, 800 |
| Boost converter | $L$ = 1 mH, $C$ = 50 μF |
| LC filter | $L$ = 10 μH, $C$ = 350 μF |
| Wind speed (m/s$^2$)<br>Wind turbine | 11, 10, 12, 13, 9, 11, 14$P_n$ = 1.5e 6, $\beta$ = 0°, $\lambda$ = 8.1, $C_p$ = 0.48, base power = 1.67e 6 VA, base wind speed = 12 m/s |
| SEIG | $P_n$ = 7.5 VA, V(L-L) = 415 V, $Rs$ = 0.9 Ω, $R_r'$ = 0.66 Ω, $L_{ls}$ = 0.00457 H, $L_{lr}'$ = 0.00457 H, $J$ = 0.1384 kgm$^2$, $F$ = 0 Nm, $p$ = 2 |

# 10

# On partial approximate controllability of semilinear systems

Agamirza E. Bashirov[1,2*] and Noushin Ghahramanlou[1]

*Corresponding author: Agamirza E. Bashirov, Department of Mathematics, Eastern Mediterranean University, Mersin 10, Turkey; Institute of Cybernetics, ANAS, Baku, Azerbaijan

E-mail: agamirza.bashirov@emu.edu.tr

Reviewing editor: James Lam, University of Hong Kong, Hong Kong

**Abstract:** In this paper, a sufficient condition for the partial approximate controllability of semilinear deterministic control systems is proved. Generally, the theorems on controllability are formulated for control systems given as a first-order differential equation, while many systems can be written in this form only by enlarging the dimension of the state space. The ordinary controllability conditions for such systems are too strong because they involve the enlarged state space. Therefore, it becomes useful to define partial controllability concepts, which assume the original state space. The method of proof, given in this paper, differs from the traditional proofs by fixed point theorems. The obtained result is demonstrated on examples.

**Subjects: Applied Mathematics, Engineering & Technology, Science, Systems & Control Engineering, Technology**

**Keywords: approximate controllability, exact controllability, partial controllability, semilinear system**

**AMS subject classification: 93B05**

## 1. Introduction

In 1960, Kalman (1960) defined the concept of controllability as a property of control systems to attain every point in the state space from every initial state point for a finite time. Further studies in this field resulted with a separation of this concept into two concepts: a stronger concept of exact controllability and a weaker concept of approximate controllability. The reason for this was the fact

## ABOUT THE AUTHORS

Agamirza E. Bashirov joined the Department of Mathematics of the Eastern Mediterranean University after defending his Doctor of Sciences dissertation in mathematics at the Kiev State University in 1991 and held the positions: associate professor from September 1992 till March 1994, and then professor till date. His research concerns stochastic systems, optimal control, estimation theory, controllability theory, and multiplicative calculus.

Noushin Houshyar Ghahramanlou has received BS degree from Urmia Payam nour University of IRAN in 2006 and MS degree from URMIA University in 2009. Presently, she is a PhD student at the Department of Mathematics of Eastern Mediterranean University. The subject of her PhD thesis is related to controllability theory for deterministic and stochastic systems.

## PUBLIC INTEREST STATEMENT

In the real life, the controllability means being able to manipulate completely or almost completely with the state of an object. For example, TVs are manipulated by remote controls and marionettes by actors. E-mail messaging is controllable if everyone is able to send a message to everyone. If some of e-mail accounts are blocked, then e-mail messaging becomes partially controllable as displayed in the cover image. In engineering, a control system is controllable if it is possible to reach its every (or almost every) state for a finite time. It has great applications, for example, in robotics. In this paper, conditions are proved under which a given system is controllable.

that many infinite dimensional control systems are approximately controllable while they are not completely controllable (see Fattorini, 1966; Russel, 1967). The controllability concepts for linear systems are discussed in Curtain and Zwart (1995), Bensoussan (1992), Bensoussan, Da Prato, Delfour, and Mitter (1993), Zabczyk (1995), Bashirov (2003), Klamka (1991), etc.

Recently, in Bashirov, Etikan, and Şemi (2010) and Bashirov, Mahmudov, Şemi, and Etikan (2007), the partial controllability concepts were introduced and the basic controllability conditions from Bashirov and Mahmudov (1999a) and Bashirov and Kerimov (1997) (see also Bashirov, 1996; Bashirov & Mahmudov, 1999b, 1999c) were extended to partial controllability concepts by a replacement of the controllability operator by its partial version. A study of partial controllability concepts is a significant part of the overall study in the area of controllability. This is motivated by the fact that the theorems on controllability are formulated for control systems given as a first-order differential equation, while many systems, such as higher order differential equations, wave equations, and delay equations, can be written in this form only by enlarging the dimension of the state space. Therefore, the ordinary controllability conditions for such systems are too strong because they involve the enlarged state space while controllability concepts require the original state space. This motivates to define the partial controllability concepts that differ from ordinary controllability concepts by an additional projection of the enlarged state to the original state. In more details, this is discussed in Section 2.

In this paper, we study the partial approximate controllability for semilinear systems. The controllability concepts for nonlinear systems are intensively studied in the literature (see, e.g. Balachandran & Sakthivel, 2001; Bashirov & Jneid, 2013; Klamka, 2000, 2001, 2002, 2008; Leiva, Merentes, & Sanchez, 2011, 2012, 2013; Ren, Dai, & Sakthivel, 2013; Sakthivel, Ganesh, & Suganya, 2012; Sakthivel, Ganesh, Ren, & Anthoni, 2013; Sakthivel, Mahmudov, & Kim, 2009; Sakthivel, Mahmudov, & Nieto, 2012; Sakthivel & Ren, 2013; Sakthivel, Suganya, & Anthoni, 2012, etc.). An underlying method of study in these works is fixed point theorems. In the present paper, we apply a different and more natural method. The idea of this method is that we divide the time interval $[0, T]$ into two subintervals $[0, T-\delta]$ and $[T-\delta, T]$. On the interval $[0, T-\delta]$, we choose any control and steer the initial state to some state at $T-\delta$. Then on the interval $[T-\delta, T]$, we choose the sequence of controls steering the state at $T-\delta$ arbitrarily close to target state at $T$ along the linear part of the semilinear system. Using the fact that on the small time interval, the nonlinearity of the semilinear system disturbs its linear part for a small value, we obtain the partial approximate controllability of the semilinear system. This simple idea is realized in this paper and a sufficient condition of partial approximate controllability for the semilinear system is proved. A significant point in the proved sufficient condition is that instead of the positiveness of the controllability operator, we assume the positiveness of the partial controllability operator. This produces a weaker condition in comparison with the similar condition for ordinary approximate controllability. So, the main contribution of the paper is the proved sufficient condition as well as the method of the proof. The obtained result is demonstrated on examples.

One major notation is that we prefer to write the arguments of functions in the subscripts, for example, $f_t$ instead of $f(t)$. $\mathbb{R}^n$ denotes an $n$-dimensional Euclidean space and $\mathbb{R}^{n\times k}$ the space of $n\times k$) -matrices. As always, $\mathbb{R}=\mathbb{R}^1$. The norm and scalar products in all considered spaces are denoted by $\|\cdot\|$ and $\langle\cdot,\cdot\rangle$, being clear from the context. $I$ and $0$ are the identity and zero matrices or operators independently on their dimensions. $A^*$ is the adjoint of the linear closed operator $A$. In the case when $A\in\mathbb{R}^{n\times m}$, $A^*$ becomes the transpose of $A$. For a linear operator $F$ on $H$, we write $F\geq 0$ (respectively, $F>0$) if $F^*=F$ and $\langle Fh,h\rangle\geq 0$ for all $h\in H$ (respectively, $\langle Fh,h\rangle>0$ for all nonzero $h\in H$). For $H$-valued functions on $[a, b]$, we use the symbols $C(a, b; H)$ for continuous functions, $PC(a, b; H)$ for piecewise continuous functions, and $L_2(a, b; H)$ for square integrable functions. In the case $H=\mathbb{R}$, $H$ is dropped in these symbols.

## 2. Setting the problem and motivation

Consider the semilinear control system

$$x_t' = Ax_t + Bu_t + f(t, x_t, u_t) \tag{1}$$

on the interval [0, T] with $T > 0$. Here, $x$ is a state process and $u$ is a control from $U_{ad}$. We assume:

(A) $X$ and $U$ are separable Hilbert spaces, $H$ is a closed subspace of $X$, and $L$ is a projection operator from $X$ to $H$.

(B) $A$ is a densely defined closed linear operator on $X$, generating a strongly continuous semigroup $e^{At}, t \geq 0$.

(C) $B$ is a bounded linear operator from $U$ to $X$.

(D) $f$ is a nonlinear function from $[0,T] \times X \times U$ to $X$ and satisfies:

- $f$ is continuous on $[0, T] \times X \times U$,
- $f$ is bounded on $[0, T] \times X \times U$,
- $f$ satisfies Lipschitz condition with respect to $x$.

(E) $U_{ad} = PC(0, T; U)$.

Under these conditions, for every $u \in U_{ad}$ and $x_0 \in X$, Equation 1 admits a unique continuous mild solution (see Li & Yong, 1995), that is, there is a unique continuous function $x^{u,x_0}$ from [0, T] to $X$ such that

$$x_t^{u,x_0} = e^{At} x_0 + \int_0^t e^{A(t-s)} \left( Bu_s + f\left(s, x_s^{u,x_0}, u_s\right) \right) ds, \quad 0 \leq t \leq T$$

Define the set

$$D_T^{x_0} = \left\{ x \in X : \exists u \in U_{ad} \text{ such that } x = x_T^{u,x_0} \right\}$$

According to Bashirov et al. (2010, 2007), the semilinear system in Equation 1 is said to be $L$-partially $A$-controllable on $U_{ad}$ if $\overline{L(D_T^{x_0})} = H$ for all $x_0 \in X$, where $\bar{D}$ is the closure of $D$ and $L(D)$ is the image of $D$ under $L$. Similarly, the semilinear system in Equation (1) is said to be $L$-partially $E$-controllable on $U_{ad}$ if $L\left(D_T^{x_0}\right) = H$ for all $x_0 \in X$. Here, $A$ and $E$ are abbreviations for the terms "approximate" and "exact," respectively. In the case $H = X$, these are just well-known approximate and exact controllability concepts, respectively. If $H$ is the zero-dimensional subspace of $X$, then the $L$-partial controllability concepts reduce to the null-controllability.

To motivate the partial controllability concepts, consider a few examples of systems which can be written in the form of Equation 1 after enlarging the state space.

**Example 1** Consider the nonlinear system

$$x_t^{(n)} = f\left(t, x_t, x_t', \ldots, x_t^{(n-1)}, u_t\right) \tag{2}$$

with the one-dimensional state space $X = \mathbb{R}$. Write this system as a first-order differential equation

$$y_t' = Ay_t + F(t, y_t, u_t) \tag{3}$$

for

$$y_t = \begin{bmatrix} x_t \\ x_t' \\ \vdots \\ x_t^{(n-2)} \\ x_t^{(n-1)} \end{bmatrix}, \quad A = \begin{bmatrix} 0 & 1 & \cdots & 0 & 0 \\ 0 & 0 & \cdots & 0 & 0 \\ \vdots & \vdots & \ddots & \vdots & \vdots \\ 0 & 0 & \cdots & 0 & 1 \\ 0 & 0 & \cdots & 0 & 0 \end{bmatrix}$$

and

$$F(t,y,u)=\begin{bmatrix} 0 \\ 0 \\ \vdots \\ 0 \\ f\left(t,x,x',\ldots,x^{(n-1)},u\right) \end{bmatrix}$$

The state space of the system in Equation 3 is the $n$-dimensional Euclidean space $\mathbb{R}^n$ and, respectively, its attainable set is a subset of $\mathbb{R}^n$. Therefore, the controllability concepts for the system in Equation 3 are stronger than the same for in Equation 2. But if we define the projection operator $L$ by

$$L=[\ 1\quad 0\quad \cdots\quad 0\quad 0\ ]:\mathbb{R}^n\to\mathbb{R}$$

then the $L$-partial controllability concepts for the system in Equation 3 become the same as the ordinary controllability concepts for the system in Equation 2.

**Example 2** Consider the nonlinear wave equation

$$\frac{\partial^2 x_{t,\theta}}{\partial t^2}=\frac{\partial^2 x_{t,\theta}}{\partial \theta^2}+b_\theta u_t+f\left(t,x_{t,\theta},\partial x_{t,\theta}/\partial t,u_t\right) \tag{4}$$

where $x$ is a real-valued function of two variables $t\geq 0$ and $0\leq\theta\leq 1$. The state space of this system is $L_2(0, 1)$. This system can also be written as the first-order abstract differential equation

$$y_t'=Ay_t+Bu_t+F(t,y_t,u_t) \tag{5}$$

if

$$y_t=\begin{bmatrix} x_{t,\theta} \\ \partial x_{t\theta}/\partial t \end{bmatrix},\quad A=\begin{bmatrix} 0 & I \\ d^2/d\theta^2 & 0 \end{bmatrix},\quad F(t,y,u)=\begin{bmatrix} 0 \\ f(t,y_1,y_2,u) \end{bmatrix},\quad B=\begin{bmatrix} 0 \\ b \end{bmatrix}$$

where

$$y=\begin{bmatrix} y_1 \\ y_2 \end{bmatrix}\in L_2(0,1)\times L_2(0,1)$$

The state space $L_2(0,1)\times L_2(0,1)$ of the system in Equation 5 is the enlargement of the state space $L_2(0, 1)$ of the system in Equation 4. This is a cost paid to bring the wave equation to a first-order differential equation. The ordinary controllability concepts for the system in Equation 5 are too strong for the system in Equation 4. But if the projection operator $L$ is defined by

$$L=[\ I\quad 0\ ]:L_2(0,1)\times L_2(0,1)\to L_2(0,1)$$

then $L$-partial controllability concepts for the system in Equation 5 become the same as the ordinary controllability concepts for the system in Equation 4.

**Example 3** Delay equations form another class of systems suitable for application of partial controllability concepts. Consider the system

$$x_t'=f\left(t,x_t,\int_{-\varepsilon}^{0} x_{t+\theta}\,d\theta,u_t\right) \tag{6}$$

which contains a simple distributed delay in the nonlinear term, assuming that $x$ is a real-valued function. The state space of this system is $\mathbb{R}$. To bring it to a system without delay, enlarge $\mathbb{R}$ to $\mathbb{R}\times L_2(-\varepsilon,0)$ and define $L_2(-\varepsilon,0)$-valued function

$$[\bar{x}_t]_\theta = x_{t+\theta},\quad t\geq 0,\quad -\varepsilon\leq\theta\leq 0$$

Then the above system can be written as an abstract system

$$y_t' = Ay_t + f(t,y_t,u_t) \tag{7}$$

if

$$y_t = \begin{bmatrix} x_t \\ \bar{x}_t \end{bmatrix},\quad A = \begin{bmatrix} 0 & 0 \\ 0 & d/d\theta \end{bmatrix},\quad F(t,y,u) = \begin{bmatrix} f(t,x,\Gamma\bar{x},u) \\ 0 \end{bmatrix}$$

where $\Gamma$ is the integral operator

$$\Gamma\bar{x} = \int_{-\varepsilon}^{0} \bar{x}_\theta d\theta,\quad \bar{x}\in L_2(-\varepsilon,0)$$

One can easily observe that the ordinary controllability concepts for the system in Equation 7 are too strong for the system in Equation 6. But for

$$L = [\,1 \quad 0\,] : \mathbb{R}\times L_2(0,1)\to\mathbb{R}$$

the $L$-partial controllability concepts for the system in Equation 7 are exactly the ordinary controllability concepts for the system in Equation 6.

For $0<\delta<T$, we will associate with the semilinear system in Equation 1 the following linear system

$$y_t' = Ay_t + Bv_t,\quad T-\delta<t\leq T \tag{8}$$

where $v\in V_{ad}^{\delta} = C(T-\delta,T;U)$. The solution of Equation 8 is again understood in the mild sense, that is, for every $v\in V_{ad}^{\delta}$ and $y_{T-\delta}\in X$, the function

$$y_t^{v,y_{T-\delta}} = e^{A(t-T+\delta)}y_{T-\delta} + \int_{T-\delta}^{t} e^{A(t-s)}Bv_s\,ds,\quad T-\delta\leq t\leq T$$

is a unique mild solution of Equation 8. The controllability operator for the linear system in Equation 8 is defined by

$$Q_\delta = \int_{T-\delta}^{T} e^{A(T-t)}BB^*e^{A^*(T-t)}dt = \int_0^{\delta} e^{At}BB^*e^{A^*t}dt$$

We let

$$\tilde{Q}_\delta = LQ_\delta L^*$$

and call it an $L$-partial controllability operator. In addition to preceding conditions (A–E), we will also assume that

(F) For all $0<\delta\leq T, \tilde{Q}_\delta>0$.

Note that $Q_\delta>0$ implies $\tilde{Q}_\delta>0$ but the converse is not true. Therefore, condition (F) is weaker that the same kind of condition involving $Q_\delta$ instead of $\tilde{Q}_\delta$. In this paper, we study the concept of $L$-partial $A$-controllability, while concerning the concept of $L$-partial $E$-controllability as well, for the semilinear system in Equation 1. We prove that under conditions (A–F), the system in Equation 1 is $L$-partially $A$-controllable.

### 3. Main result

The resolvent of $-\tilde{Q}_\delta$ is defined by $R(\lambda, -\tilde{Q}_\delta) = (\lambda I, +\tilde{Q}_\delta)^{-1}$. Obviously, $R(\lambda, -\tilde{Q}_\delta)$ exists for all $\lambda > 0$ since $\lambda I + \tilde{Q}_\delta$ is coercive.

LEMMA **1** *Under the above conditions and notation, for given $\lambda > 0$ and $h \in H$, there exists a unique optimal control $v^\lambda \in V_{ad}^\delta$ at which the functional*

$$J^\lambda(v) = \left\| Ly_T^{v,y_{T-\delta}} - h \right\|^2 + \lambda \int_{T-\delta}^{T} \|v_t\|^2 \, dt$$

*along the linear system in Equation 8 takes its minimal value on $V_{ad}^\delta$. Moreover,*

$$v_t^\lambda = -\lambda^{-1} B^* e^{A^*(T-t)} L^* \left( Ly_T^{v^\lambda, y_{T-\delta}} - h \right), \quad T-\delta \le t \le T \tag{9}$$

*and*

$$Ly_T^{v^\lambda, y_{T-\delta}} - h = \lambda R(\lambda, -\tilde{Q}_\delta) \left( Le^{AT} y_{T-\delta} - h \right) \tag{10}$$

*Proof* This lemma is a restatement of Lemma 1 for the case of the interval $[T - \delta, T]$, proved in Bashirov et al. (2007). □

LEMMA **2** *Under the above conditions and notation, assume that the linear system in Equation 8 is L-partially A-controllable on $V_{ad}^\delta$. Then for every initial value $y_{T-\delta} \in X$ and $h \in H$,*

$$\left\| Ly_T^{v^\lambda, y_{T-\delta}} - h \right\| \to 0 \text{ as } \lambda \to 0 \tag{11}$$

*where $v^\lambda$ is a control in $V_{ad}^\delta$ defined by Equations 9–10.*

*Proof* By the resolvent condition for the $L$-partial $A$-controllability from Bashirov et al. (2007), $\lambda R(\lambda, -\tilde{Q}_\delta)$ converges to the zero operator as $\lambda \to 0$ in the strong operator topology. Therefore, by Equation 10, the convergence in Equation 11 takes place. □

LEMMA **3** *Under the above conditions and notation, let the linear system in Equation 8 be L-partially E-controllable on $V_{ad}^\delta$. Then for every $[T - \delta \le t \le T]$ and $0 < \lambda \le \lambda_0$ with some $\lambda_0 > 0$,*

$$\left\| v_t^\lambda \right\| \le c_1 \|y_{T-\delta}\| + c_2 \|h\| \tag{12}$$

*where $c_1 \ge 0$ and $c_2 \ge 0$ are constants and $v^\lambda$ is a control in $V_{ad}^\delta$ defined by Equations 9–10.*

*Proof* From Equations 9–10,

$$v_t^\lambda = -B^* e^{A^*(T-t)} L^* R(\lambda, -\tilde{Q}_\delta) \left( Le^{AT} y_{T-\delta} - h \right)$$

Denote $M = \sup_{[0,T]} \left\| e^{At} \right\|$. By the resolvent condition for the $L$-partial $E$-controllability from Bashirov et al. (2007), $R(\lambda, -\tilde{Q}_\delta)$ converges as $\lambda \to 0$ in the uniform operator topology. Hence, $\left\| R(\lambda, -\tilde{Q}_\delta) \right\| \le K$ for some $K \ge 0$. Also, $\|L\| \le 1$ since $L$ is a projection operator. Therefore, for $c_1 = M^2 K \|B\|$ and $c_2 = MK\|B\|$, the inequality in Equation 12 holds.

THEOREM 1 *Under conditions* (A–F) *the semilinear system in* Equation 1 *is L-partially A-controllable on $U_{ad}$.*

*Proof* Give arbitrary $\varepsilon > 0$. Take any $x_0 \in X$ and $h \in H$. Let $0 < \delta < T$. Consider any function $u \in C(0, T; U)$. For example, it may be zero function. Let $x_t^{u,x_0}$ be the value of the mild solution of the semilinear system in Equation 1 at $t$, corresponding $u$ and $x_0$. Define the control $u^{\lambda,\delta}$ by letting it to be $u_t^{\lambda,\delta}=u_t$ if $0\le t\le T-\delta$ and

$$u_t^{\lambda,\delta}=-B^*e^{A^*(T-t)}L^*R\left(\lambda,-\tilde{Q}_\delta\right)\left(Le^{AT}x_{T-\delta}^{u,x_0}-h\right) \text{ if } T-\delta<t\le T \tag{13}$$

Obviously, $u^{\lambda,\delta}$ is piecewise continuous for all $\lambda>0$ and $0<\delta<T$. So, $u^{\lambda,\delta}\in U_{ad}$. We can write $x_T^{u^{\lambda,\delta},x_0}$ as

$$x_T^{u^{\lambda,\delta},x_0}=e^{A\delta}x_{T-\delta}^{u,x_0}+\int_{T-\delta}^{T}e^{A(T-s)}\left(Bu_s^{\lambda,\delta}+f\left(s\,x_s^{u^{\lambda,\delta},x_0},u_s^{\lambda,\delta}\right)\right)ds$$

Also, for the mild solution of the linear system in Equation 8, we have

$$y_T^{u^{\lambda,\delta},x_{T-\delta}^{u,x_0}}=e^{A\delta}x_{T-\delta}^{u,x_0}+\int_{T-\delta}^{T}e^{A(T-s)}Bu_s^{\lambda,\delta}\,ds$$

Therefore,

$$\left\|x_T^{u^{\lambda,\delta},x_0}-y_T^{u^{\lambda,\delta},x_{T-\delta}^{u,x_0}}\right\|\le\int_{T-\delta}^{T}\left\|e^{A(T-s)}\right\|\left\|f\left(s,x_s^{u^{\lambda,\delta},x_0},u_s^{\lambda,\delta}\right)\right\|ds$$

Letting $M=\sup_{[0,T]}\|e^{At}\|$ and $K=\sup_{[0,T]\times X\times U}\|f(t,x,u)\|$, we obtain

$$\left\|x_T^{u^{\lambda,\delta},x_0}-y_T^{u^{\lambda,\delta},x_{T-\delta}^{u,x_0}}\right\|\le MK\delta$$

This yields

$$\left\|Lx_T^{u^{\lambda,\delta},x_0}-h\right\|\le\left\|Lx_T^{u^{\lambda,\delta},x_0}-Ly_T^{u^{\lambda,\delta},x_{T-\delta}^{u,x_0}}\right\|+\left\|Ly_T^{u^{\lambda,\delta},x_{T-\delta}^{u,x_0}}-h\right\|\le MK\delta+\left\|Ly_T^{u^{\lambda,\delta},x_{T-\delta}^{u,x_0}}-h\right\|$$

Now let $0<\delta<\min\{T,\varepsilon/2MK\}$. The condition $\tilde{Q}_\delta>0$ implies that the linear system in Equation 8 is $L$-partially $A$-controllable. Then, by Lemma 2, we can find sufficiently small $\lambda>0$ such that

$$\left\|Ly_T^{u^{\lambda,\delta},x_{T-\delta}^{u,x_0}}-h\right\|<\frac{\varepsilon}{2}$$

For these $\delta$ and $\lambda$, the control $u^{\lambda,\delta}$ satisfies

$$\left\|Lx_T^{u^{\lambda,\delta},x_0}-h\right\|<MK\frac{\varepsilon}{2MK}+\frac{\varepsilon}{2}=\varepsilon$$

From the arbitrariness of $x_0\in X$, $h\in H$ and $\varepsilon>0$, we arrive to the $L$-partial $A$-controllability of the semilinear system in Equation 1. □

*Remark* 1 According to the proof of Theorem 1, the control $u^{\lambda,\delta}$ from Equation 13 does not completely determine a sequence of controls, steering the initial state $x_0$ arbitrarily close to $h\in H$ because the selection of $\lambda$ depends on $\delta$, that is, $\lambda=\lambda_\delta$. This means that the rate of the convergence $\lambda\to 0$ is subject to the rate of the convergence $\delta\to 0$.

*Remark* 2 The condition on boundedness of $f$ in (D) can be dropped if $X=\mathbb{R}^n$ and $U=\mathbb{R}^m$. Indeed, the concepts of $L$-partial $A$- and $E$-controllability for the linear systems in finite dimensions coincide. Therefore, we can use Lemma 3. Fix $x_0\in X$ and $h\in H$. For a fixed function $u\in PC\left(0,T;\mathbb{R}^m\right)$, the mild solution $x^{u,x_0}$ of Equation 1 is a continuous function on the compact interval [0, $T$]. Therefore, it is bounded. Let $\|x_t^{u,x_0}\|\le c$ for some $c\ge 0$. Then from Lemma 3,

$$\left\|u_t^{\lambda,\delta}\right\|\le c_1c+c_2\|h\|=r_1$$

This and the Lipschitz condition in (D) implies that all $x^{u^{\lambda,\delta},x_0}$ range in a bounded set, that is,

$$\left\|x_t^{u^{\lambda,\delta},x_0}\right\|\le r_2$$

for some $r_2 \geq 0$. Therefore, we can restrict the function $f$ into the compact set $[0,T]\times B^n(r_2)\times B^m(r_1)$, where $B^n(r_2)$ and $B^m(r_1)$ are closed balls in $\mathbb{R}^n$ and $\mathbb{R}^m$ with radii $r_2$ and $r_1$, respectively, centered at the origin. Then the continuity condition on $f$ in (D) implies the boundedness of $f$ on $[0,T]\times B^n(r_2)\times B^m(r_1)$. So, the boundedness of $f$ becomes a consequence from the other conditions.

*Remark* 3 The system in Equation 1 can also be written as

$$x_t' = (A+A_1)x_t + (B+B_1)u_t + (f(t,x_t,u_t) - A_1x_t - B_1u_t), \quad 0 < t \leq T$$

for some linear bounded operators $A_1$ and $B_1$. Therefore, in some circumstances, the boundedness of the nonlinear function $f(t, x, u)$ in the condition (D) can be replaced by the boundedness of $f(t, x, u) - A_1x - B_1u$ for suitable $A_1$ and $B_1$.

*Remark* 4 The used method of proof has three advantages in comparison to the method by fixed point theorems:

- There is no need to consider larger space $L_2(0, T; U)$ as a set of admissible controls because the sequence of controls, used in the proof, are piecewise continuous.
- It suffices the Lipschitz continuity of $f(t, x, u)$ just in $x$ for the existence and uniqueness of the mild solution of Equation 1, that is, the condition on Lipschitz continuity in $u$ can be removed.
- No need in unusual inequality, required in the method by fixed point theorems.

At the same time its disadvantage is that it is not applicable for study of exact controllability.

## 4. Examples

We demonstrate the features of Theorem 1 in the following examples of control systems.

**Example 4** To demonstrate that the conditions of Theorem 1 are just sufficient for $L$-partial $A$-controllability but not necessary, let $L = I$, reducing the $L$-partial $A$-controllability to approximate controllability. Consider any infinite-dimensional control system of the form

$$x_t' = 2Ax_t + Bu_t, \quad x_0 \in \mathbb{R} \tag{14}$$

where $A$ is closed but not bounded and $B$ is bounded operators. Assume that

$$Q_\delta = \int_0^\delta e^{2At}BB^*e^{2A^*t}\,dt > 0$$

for all $\delta > 0$. Then the system in Equation 14 is approximately controllable. For example, such a system may be controllable heat equation studied in Bashirov and Mahmudov (1999a). Have another look to the system in Equation 14 by writing it as

$$x_t' = Ax_t + Bu_t + f(x_t), \quad x_0 \in \mathbb{R} \tag{15}$$

with $f(x) = Ax$. Here, the function $f$ is neither continuous nor bounded since $A$ is an unbounded operator. Therefore, it does not satisfy the conditions of Theorem 1 while it is approximately controllable.

**Example 5** To demonstrate that a system may not be approximately controllable while being $L$-partially $A$-controllable, consider the control system consisting of two one-dimensional differential equations

$$\begin{cases} x_t' = y_t + bu_t, & x_0 \in \mathbb{R} \\ y_t' = f(t,x_t,y_t,u_t), & y_0 \in \mathbb{R} \end{cases} \tag{16}$$

on $[0, T]$, where $u \in U_{ad} = PC(0,T;\mathbb{R})$. We can write the system in Equation 16 as the following semilinear system

$$z_t' = Az_t + Bu_t + F(t, z_t, u_t) \tag{17}$$

where

$$z_t = \begin{bmatrix} x_t \\ y_t \end{bmatrix}, \quad A = \begin{bmatrix} 0 & 1 \\ 0 & 0 \end{bmatrix} \quad B = \begin{bmatrix} b \\ 0 \end{bmatrix}, \quad F(t,z,u) = \begin{bmatrix} 0 \\ f(t,x,y,u) \end{bmatrix}$$

and $z \in \mathbb{R}^2$ is the vector

$$z = \begin{bmatrix} x \\ y \end{bmatrix}$$

Solving the system of equations

$$\begin{cases} x_t' = y_t, & x_0 = c_1 \\ y_t' = 0, & y_0 = c_2 \end{cases}$$

we find $x_t = c_2 t + c_1$ and $y_t = c_2$. Therefore,

$$\begin{bmatrix} x_t \\ y_t \end{bmatrix} = \begin{bmatrix} 1 & t \\ 0 & 1 \end{bmatrix} \begin{bmatrix} c_1 \\ c_2 \end{bmatrix}$$

implying

$$e^{At} = \begin{bmatrix} 1 & t \\ 0 & 1 \end{bmatrix}$$

Hence, the controllability operator of the system in Equation 17 is

$$Q_\delta = \int_0^\delta e^{At} BB^* e^{A^* t} dt = b^2 \delta \begin{bmatrix} 1 & 0 \\ 0 & 0 \end{bmatrix}$$

Thus, the condition $Q_\delta > 0$ fails for this example. Therefore, the linear part of the system in Equation 17 is not approximately controllable. Respectively, all approximate controllability results for the system in Equation 17 (or 16) that are based on approximate controllability of its linear part fail. Instead, we can investigate the *L*-partial *A*-controllability of the system in Equation 17 (or 16) related to the first component $x_t$ of $z_t$ Letting $L = [1 \quad 0]$, we have

$$\tilde{Q}_\delta = LQ_T L^* = b^2 \delta > 0, \quad 0 < \delta \le T$$

Therefore, by Theorem 1 and Remark 2, the semilinear system in Equation 17 (or 16) is *L*-partially *A*-controllable if *f* is continuous and satisfies the Lipschitz condition in *x* and *y*

**Example 6** Although the wave equation from Example 2 looks like suitable for application of partial controllability concepts, indeed, the system in 11 does not satisfy condition (F) of Theorem 1 since its linear part is *L*-partially *A*-controllable only for the time $T \ge 2$ (if the Fourier sine coefficients of *b* are nonzero), and, respectively, $\tilde{Q}_\delta > 0$ only for $\delta > 2$. For this result, we refer to Zabczyk (1995) and Bashirov and Mahmudov (1999a). Making the nonlinear part of this system to be zero, we obtain another example of approximately controllable system (for the time $T \ge 2$ which does not satisfy condition (F). This demonstrates that the condition of Theorem 1 is sufficient but not necessary.

**Example 7** Delay equations are typical for demonstration of partial controllability concepts. Consider a semilinear delay equation with distributed delays in the linear and nonlinear terms:

$$\begin{cases} x_t' = Ax_t + \int_{-\varepsilon}^0 M_\theta x_{t+\theta} d\theta + Bu_t + f\left(t, x_t, \int_{-\varepsilon}^0 N_\theta x_{t+\theta} d\theta, u_t\right) \\ x_0 = \xi, \quad x_\theta = \eta_\theta, \quad -\varepsilon \le \theta \le 0 \end{cases} \tag{18}$$

on [0, T], where $\varepsilon > 0$, $A \in \mathbb{R}^{n\times n}, B \in \mathbb{R}^{n\times m}, M, N \in C(-\varepsilon, 0, \mathbb{R}^{n\times n}), \xi \in \mathbb{R}^n, \eta \in L_2(-\varepsilon, 0; \mathbb{R}^n)$ and $u \in U_{ad} = PC(0, T; \mathbb{R}^m)$.

Let $\bar{x} : [0, T] \to L_2(-\varepsilon, 0; \mathbb{R}^n)$ be a function defined by

$$[\bar{x}_t]_\theta = x_{t+\theta}, \quad 0 \le t \le T, \quad -\varepsilon \le \theta \le 0$$

Then

$$\bar{x}_t' = (d/d\theta)\bar{x}_t, \quad \bar{x}_0 = \eta, \quad 0 < t \le T$$

Let $\mathcal{T}_t, t \ge 0$, be the semigroup generated by the differential operator $d/d\theta$ and let $\Gamma_1$ and $\Gamma_2$ be the integral operators from $L_2(-\varepsilon, 0; \mathbb{R}^n)$ to $\mathbb{R}^n$, defined by

$$\Gamma_1 h = \int_{-\varepsilon}^{0} M_\theta h_\theta d\theta, \quad \Gamma_2 h = \int_{-\varepsilon}^{0} N_\theta h_\theta d\theta, \quad h \in L_2(-\varepsilon, 0; \mathbb{R}^n)$$

Then for

$$z_t = \begin{bmatrix} x_t \\ \bar{x}_t \end{bmatrix}, \quad \zeta = \begin{bmatrix} \xi \\ \eta \end{bmatrix} \in \mathbb{R}^n \times L_2(-\varepsilon, 0; \mathbb{R}^n)$$

we can write the system in Equation 18 as a semilinear system

$$z_t' = \tilde{A} z_t + F(t, z_t, u_t) + \tilde{B} u_t, \quad z_0 = \zeta \tag{19}$$

where

$$\tilde{A} = \begin{bmatrix} A & \Gamma_1 \\ 0 & \partial/\partial\theta \end{bmatrix}, \quad F(t, z, u) = \begin{bmatrix} f(t, x, \Gamma_2 \bar{x}, u) \\ 0 \end{bmatrix}, \quad \tilde{B} = \begin{bmatrix} B \\ 0 \end{bmatrix}$$

and

$$z = \begin{bmatrix} x \\ \bar{x} \end{bmatrix} \in \mathbb{R}^n \times L_2(-\varepsilon, 0; \mathbb{R}^n)$$

Define

$$L = [\ I \quad 0\ ] : \mathbb{R}^n \times L_2(\varepsilon, 0; \mathbb{R}^n) \to \mathbb{R}^n$$

Then the approximate controllability of the system in Equation 18 is the same as the $L$-partial $A$-controllability of the system in Equation 19. The $L$-partial controllability operator of the system in Equation 19 is calculated in Bashirov et al. (2007) in the form

$$\tilde{Q}_\delta = \int_0^\delta \mathcal{Y}_s B B^* \mathcal{Y}_s^* \, ds$$

where $\mathcal{Y}$ is a unique operator solution of the equation

$$\mathcal{Y}_t = e^{At} + \int_0^{\max(0, t-\varepsilon)} \int_{-\varepsilon}^{0} e^{Ar} M_\theta \mathcal{Y}_{t-r+\theta} \, d\theta dr$$

Hence, by Theorem 1, the system in Equation 19 is $L$-partially $A$-controllable and, respectively, the system in Equation 18 is approximately controllable if

$$\int_0^\delta \mathcal{Y}_s B B^* \mathcal{Y}_s^* ds > 0 \quad \text{for all} \quad \delta > 0$$

and the function $f$ is continuous, bounded, and satisfies Lipschitz condition in its second and third variables.

In particular, if $n = m = 1$, $A = a$, $B = b$ and $M_\theta \equiv 0$, then in Bashirov and Jneid (2013) (see Equation 65) the $L$-partial controllability operator is calculated in the form

$$\tilde{Q}_\delta = \frac{b^2(e^{2a\delta} - 1)}{2a} > 0$$

for all $\delta > 0$. So it just remains to assume the preceding conditions on $f$ to obtain the approximate controllability of the system in Equation 18.

## 5. Conclusion

The basic contribution of this paper can be summarized in two items: (1) finding a sufficient condition for partial approximate controllability of a semilinear system and (2) proposing an alternative method for study of the controllability concepts.

The sufficient condition, given in Theorem 1, allows to get approximate controllability of one or several components of the state vector and becomes useful in the case when the total of the state vector is not approximately controllable. It is especially useful for systems which can be written as a first-order differential equation by enlarging the state space. Unfortunately, the important in applications wave equation does not fit to the frame of this sufficient condition (Example 6). At the same time, delay equations well suit to this frame (Example 7). Another kind of systems, for which partial controllability can be suitable are stochastic systems driven by wide band noises. In the linear case, these systems are investigated in Bashirov et al. (2007, 2010). This issue is not yet investigated for nonlinear stochastic systems and can be considered as a subject for a separate paper.

The proof method of Theorem 1 differs from the traditional method by fixed point theorems. We find this method natural and less complicated, although it has also disadvantages (Remarks 1 and 4). This method requires a separate consideration of the linear and nonlinear parts of a control system while the method by fixed point theorems combines the linear and nonlinear parts into one total. An interesting development may be a combination of these methods in the form: application of fixed point theorems on small intervals $[T - \delta, T]$ rather than on the total interval $[0, T]$. It seems a sufficient condition for the partial (or not) exact controllability can be proved by this combined method, in which the conditions on $f$ can be relaxed.

**Funding**
The authors received no direct funding for this research.

**Author details**
Agamirza E. Bashirov[1,2]
E-mail: agamirza.bashirov@emu.edu.tr
Noushin Ghahramanlou[1]
E-mail: noushin.ghahramanlou@emu.edu.tr
[1] Department of Mathematics, Eastern Mediterranean University, Mersin 10, Turkey.
[2] Institute of Cybernetics, ANAS, Baku, Azerbaijan.

**References**

Balachandran, K., & Sakthivel, R. (2001). Controllability of integrodifferential systems in Banach spaces. *Applied Mathematics and Computation, 118*, 63–71. http://dx.doi.org/10.1016/S0096-3003(00)00040-0

Bashirov, A. E. (1996). On weakening of the controllability concepts. In *Proceedings of the 35th Conference on Decision and Control* (pp. 640–645). Kobe, Japan.

Bashirov, A. E. (2003). Partially observable linear systems under dependent noises. *Systems & Control: Foundations & Applications*. Basel: Birkhäuser.

Bashirov, A. E., Etikan, H. ,& Şemi, N. (2010). Partial controllability of stochastic linear systems. *International Journal of Control, 83*, 2564–2572. http://dx.doi.org/10.1080/00207179.2010.532570

Bashirov, A. E., & Jneid, M. (2013). On partial complete controllability of semilinear systems. *Abstract and Applied Analysis, 2013*, 8 p. doi:10.1155/2013/521052

Bashirov, A. E., & Kerimov, K. R. (1997). On controllability conception for stochastic systems. *SIAM Journal on Control and Optimization, 35*, 384–398. http://dx.doi.org/10.1137/S0363012994260970

Bashirov, A. E., & Mahmudov, N. I. (1999a). On concepts of controllability for deterministic and stochastic systems. *SIAM Journal on Control and Optimization, 37*, 1808–1821. http://dx.doi.org/10.1137/S036301299732184X

Bashirov, A. E., & Mahmudov, N. I. (1999b). Controllability of linear deterministic and stochastic systems. In

*Proceedings of the 38th Conference on Decision and Control* (pp. 3196–3201). Phoenix, AZ, USA.

Bashirov, A. E., & Mahmudov, N. I. (1999c). Some new results in the theory of controllability. In *Proceedings of the 7th Mediterranean Conference on Control and Automation* (pp. 323–343). Haifa, Israel.

Bashirov, A. E., Mahmudov, N. I., Şemi, N., & Etikan, H. (2007). Partial controllability concepts. *International Journal of Control, 80*(1), 1–7. http://dx.doi.org/10.1080/00207170600885489

Bensoussan, A. (1992). *Stochastic control of partially observable systems*. Cambridge: Cambridge University Press. http://dx.doi.org/10.1017/CBO9780511526503

Bensoussan, A., Da Prato, G., Delfour, M. S., Mitter, S. K. (1993). Representation and control of infinite dimensional systems. *Systems & Control: Foundations & Applications* (Vol. 2). Boston, MA: Birkhauser.

Curtain, R. F., & Zwart, H. J. (1995). *An introduction to infinite dimensional linear systems theory*. Berlin: Springer-Verlag. http://dx.doi.org/10.1007/978-1-4612-4224-6

Fattorini, H. O. (1966). Some remarks on complete controllability. *SIAM Journal on Control, 4*, 686–694. http://dx.doi.org/10.1137/0304048

Kalman, R. E. (1960). A new approach to linear filtering and prediction problems. *Transactions ASME, Series D (Journal of Basic Engineering), 82*, 35–45. http://dx.doi.org/10.1115/1.3662552

Klamka, J. (1991). *Controllability of dynamical systems*. Dordrecht: Kluwer.

Klamka, J. (2000). Shauder's fixed point theorem in nonlinear controllability problems. *Control and Cybernetics, 29*, 153–165.

Klamka, J. (2001). Constrained controllability of semilinear systems. *Nonlinear Analysis: Theory, Methods & Applications, 47*, 2939–2949.

Klamka, J. (2002). Constrained exact controllability of semilinear systems. *Systems and Control Letters, 47*, 139–147. http://dx.doi.org/10.1016/S0167-6911(02)00184-6

Klamka, J. (2008). Constrained controllability of semilinear systems with delayed controls. *Bulletin of the Polish Academy of Sciences, Technical Sciences, 56*, 333–337.

Leiva, H., Merentes, N., & Sanchez, J. L. (2011). Interior controllability of the *n*-dimensional semilinear heat equation. *African Diaspora Journal of Mathematics, 12* (2), 1–12.

Leiva, H., Merentes, N., & Sanchez, J. L. (2012). Approximate controllability of semilinear reaction diffusion equations. *Mathematical Control and Related Fields, 2*, 171–182. http://dx.doi.org/10.3934/mcrf

Leiva, H., Merentes, N., & Sanchez, J. L. (2013). A characterization of semilinear dense range operators and applications. *Abstract and Applied Analysis, 2013*, 11 p. doi:10.1155/2013/729093

Li, X., & Yong, J. (1995). Optimal control theory for infinite dimensional systems. *Systems & Control: Foundations & Applications*. Boston, MA: Birkhäuser.

Ren, Y., Dai, H., & Sakthivel, R. (2013). Approximate controllability of stochastic differential systems driven by a Lévy process. *International Journal of Control, 86*, 1158–1164. http://dx.doi.org/10.1080/00207179.2013.786188

Russell, D. L. (1967). Nonharmonic Fourier series in the control theory of distributed parameter systems. *Journal of Mathematical Analysis and Applications, 18*, 542–560. http://dx.doi.org/10.1016/0022-247X(67)90045-5

Sakthivel, R., Ganesh, R., Ren, Y., & Anthoni, S. M. (2013). Approximate controllability of nonlinear fractional dynamical systems. *Communications in Nonlinear Science and Numerical Simulation, 18*, 3498–3508. http://dx.doi.org/10.1016/j.cnsns.2013.05.015

Sakthivel, R., Ganesh, R., & Suganya, S. (2012). Approximate controllability of fractional neutral stochastic system with infinite delay. *Reports on Mathematical Physics, 70*, 291–311. http://dx.doi.org/10.1016/S0034-4877(12)60047-0

Sakthivel, R., Mahmudov, N. I., & Kim, J. H. (2009). On controllability of second order nonlinear impulsive differential systems. *Nonlinear Analysis: Theory, Methods & Applications, 71*, 45–52.

Sakthivel, R., Mahmudov, N. I., & Nieto, J. J. (2012). Controllability for a class of fractional-order neutral evolution control systems. *Applied Mathematics and Computation, 218*, 10334–10340. http://dx.doi.org/10.1016/j.amc.2012.03.093

Sakthivel, R., & Ren, Y. (2013). Approximate controllability of fractional differential equations with state-dependent delay. *Results in Mathematics, 63*, 949–963.

Sakthivel, R., Suganya, S., & Anthoni, S. M. (2012). Approximate controllability of fractional stochastic evolution equations. *Computers and Mathematics with Applications, 63*, 660–668. http://dx.doi.org/10.1016/j.camwa.2011.11.024

Zabczyk, J. (1995). Mathematical control theory: An Introduction. *Systems & Control: Foundations & Applications*. Berlin: Birkhäuser.

# 11

# Event-triggered networked predictive control of system with data loss

Quan Wang[1], Yuanyuan Zou[1]* and Shaobo Wang[2]

*Corresponding author: Yuanyuan Zou, Key Laboratory of Advanced Control and Optimization for Chemical Processes, Ministry of Education, East China University of Science and Technology, Shanghai, China, 200237

E-mail: yyzou@ecust.edu.cn

Reviewing editor: James Lam, University of Hong Kong, Hong Kong

**Abstract:** This paper investigates the problem of event-triggered networked predictive control for systems with data loss. An event-triggered networked predictive control system is proposed. Based on predictive control model, a data loss compensation strategy is presented and an extended event-triggered transmission mechanism is developed. The closed-loop event-triggered predictive control system is described as a switched system and sufficient closed-loop stability conditions related to event-triggered mechanism are established. Under the event-triggered networked predictive control scheme, the consumption of the communication resources is reduced. Finally, an example is provided to illustrate the effectiveness of the proposed method.

**Subjects: Automation Control; Control Engineering; Dynamical Control Systems; General Systems; Intelligent Systems; Robotics Cybernetics; Systems Engineering**

**Keywords: Networked control systems (NCSs); predictive control; event-triggered; data loss**

## 1. Introduction

In recent years, with the development of computer science and network communication technique, networked control systems (NCSs), where the network is inserted in the feedback control loop, have received widespread attention. Compared with traditional control systems, NCSs have many advantages, such as easy installation and service, suitable for long-distance operation and control, and low cost. So they have been widely used in industrial automation, underwater detection, highway

### ABOUT THE AUTHOR

**Quan Wang is currently pursuing her master's degree in East China University of Science and Technology. Her research interests include model predictive control and networked control systems and their applications.**

### PUBLIC INTEREST STATEMENT

Model predictive control (MPC) is a popular technique for industrial process control. The essence of MPC is to obtain a sequence of control actions by online solving a finite-horizon optimization problem. Event-triggered scheme is an effective scheme to deal with the limited network resources by reducing the information transmission while guaranteeing the closed-loop stability.

In networked control systems (NCSs), data losses can result in severe performance deterioration or may even cause instability. To overcome this problem and reduce the consumption of communication resources in NCSs, the event-triggered networked predictive controller is designed to compensate for data losses, save the communication resources consumption, and achieve a desired control performance.

systems, and so on (Ge, Yang, & Han, Ge2015; Gupta, & Chow, 2010; Hespanha, Naghshtabrizi, & Xu, 2007; Leung et al., 2005; Lu, Li, & Xi, 2013; Yang, 2006; Zhang, Gao, & Kaynak, 2013).

Despite the advantages, NCSs also have some inevitable disadvantages, such as data loss. When data are transmitted through the networks, it may be lost, and this phenomenon can deteriorate the control performance of NCSs. To compensate for the effect of data loss, model predictive control method has been proposed (Li & Shi, 2013; Liu, 2010; Song & Fang, 2014; Zou, Lam, Niu, & Li, 2015; Zou & Niu, 2013). In Zou and Niu (2013), the data loss was described as a Bernoulli process, and the predictive control with data loss compensation strategy was proposed to guarantee the closed-loop stability. In Zou et al. (2015), the data loss process was defined as a discrete-time homogeneous Markov chain, and the model predictive control synthesis approach for quantized systems was presented to guarantee the satisfaction of system constraints and the closed-loop stability. In Liu (2010), a networked predictive controller with data losses and network delays was designed, where the control prediction was adopted to compensate for the data losses and network delays actively. In Li and Shi (2013), the min-max model predictive control method was investigated for a constrained nonlinear networked control system, and the proposed method can effectively compensate for the data loss while guaranteeing the input-to-state practical stability. In Song and Fang (2014), a distributed model predictive control was proposed for a linear uncertain system with a polytopic description and subject to randomly occurring packet loss. It is worth noting that the time-triggered communication strategy is adopted in the aforementioned literatures, which may result in a waste of the communication resources.

In order to reduce the consumption of communication resources, the event-triggered strategy has been proposed (Åström & Bernhardsson, 1999). The main idea of the event-triggered strategy is to reduce the communication among the sensor, controller, and actuator by introducing an event-triggered condition, that is the data are transmitted only when the event-triggered condition is satisfied. Many results have been investigated in event-triggered control strategy (Dong, Wang, Alsaadi, & Ahmad, 2015; Li, & Shi, 2014; Liu, Wang, He, & Zhou, 2015; Peng, & Yang, 2013; Yin, Yue, & Hu, 2014). In Peng and Yang (2013), an event-triggered strategy and $H_\infty$ control method were designed for networked control systems, and the stability was analyzed using the Lyapunov–Krasovskii functional theory. In order to utilize the limited resources of wireless sensor networks efficiently, the event-based distributed filtering and state estimation problems were investigated in Liu et al. (2015) and Dong et al. (2015). In Yin et al. (2014), the problem of the model-based event-triggered predictive control was studied for networked systems with network delays, and the control gain and the parameter of the event-triggered condition were co-designed. In Li and Shi (2014), the event-triggered model predictive control for continuous-time nonlinear system was studied, and the computational efficiency of the proposed scheme has been proved to be higher than the conventional model predictive control. To the best of the authors' knowledge, few works have investigated the problem of event-triggered networked predictive control for NCSs under the effect of data loss. How to design an event-triggered networked predictive controller to compensate for the data loss actively and reduce the consumption of communication resources while guaranteeing the closed-loop stability still remains challenging. This motivates the present study.

In this paper, an architecture of the event-triggered networked predictive control system is established, and the design of event-triggered mechanism and the compensation of data loss are considered simultaneously. In order to avoid an undesired performance degradation caused by data loss, a data loss compensation strategy is presented based on event-triggered predictive control model and an extended event-triggered mechanism is proposed using ACK. The closed-loop system is described as a switched system and the closed-loop stability conditions related to event-triggered parameters are established.

The organization of this paper is as follows. Section 2 introduces the basic setup of event-triggered networked predictive control system. In Section 3, the event-triggered predictive controller is designed, the data loss compensation strategy is proposed, and the closed-loop system is modeled. In

Section 4, the closed-loop stability is analyzed and stability conditions are derived. Simulation results are given to show the effectiveness of the proposed methods in Section 5, and conclusions are made in Section 6.

**Notation.** Throughout this paper, $\mathbb{R}^m$ denotes the $m$-dimensional Euclidean space and $\mathbb{R}^+$ stands for the set of positive real numbers, respectively. Superscript $T$ denotes matrix transposition and $I$ is the identity matrix with appropriate dimension. In symmetric block matrices, "*" is used as an ellipsis for terms induced by symmetry. $\mathrm{diag}\{\cdot\}$ denotes the block diagonal matrix.

## 2. Problem formulation

We consider an event-triggered networked predictive control system shown in Figure 1, where the network exists between the event generator and predictive controller. The plant is given by the following discrete-time linear time-invariant (LTI) system

$$x(k+1) = Ax(k) + Bu(k), \tag{1}$$

where $x(k) \in \mathbb{R}^n$ is the state of the plant; $u(k) \in \mathbb{R}^m$ is the control input; and $A \in \mathbb{R}^{n\times n}$ and $B \in \mathbb{R}^{n\times m}$ are known matrices with appropriate dimensions.

In order to reduce the communication resources consumption in the NCS, we introduce an event generator to limit the number of information transmission, that is the state information will be transmitted to the controller only when a certain event-triggered condition is satisfied. We denote the event-triggered instant as $k_i (i = 0, 1, 2, \dots)$. If the current state $x(k_i)$ is transmitted to the controller at time $k_i$, the next event-triggered instant $k_{i+1}$ is decided by the following condition

$$k_{i+1} = k_i + \min_r\{r|[\hat{x}(k_i + r) - x(k_i)]^T \Phi[\hat{x}(k_i + r) - x(k_i)] > \mu\hat{x}(k_i + r)^T \Phi\hat{x}(k_i + r)\}. \tag{2}$$

In (2), $\hat{x}(k_i + r)$ is the estimated state at time $k_i + r$, which will be derived from Section 3.2; $r$ is a positive integer; $\mu \in (0, 1)$ is a given parameter; and $\Phi$ is a positive definite symmetric matrix to be designed in Section 4.

In practical NCS, data loss may happen when the data are transmitted over networks. At the event-triggered instant $k_i$, when the state $x(k_i)$ is transmitted to the controller, it may be lost, that is $x(k_i)$ cannot arrive at the controller. In the sequel, the instant when the state is successfully transmitted to the controller is denoted as $t_j$ $(j = 0, 1, 2, \dots)$.

*Remark 1* According to the above description, not every sampling instant is event-triggered instant and the state may be lost when it is transmitted through the networks at the event-triggered instant. So we have $\{t_0, t_1, t_2, \dots\} \subset \{k_0, k_1, k_2 \dots\} \subset \{0, 1, 2, \dots\}$, and an example is shown in Figure 2.

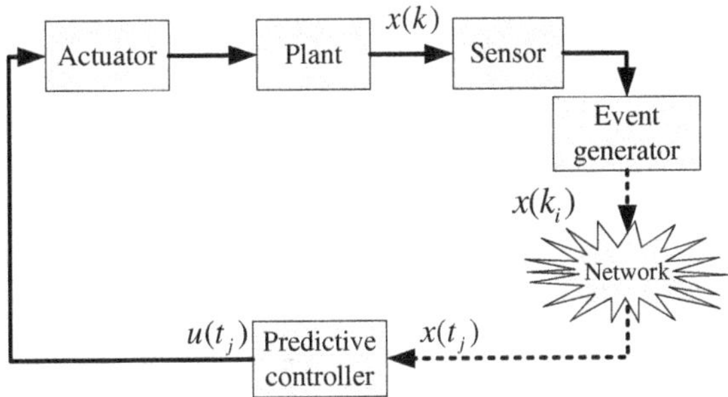


**Figure 1. Architecture of the event-triggered networked predictive control system.**

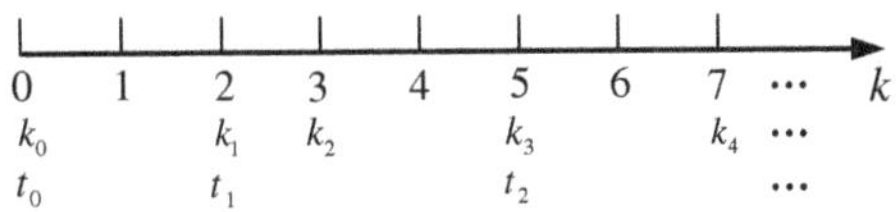


**Figure 2. An example for various sequences.**

The main objective of this paper is to design an event-triggered networked predictive control scheme with a data loss compensation strategy, such that the closed-loop stability is guaranteed and the communication burden of the networks is relieved.

## 3. Event-triggered networked predictive control with data loss compensation strategy

In the traditional time-triggered networked predictive control scheme, the state $x(k)$ is sent to the controller at each sampling instant and the control law is updated periodically. In the event-triggered setup, the state is sent to the controller only when the event-triggered condition (2) is satisfied. Furthermore, in the process of event-triggered transmission, the presence of data loss results in the decrease of the control performance. In the following, the calculation of the event-triggered predictive control law and the data loss compensation strategy will be presented.

### 3.1. Event-triggered predictive control law

At time instant $t_j$, the data are successfully transmitted to the controller and the following optimization problem of predictive control is solved:

$$\min J(t_j) = \sum_{l=1}^{N_p} x(t_j + l \mid t_j)^T Q x(t_j + l \mid t_j) + \sum_{l=0}^{N_u - 1} u(t_j + l \mid t_j)^T R u(t_j + l \mid t_j), \tag{3}$$

s.t.

$$x(t_j + l + 1 \mid t_j) = Ax(t_j + l \mid t_j) + Bu(t_j + l \mid t_j), \tag{4}$$

where $Q$ and $R$ are symmetrical positive-definite weighting matrices; $N_p$ and $N_u$ are predictive horizon and control horizon, respectively; and $x(t_j + l \mid t_j)$ and $u(t_j + l \mid t_j)$ are the state and control inputs predicted at time $t_j + l$ based on the measurement at time $t_j$. According to (4), the predictive equation of the plant can be described as

$$X(t_j + 1) = A_p x(t_j) + B_p U(t_j), \tag{5}$$

where

$$X(t_j + 1) = [x(t_j + 1|t_j)^T \cdots x(t_j + N_p|t_j)^T]^T,$$
$$U(t_j) = [u(t_j|t_j)^T \cdots u(t_j + N_u - 1|t_j)^T]^T,$$
$$A_p = [A^T \ (A^2)^T \cdots (A^{N_p})^T]^T,$$

$$B_p = \begin{bmatrix} B & 0 & \cdots & 0 & 0 \\ AB & B & \cdots & 0 & 0 \\ \vdots & \vdots & \cdots & \vdots & \vdots \\ ??A^{N_u-1}B & A^{N_u-2}B & \cdots & AB & B \\ \vdots & \vdots & \cdots & \vdots & \vdots \\ A^{N_p-1}B & A^{N_p-2}B & \cdots & A^{N_p-N_u+1}B & \sum_{i=0}^{N_p-N_u} A^i B \end{bmatrix}$$

Further, the performance objective (3) can be written as

$$J(t_j) = X(t_j + 1)^T \bar{Q} X(t_j + 1) + U(t_j)^T \bar{R} U(t_j), \tag{6}$$

where $\bar{Q} = diag\{Q, \ldots, Q\}, \bar{R} = diag\{R, \ldots, R\}$. Then, the predictive control optimization problem (3 and 4) can be reformulated as

$$\min_{U(t_j)} J(t_j) \tag{7}$$

subject to (5). According to $\partial J(t_j)/\partial U(t_j) = 0$, the optimal solution can be obtained as follows:

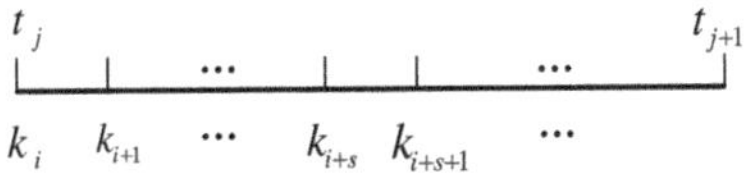

$t_j$ $t_{j+1}$

$k_i$ $k_{i+1}$ ⋯ $k_{i+s}$ $k_{i+s+1}$ ⋯

**Figure 3. Timeline.**

$$U^*(t_j) = -(B_p^T\bar{Q}B_p + \bar{R})^{-1}B_p^T\bar{Q}A_p x(t_j). \tag{8}$$

Then, the control law at event-triggered time instant $t_j$ is $u(t_j) = u(t_j|t_j) = Fx(t_j)$, where the predictive control feedback gain is $F = -(I\ 0 \cdots 0)(B_p^T\bar{Q}B_p + \bar{R})^{-1}B_p^T\bar{Q}A_p$.

### 3.2. Data loss compensation strategy under event-triggered transmission mechanism

Assume that the state $x(k_i)$ is transmitted successfully at time $k_i$, that is we have $t_j = k_i$. Between the two successfully transmitted instants $[t_j, t_{j+1})$ (see Figure 3), the data loss compensation strategy is proposed as follows: at the event-triggered time instant $k_{i+s}$ $(s = 1, 2, \ldots, d)$, the following control law is used

$$u(k_{i+s}) = F\hat{x}(k_{i+s}), \tag{9}$$

and between the two event-triggered instants $[k_{i+s}, k_{i+s+1})$, the control law remains constant, i.e.

$$u(k) = F\hat{x}(k_{i+s}), k \in [k_{i+s}, k_{i+s+1}) \subset [t_j, t_{j+1}). \tag{10}$$

Then, the corresponding states between the two successfully transmitted instants $[t_j, t_{j+1})$ can be estimated:

$$x(k_i) = x(t_j) \tag{11}$$
$$\hat{x}(k_i + 1) = Ax(k_i) + BFx(k_i) \tag{12}$$
$$\hat{x}(k_i + j + 1) = A\hat{x}(k_i + j) + BFx(k_i) \quad j = 1, 2, \ldots, k_{i+1} - k_i - 1 \tag{13}$$
$$\hat{x}(k_{i+s} + j + 1) = A\hat{x}(k_{i+s} + j) + BF\hat{x}(k_{i+s}) \quad j = 0, 1, 2, \ldots, k_{i+s+1} - k_{i+s} - 1,\ s = 1, 2, \ldots, d, \tag{14}$$

where $k_{i+s}$ $(s = 1, 2, \ldots, d)$ denotes the event-triggered instant triggered by the condition (2), and the data may be lost at these instants; $d$ denotes the maximal allowable number of successive data losses; and $\hat{x}(k)$, $k \in [t_j, t_{j+1})$ denotes the estimated state at time $k$ under the action of the control laws (9) and (10).

### 3.3. Closed-loop model of the event-triggered networked predictive control with data loss

If there is no restriction on the number of successive data losses, the measurement difference $\hat{x}(k_i + r) - x(k_i)$ can be arbitrarily large. In order to avoid an undesired performance degradation in the event-triggered NCS, we propose an extended event-triggered strategy using acknowledgment signals (ACKs). The transmission mechanism is described as follows: at the event-triggered instant $k_{i+s}$, the data $\hat{x}(k_{i+s})$ are sent from the event generator to the controller. If $\hat{x}(k_{i+s})$ is successfully transmitted to the controller, the controller sends an ACK to the event generator over a reliable channel. If $\hat{x}(k_{i+s})$ is lost, the controller will not send the ACK to the event generator. After a predefined waiting time $T_w \in \mathbb{R}^+$, if the event generator does not get any ACK, it resends the data $\hat{x}(k_{i+s} + T_w)$ to the controller at time $k_{i+s} + T_w$, which is considered as the new event-triggered instant $k_{i+s+1}$. $\hat{x}(k_{i+s} + T_w)$ can be obtained from the predictive state (14), i.e.

$$\hat{x}(k_{i+s} + T_w) = A^{T_w}\hat{x}(k_{i+s}) + \sum_{j=1}^{T_w} A^{T_w - j}BF\hat{x}(k_{i+s}).$$

The extended event-triggered mechanism becomes

$$\begin{cases} k_{i+s+1} = k_{i+s} + T_w, \text{ if the transmission has failed at time } k_{i+s}, & (15) \\ k_{i+s+1} = k_{i+s} + \min_r \{r|[\hat{x}(k_{i+s}+r) - x(k_i)]^T \Phi[\hat{x}(k_{i+s}+r) - x(k_i)] \\ \quad > \mu \hat{x}(k_{i+s}+r)^T \Phi \hat{x}(k_{i+s}+r)\}, \text{ else.} & (16) \end{cases}$$

At the next event-triggered instant $k_{i+s+1}$, the state $x(k_{i+s+1})$ is updated according to

$$x(k_{i+s+1}) = \begin{cases} \hat{x}(k_{i+s}+r), \text{ if } [\hat{x}(k_{i+s}+r) - x(k_i)]^T \Phi[\hat{x}(k_{i+s}+r) - x(k_i)] \\ > \mu \hat{x}(k_{i+s}+r)^T \Phi \hat{x}(k_{i+s}+r), \\ \hat{x}(k_{i+s}+T_w), \text{ else} \end{cases} \tag{17}$$

where $\hat{x}(k_{i+s}+r)$ and $\hat{x}(k_{i+s}+T_w)$ can be obtained from the predictive state (14), and then the data are transmitted toward the predictive controller.

Under the event-triggered mechanism (15–16), system (1) with data loss compensation (9–10) can be described as the following closed-loop system:

$$\begin{aligned} x(k+1) &= Ax(k) + Bu(k_{i+s}) = Ax(k) + BF\hat{x}(k_{i+s}) \\ & k \in [k_{i+s}, k_{i+s+1}) \subset [t_j, t_{j+1}). \end{aligned} \tag{18}$$

According to the event-triggered predictive states (11–14), we have

$$\hat{x}(k_{i+s}) = [A^{k_{i+s}-k_i} + \sum_{j=1}^{k_{i+s}-k_i} A^{k_{i+s}-k_i-j} BF] x(k_i). \tag{19}$$

So the closed-loop system can be rewritten as:

$$\begin{aligned} x(k+1) &= Ax(k) + BF[A^{k_{i+s}-k_i} + \sum_{j=1}^{k_{i+s}-k_i} A^{k_{i+s}-k_i-j} BF] x(k_i) \\ & k \in [k_{i+s}, k_{i+s+1}) \subset [t_j, t_{j+1}), \ s = 0, 1, 2, \ldots d. \end{aligned} \tag{20}$$

Define the state measurement difference $e(k) = \hat{x}(k) - x(k_i)$ for $k \in [k_{i+s}, k_{i+s+1}) \subset [t_j, t_{j+1}), s = 0, 1, 2, \ldots d$. Then, the closed-loop system can be further described as the following switched system:

$$\begin{aligned} x(k+1) &= \Pi_{\sigma_s} x(k) - \Xi_{\sigma_s} e(k), \\ & \sigma_s \in S = \{\sigma_0, \sigma_1, \ldots, \sigma_d\} \end{aligned} \tag{21}$$

where

$$\Pi_{\sigma_s} = A + BF[A^{\sigma_s} + \sum_{j=1}^{\sigma_s} A^{\sigma_s - j} BF], \ \Xi_{\sigma_s} = BF[A^{\sigma_s} + \sum_{j=1}^{\sigma_s} A^{\sigma_s - j} BF],$$

$$\sigma_s = k_{i+s} - k_i, \ k \in [k_{i+s}, k_{i+s+1}) \subset [t_j, t_{j+1}), \ s = 0, 1, 2, \ldots d.$$

*Remark 2* In this paper, it can be seen that $x(k) = \hat{x}(k)$ for $k \in [t_j, t_{j+1})$, $j = 0, 1, 2, \ldots$. It can be proved by means of the mathematical induction. For the successfully transmitted state $x(t_j) = x(k_i)$, according to (1) and (12), we have $x(k_i+1) = Ax(k_i) + BFx(k_i)$ and $\hat{x}(k_i+1) = Ax(k_i) + BFx(k_i)$, so $x(k_i+1) = \hat{x}(k_i+1)$. For $\forall k \in (t_j, t_{j+1})$, we assume $x(k) = \hat{x}(k)$. Then, we just need to prove that $x(k+1) = \hat{x}(k+1)$. For $k \in [k_{i+s}, k_{i+s+1}) \subset (t_j, t_{j+1})$, $s = 0, 1, 2, \ldots d$, we have $x(k+1) = Ax(k) + BF\hat{x}(k_{i+s})$. According to (11)-(14), it can be concluded that $\hat{x}(k+1) = A\hat{x}(k) + BF\hat{x}(k_{i+s})$. So, we have $x(k+1) = \hat{x}(k+1)$.

*Remark 3* The switching modes $\sigma_0, \sigma_1, \dots, \sigma_d$ are related to the number of successive data losses, and the closed-loop system is switched among the $d+1$ subsystems according to the number of successive data losses.

*Remark 4* According to the event-triggered conditions (15–16), no event is triggered at time $k \in [k_{i+s}, k_{i+s+1}) \subset [t_j, t_{j+1})$. So we have

$$e(k)^T \Phi e(k) \le \mu x(k)^T \Phi x(k),\ k \in [k_{i+s}, k_{i+s+1}) \subset [t_j, t_{j+1}). \tag{22}$$

## 4. Stability analysis

In this section, based on the closed-loop system (21), we discuss the design of the event-triggered parameter $\Phi$ using the Lyapunov stability theory and the LMI technique. The stability condition is provided in the following theorem.

THEOREM 1 *Consider the proposed event-triggered networked control system with data loss. For given system matrices A and B, the constant parameter $\mu \in (0,1)$, and the predictive control feedback gain F if there exist matrices $P > 0$ and $\Phi > 0$ with appropriate dimensions such that the following LMIs hold for all $\sigma_s \in S$:*

$$\begin{bmatrix} -P+\mu\Phi & * & * \\ 0 & -\Phi & * \\ P\Pi_{\sigma_s} & -P\Xi_{\sigma_s} & -P \end{bmatrix} < 0,\ \sigma_s = k_{i+s} - k_i,\ s = 0,1,2,\dots d \tag{23}$$

then the closed-loop switched system in (21) is asymptotically stable.

*Proof* Choose the Lyapunov function of the closed-loop switched system as

$$V(x(k)) = x(k)^T P x(k), \tag{24}$$

where $P$ is a positive definite symmetric matrix. For $k \in [k_{i+s}, k_{i+s+1}) \subset [t_j, t_{j+1})$, calculating the difference of the Lyapunov function according to the closed-loop system (21) and taking the inequality (22) into account, we have

$$\begin{aligned} \Delta V &= V(x(k+1)) - V(x(k)) \\ &= x(k+1)^T P x(k+1) - x(k)^T P x(k) \\ &\le x(k+1)^T P x(k+1) - x(k)^T P x(k) - e(k)^T \Phi e(k) + \mu x(k)^T \Phi x(k) \\ &= [\Pi_{\sigma_s} x(k) - \Xi_{\sigma_s} e(k)]^T P [\Pi_{\sigma_s} x(k) - \Xi_{\sigma_s} e(k)] - x(k)^T P x(k) - e(k)^T \Phi e(k) + \mu x(k)^T \Phi x(k) \\ &= \begin{bmatrix} x(k) \\ e(k) \end{bmatrix}^T \Omega \begin{bmatrix} x(k) \\ e(k) \end{bmatrix} \end{aligned}$$

where

$$\Omega = \begin{bmatrix} \Pi_{\sigma_s}^T P \Pi_{\sigma_s} - P + \mu\Phi & * \\ -\Xi_{\sigma_s}^T P \Pi_{\sigma_s} & \Xi_{\sigma_s}^T P \Xi_{\sigma_s} - \Phi \end{bmatrix}.$$

$\Omega$ can be decomposed as

$$\Omega = \begin{bmatrix} -P+\mu\Phi & * \\ 0 & -\Phi \end{bmatrix} + \begin{bmatrix} \Pi_{\sigma_s}^T P \\ -\Xi_{\sigma_s}^T P \end{bmatrix} P^{-1} \begin{bmatrix} P\Pi_{\sigma_s} & -P\Xi_{\sigma_s} \end{bmatrix}.$$

Using the Schur complement, $\Omega < 0$ is equivalent to the LMIs (23). Thus, if the LMIs (23) hold, then $\triangle V < 0$. According to the Lyapunov stability theory, we can conclude that the closed-loop system is asymptotically stable. This completes the proof. □

The algorithm of the event-triggered networked predictive control with data loss is given as follows:

**Offline:**

**Step 1.1** Given $x(t_0) = x(k_0) = x(0)$, $N_p$, $N_u$, $Q$, $R$, compute the predictive control feedback gain $F$.

**Step 1.2** Given $\mu$, solve the LMIs (23) to get the parameter $\Phi$ in the event-triggered condition (16).

**Online:**

**Step 2.1** Compute the predictive control law $u(0) = Fx(0)$. Set $i = i + 1$.

**Step 2.2** At time instant $k > 0$, measure the system state $x(k)$. *Check the event-triggered condition (16), if (16) is satisfied, go to* **Step 2.3**. *Otherwise, go to* **Step 2.8**.

**Step 2.3** Set $x(k_i) = x(k)$. Transmit the state $x(k_i)$ to the controller over the networks. If the data are successfully transmitted, go to **Step 2.4**. Otherwise, go to **Step 2.5**.

**Step 2.4** The controller sends an ACK to the event generator, and set $j = j + 1$. Compute the control law $u(k) = Fx(t_j)$. Set $k = k + 1$, $i = i + 1$, and repeat **Step 2.2**.

**Step 2.5** The event generator does not get any ACK, then feed $u(k) = F\hat{x}(k_i)$ to the plant. Set $k = k + 1$, and go to **Step 2.6**.

**Step 2.6** *If the event-triggered condition (15) is satisfied, that is* $k = k_i + T_w$, *set* $i = i + 1$, *and go to* **Step 2.7**. *Otherwise, go to* **Step 2.5**.

**Step 2.7** Set $x(k_i) = x(k)$. Transmit the state $x(k_i)$ to the controller over the networks. If the data are successfully transmitted, go to **Step 2.4**. Otherwise, go to **Step 2.5**.

**Step 2.8** Feed $u(k) = F\hat{x}(k_i)$ to the plant. Set $k = k + 1$, and repeat **Step 2.2**.

## 5. Simulation example

In this section, an example is given to illustrate the effectiveness of the proposed algorithm of the event-triggered networked predictive control with data loss for NCS. Consider the following discrete-time system matrices:

$$A = \begin{bmatrix} 1 & 0.1 & -0.0124 & -0.0004 \\ 0 & 1 & -0.25 & -0.0124 \\ 0 & 0 & 0.0619 & 0.1021 \\ 0 & 0 & 1.2502 & 0.0619 \end{bmatrix}, B = \begin{bmatrix} 0.0013 \\ 0.0251 \\ -0.0013 \\ -0.0255 \end{bmatrix}.$$

The initial state is $x_0 = [0.05, -0.1, 0, 0.02]^T$; the predictive control parameters are $Q = 4 \times I_{4\times4}$, $R = 0.01$, $N_p = 5$, and $N_u = 3$ and the parameter of the event-triggered condition provided by (16) is $\mu = 0.1$; the waiting time $T_w = 2$; and the maximal allowable number of successive data losses is $d = 2$. According to Equation (8), the predictive control feedback gain matrix is $F = \begin{bmatrix} -4.3617 & -14.3952 & 10.1882 & 0.5811 \end{bmatrix}$.

The state trajectories and control input of the closed-loop system are depicted in Figures 4 and 5, respectively, by which we can see that the closed-loop system with data loss is stable. Under the event-triggered conditions (15–16), the event-triggered instants of the system (1) with data loss are

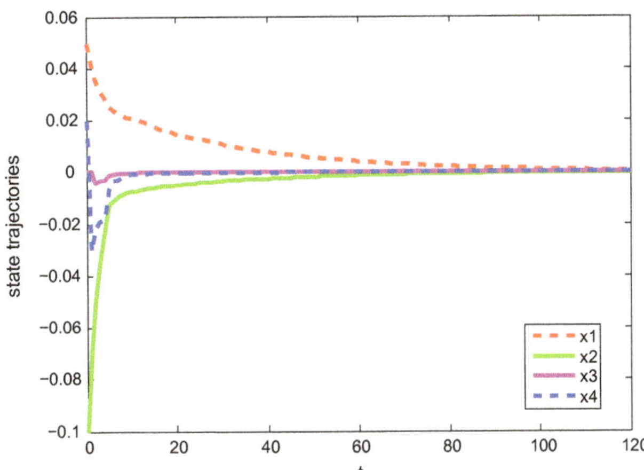

**Figure 4. The state trajectories of the event-triggered networked predictive control.**

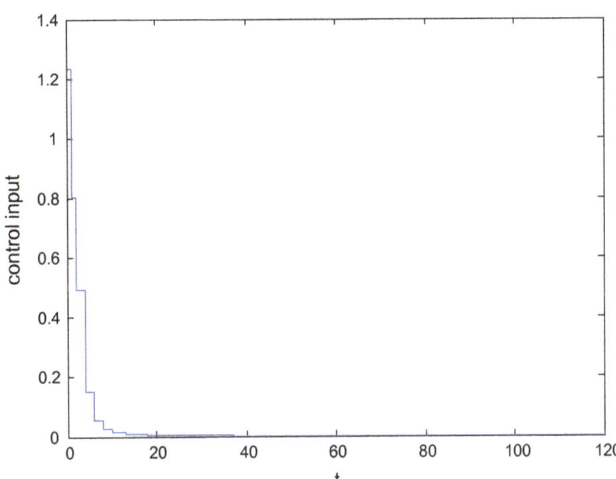

**Figure 5. The control input of the event-triggered networked predictive control.**

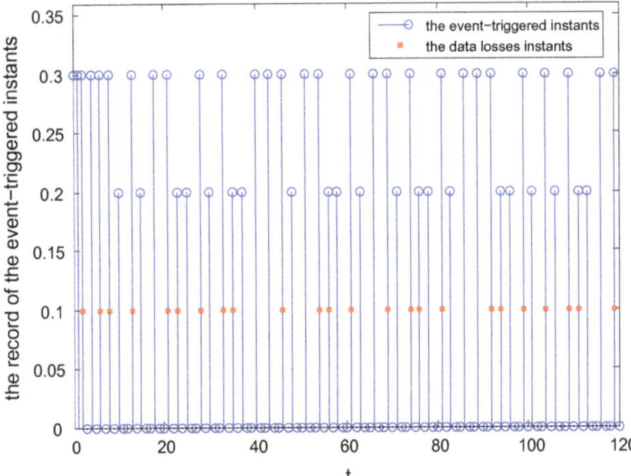

**Figure 6. The record of the event-triggered instants and the data losses instants.**

described in Figure 6. In this figure, the value 0.3 represents that the event is triggered by the condition (16), the value 0.2 represents that the event is triggered by the condition (15), and the value 0 represents that no event is triggered. The instants that the event is triggered but the data are lost are denoted by the symbol of red "x." It can be shown that the proposed algorithm of the event-triggered networked predictive control with data loss can reduce the communication resources' consumption and guarantee the closed-loop stability.

## 6. Conclusion

In this paper, the event-triggered networked predictive control for NCS with data loss has been studied. The data loss compensation strategy and the extended event-triggered mechanism have been proposed. The stability conditions of the closed-loop switched system have been given based on the Lyapunov stability theory.

**Funding**

This work is supported by National Nature Science Foundation of China [61374107], the Fundamental Research Funds for the Central Universities, and the Program of Shanghai Subject Chief Scientist [14XD1420900].

**Author details**

Quan Wang[1]

E-mail: quanwang2014@gmail.com

Yuanyuan Zou[1]

E-mail: yyzou@ecust.edu.cn

Shaobo Wang[2]

E-mail: wangshaobo@tsinghua.org.cn

[1] Key Laboratory of Advanced Control and Optimization for Chemical Processes, Ministry of Education, East China University of Science and Technology, Shanghai, 200237, China.

[2] Automation Division of Shanghai Electric Group Co., Ltd., Shanghai, China.

**References**

Åström, K., & Bernhardsson, B. (1999). Comparison of periodic and event based sampling for first-order stochastic systems.. *Proceedings of the 14th IFAC World Congress*, 301–306.

Dong, H., Wang, Z., Alsaadi, F., & Ahmad, B. (2015). Event-triggered robust distributed state estimation for sensor networks with state-dependent noises. *International Journal of General Systems, 44*, 254–266. doi: http://dx.doi.org/10.1080/03081079.2014.973726

Ge, X., Yang, F., & Han, Q. (2015). Distributed networked control systems: A brief overview. *Information Sciences*. doi:10.1016/j.ins.2015.07.047

Gupta, R., & Chow, M. (2010). Networked control system: Overview and research trends. *IEEE Transactions on Industrial Electronics, 57*, 2527–2535. doi:10.1109/TIE.2009.2035462

Hespanha, J., Naghshtabrizi, P., & Xu, Y. (2007). A survey of recent results in networked control systems. *Proceedings of the IEEE, 95*, 138–162. doi:10.1109/JPROC.2006.887288

Leung, W., Vanijjirattikhan, R., Zheng, L., Le, X., Richards, T., Ayhan, B., & Chow, M. (2005). Intelligent space with time sensitive applications. *Proceedings of IEEE/ASME International Conference on Advanced Intelligent Mechatronics*, 1413–1418, doi:10.1109/AIM.2005.1511209

Li, H., & Shi, Y. (2014). Event-triggered robust model predictive control of continuous-time nonlinear systems. *Automatica, 50*, 1507–1513. doi:10.1016/j.automatica.2014.03.015

Li, H., & Shi, Y. (2013). Networked min-max model predictive control of constrained nonlinear systems with delays and packet dropouts. *International Journal of Control, 86*, 610–624. doi:10.1080/00207179.2012.751628

Liu, G. (2010). Predictive controller design of networked systems with communication delays and data loss. *IEEE Transactions on Circuits and Systems II: Express Briefs, 57*, 481–485. doi:10.1109/TCSII.2010.2048377

Liu, Q., Wang, Z., He, X., & Zhou, D. (2015). Event-based distributed filtering with stochastic measurement fading. *IEEE Transactions on Industrial Informatics*. doi: http://dx.doi.org/10.1109/TII.2015.2444355

Lu, J., Li, D., & Xi, Y. (2013). Probability-based constrained MPC for structured uncertain systems with state and random input delays. *International Journal of Systems Science, 44*, 1354–1365. doi:10.1080/00207721.2012.659711

Peng, C., & Yang, T. (2013). Event-triggered communication and $H_\infty$ control co-design for networked control systems. *Automatica, 49*, 1326–1332. doi:10.1016/j.automatica.2013.01.038

Song, Y., & Fang, X. (2014). Distributed model predictive control for polytopic uncertain systems with randomly occurring actuator saturation and packet loss. *IET Control Theory & Applications, 8*, 297–310. doi:10.1049/iet-cta.2013.0376

Yang, T. (2006). Networked control system: A brief survey. *IEE Proceedings-Control Theory and Applications, 153*, 403–412. doi:10.1049/ip-cta:20050178

Yin, X., Yue, D., & Hu, S. (2014). Model-based event-triggered predictive control for networked systems with communication delays compensation. *International Journal of Robust and Nonlinear Control*. doi:10.1002/rnc.3281.

Zhang, L., Gao, H., & Kaynak, O. (2013). Network-induced constraints in networked control systems–A survey. *IEEE Transactions on Industrial Informatics, 9*, 403–416. doi:10.1109/TII.2012.2219540

Zou, Y., Lam, J., Niu, Y., & Li, D. (2015). Constrained predictive control synthesis for quantized systems with Markovian data loss. *Automatica, 55*, 217–225. doi:10.1016/j.automatica.2015.03.016

Zou, Y., & Niu, Y. (2013). Predictive control of constrained linear systems with multiple missing measurements. *Circuits, Systems, and Signal Processing, 32*, 615–630. doi:10.1007/s00034-012-9482-2

# 12

# Vector control of asymmetrical six-phase synchronous motor

Arif Iqbal[1]*, G.K. Singh[1] and Vinay Pant[1]

*Corresponding author: Arif Iqbal, Department of Electrical Engineering, Indian Institute of Technology Roorkee, Roorkee 247667, Uttarakhand, India

E-mails: arif.iqbal.in@gmail.com, arif0548@gmail.com

Reviewing editor: Wei Meng, Wuhan University of Technology, China

**Abstract:** Vector control scheme has been well adopted for higher performance applications of AC motor. Therefore, this paper presents an extensive development and investigation of vector control scheme for asymmetrical six-phase synchronous motor in a new two-axis (M–T) coordinate system. Phasor diagram has also been developed in a simplified way, followed by its implementation technique. Analytical results have been presented for four-quadrant operation of synchronous motor, employing the developed vector control scheme. In analytical control model, common mutual leakage reactance between the two winding sets occupying same stator slot has been considered.

**Subjects: Automation Control; Mechatronics; Systems & Controls**

**Keywords: six-phase synchronous motor; vector control; motor drive**

## 1. Introduction

The multiphase (more than three phase) AC motor drives are used as a substitute of conventional three-phase motor in different applications particularly, in propulsion system of ship and vehicle, textile and steel industries, rolling mills, power plants, etc. This is because it offers certain potential advantages when compared with its three-phase counterpart, like reduction in space and time harmonics, reduced torque pulsation, increased power handling capability, higher reliability, etc. (Levi, 2008; Singh, 2002).

Field-oriented control (i.e. vector control) technique has been widely used in high performance of AC motor drives. In this regard, an abundant number of literatures are available for three-phase motor (Bose, 2002; Das & Chattopadhyay, 1997; Jain & Ranganathan, 2011; Krause, Wasynczuk, & Sudhoff, 2004), but a few for six-phase induction motor (Bojoi, Lazzari, Profumo, & Tenconi, 2003; Singh, Nam, & Lim, 2005), wherein *d–q* model of machine has been used in synchronously rotating

### ABOUT THE AUTHOR

Arif Iqbal received his BTech and MTech degrees both in electrical engineering in 2005 and 2007, respectively, from Aligarh Muslim University, Aligarh, India. After having experience in industry and teaching in the field AC drives and power system for a few years, he is currently pursuing his PhD from Indian Institute of Technology Roorkee, India. His area of interest is multiphase AC machine and drives, power electronics, and renewable energy system.

### PUBLIC INTEREST STATEMENT

It is always advantageous to use the electrical machine with its performance at higher level using a suitable technique. Among different techniques, vector control is widely used as it provides the decoupled control of torque and flux component of machine current. Therefore, the paper explores a detailed asymmetrical six-phase synchronous motor operation employing this technique. Analysis and simulation of complete drive system has been carried out in Matlab/Simulink environment, where dynamic behavior of motor was found to be substantially improved. It is believed that the paper will serve as a source toward higher performance of six-phase synchronous motor.

reference frame. But the utilization of this scheme has not been reported for field excited six-phase synchronous motor. Therefore, the paper is dedicated to explore and develop the control technique for asymmetrical six-phase synchronous motor. In the control scheme, a new two axis (*M–T* coordinate) has been introduced along which the decoupled control of flux and torque is achieved. Following the inclusion of control technique, a detailed analytical results have been presented for motor operation in four quadrants.

## 2. Mathematical modeling

For the purpose of realizing the six-phase motor, it is a common practice to split the existing three-phase stator winding into two, namely *a b c* and *a′b′c′*. Both splitted stator winding sets (*a b c* and *a′b′c′*) are physically displaced 30° apart to realize asymmetrical six-phase winding configuration. Asymmetrical six-phase winding configuration yields the reduced torque pulsation (Singh, 2002, 2011) due to the elimination of lower order harmonics. On the rotor side, it is equipped with field winding *fr* together with the damper windings $K_d$ and $K_q$ along *d* and *q*-axis, respectively.

The equation of the motor can be written using machine variables. But this will yield a set of nonlinear differential equations. Nonlinearity is due to the existence of inductance term which is time varying in nature. Such equations are computationally complex and time-consuming. Therefore, to simplify the motor equations with constant inductance term, it will be conveniently written in rotor reference frame using Park's variables (Iqbal, Singh, & Pant, 2014, in press; Schiferl & Ong, 1983; Singh, 2011).

$$\mathbf{v}_{dqk} = \mathbf{r}_{sk}\mathbf{i}_{dqk} + \frac{\omega_r}{\omega_b}\boldsymbol{\psi}_{dqk} + \frac{p}{\omega_b}\boldsymbol{\psi}_{dqk} \tag{1}$$

$$\mathbf{v}_{dqr} = \mathbf{r}_{dqr}\mathbf{i}_{dqk} + \frac{p}{\omega_b}\boldsymbol{\psi}_{dqr} \tag{2}$$

$$v_{dqk} = \begin{bmatrix} v_{dk} & v_{qk} \end{bmatrix}^T \tag{3}$$

$$\psi_{dqk} = \begin{bmatrix} \psi_{dk} & -\psi_{qk} \end{bmatrix}^T \tag{4}$$

$$r_{dqk} = \begin{bmatrix} r_{dk} & r_{qk} \end{bmatrix}^T \tag{5}$$

$$\text{for } k = \begin{cases} 1, & \text{for winding set } a\ b\ c \\ 2, & \text{for winding set } a'b'c' \end{cases}$$

$$v_{dqr} = \begin{bmatrix} v_{Kd} & v_{Kq} & v_{fr} \end{bmatrix}^T \tag{6}$$

$$v_{fr} = e_{xfd}\frac{r_{fr}}{x_{md}}\ [3,\ 12] \tag{7}$$

$$\psi_{dqr} = \begin{bmatrix} \psi_{Kd} & \psi_{Kq} & \psi_{fr} \end{bmatrix}^T \tag{8}$$

$$r_{dqk} = \begin{bmatrix} r_{Kd} & r_{Kq} & r_{fr} \end{bmatrix}^T \tag{9}$$

Equations of motor flux linkage per second may be conveniently written as the function of currents,

$$\psi = xi \tag{10}$$

where $i = \begin{bmatrix} i_{dqk} & i_{dqk} \end{bmatrix}^T, \psi = \begin{bmatrix} \psi_{dqk} & \psi_{dqk} \end{bmatrix}^T$

$x$ is defined in Appendix 1.

The developed motor torque is expressed as

$$\tau_e = \tau_{e1} + \tau_{e2} \tag{11}$$

where $\tau_{e1}$ and $\tau_{e2}$ are the developed motor torque associated with winding sets *a b c* and *a′b′c′*, respectively, and expressed as

$$\tau_{e1} = c(i_{q1}\psi_{d1} - i_{d1}\psi_{q1}) \tag{12}$$

$$\tau_{e2} = c(i_{q2}\psi_{d2} - i_{d2}\psi_{q2}) \tag{13}$$

with $c = \frac{3}{2}\frac{P}{2}\frac{1}{\omega_b}$

The rotor dynamics having $P$ number of poles is expressed as

$$\frac{\omega_r}{\omega_b} = \frac{1}{p}\left[\frac{1}{\omega_b}\frac{P}{2}\frac{1}{J}(\tau_e - \tau_l)\right] \tag{14}$$

wherein, $\tau_l$ is the load torque, $p$ represents the derivative function w.r.t. time and all symbols stand to their usual meaning (Iqbal et al., in press). Evaluation of motor parameters is determined from the standard test procedure (Aghamohammadi & Pourgholi, 2008; Alger, 1970; Jones, 1967).

### 3. Vector control scheme

The operating performance of vector-controlled motor is greatly improved and similar to that of a separately excited DC motor (Bose, 2002). This is because of decoupled control of both flux component and torque component of stator current. The inclusion of vector control scheme is not similar to that of induction motor drive. The main difference lies on the fact that the air gap flux is attributed by both stator flux as well as field flux. Therefore, the resultant air gap flux will align along the axis which is different from conventional *d*-axis. Hence, a new (*M–T*) coordinate axis has been introduced wherein, the resultant flux vector and torque current component will align along *M* and *T* axis, respectively. Machine variables in newly defined coordinate axis (*M–T* axis) may be readily transformed to its equivalent *d–q* or vice versa by relation

$$\begin{bmatrix} M_k \\ T_k \end{bmatrix} = \begin{bmatrix} \cos\delta_k & -\sin\delta_k \\ \sin\delta_k & \cos\delta_k \end{bmatrix}\begin{bmatrix} d_k \\ q_k \end{bmatrix} \tag{15}$$

In above relation, the torque angle $\delta_1$ and $\delta_2$ are associated with winding sets *a b c* and *a′b′c′*, respectively, and defined as

$$\delta_1 = \delta_0 \tag{16}$$

$$\delta_1 = \delta_0 + \emptyset + \xi \tag{17}$$

$\delta_0$ is the initial value of load angle, whereas $\emptyset$ is the phase difference between voltage fed to phase *a* and *a′*. Hence, torque attributed by each stator winding sets *a b c* and *a′b′c′* is given by Equations (12) and (13) may be readily written in *M–T* axes,

$$\tau_{e1} = c(\psi_{M1}I_{T1} - \psi_{T1}I_{M1}) \tag{18}$$

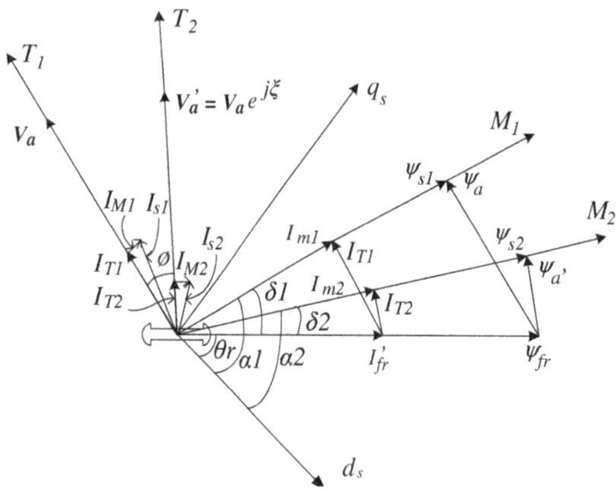


**Figure 1. Phasor diagram of vector-controlled six-phase synchronous motor.**

$$\tau_{e2} = c(\psi_{M2}I_{T2} - \psi_{T2}I_{M2}) \tag{19}$$

where, the flux linkage $\psi_{Mk}$ and $\psi_{Tk}$, and stator current $I_{Mk}$ and $I_{Tk}$ are aligned along $M_k$–$T_k$ axes respectively, for winding sets $a\ b\ c$ (for $k = 1$) and $a'b'c'$ (for $k = 2$). Since, the resultant armature air gap flux is only aligned along flux axis (i.e. $M_k$ axis). Therefore,

$\psi_{T1} = 0$ and $\psi_{M1} = \psi_{s1}$, where

$$\psi_{s1} = \sqrt{\psi_{M1}^2 + \psi_{T1}^2} \tag{20}$$

$\psi_{T2} = 0$ and $\psi_{M2} = \psi_{s2}$, where

$$\psi_{s2} = \sqrt{\psi_{M2}^2 + \psi_{T2}^2} \tag{21}$$

Hence, motor torque equation may be simplified as

$$\tau_e = \tau_{e1} + \tau_{e2} = c(\psi_{s1}I_{T1} + \psi_{s2}I_{T2}) \tag{22}$$

The developed motor torque is dependent on flux linkage $\psi_{s1}$ (and $\psi_{s2}$) and current $I_{T1}$ (and $I_{T2}$) which are orthogonal. This is the introduction of vector-controlled six-phase synchronous motor. The motor operation during steady state has been shown in the developed phasor diagram in Figure 1.

## 4. Implementation of vector control scheme

The implementation of developed vector-controlled synchronous motor drive system has been shown in Figure 2. In this paper, motor operation has been investigated in constant torque region up to base speed, but same may be extended in field weakening region above base speed. In the figure, the outer speed loop is used to generate the reference value torque component of stator current $I_{Tk}^*$ through a speed controller (PI controller), whereas the reference value magnetizing current $I_{mk}^*$ is generated by the flux controller (PI controller) associated with each winding sets $a\ b\ c$ (for $k = 1$) and $a'b'c'$ (for $k = 2$). The reference magnetizing current is used to establish the required flux $\psi_{sk}$ in air gap, which related to field current by the relation,

$$\left.\begin{aligned} I_{m1} &= I_{fr}' \cos\delta_1 \\ I_{m2} &= I_{fr}' \cos\delta_2 \end{aligned}\right\} \tag{23}$$

In phasor diagram, current component $I_{T1}$ (and $I_{T2}$) is in the direction of $T_1$ (and $T_2$) axis along which the voltage vector is also aligned. Further, the magnetizing component of current $I_{m1}$ (and $I_{m2}$) is

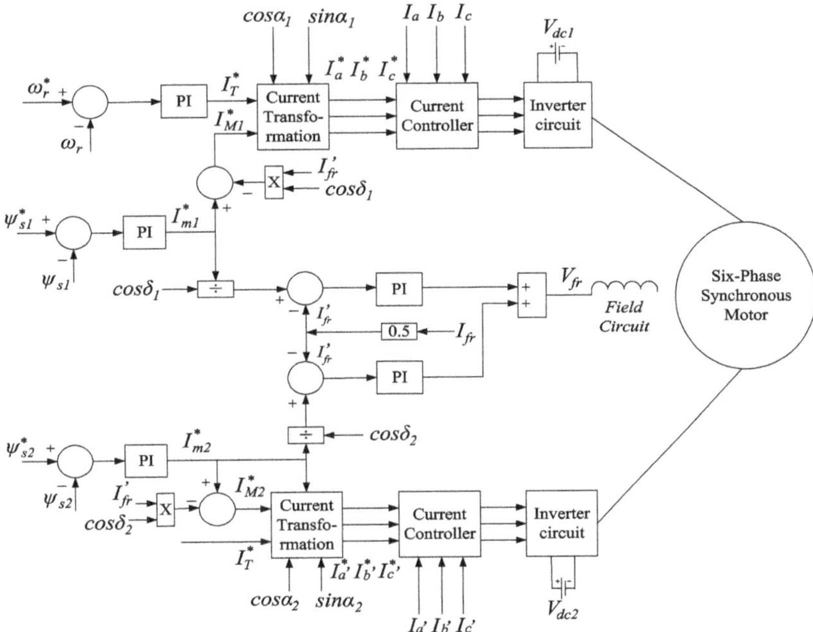


**Figure 2. Schematic diagram of vector-controlled six-phase synchronous motor drive.**

aligned along $M_1$ (and $M_2$) axis which is used to establish the flux vector $\psi_{s1}$ (and $\psi_{s2}$). At steady state, both the stator flux and armature flux vectors are orthogonal to each other, i.e. $\psi_{s1}$ (and $\psi_{s2}$) is perpendicular to $\psi_{o1}$ (and $\psi_{o2}$) as shown in phasor diagram. Therefore, at steady state, both the vectors $I_{Tk}$ and $I_{sk}$ are equal i.e. $I_{T1} = I_{s1}$ for winding set *a b c* (and $I_{T2} = I_{s2}$ for winding set *a′b′c′*) and becomes in phase with voltage vector, signifying the motor operation at unity power factor. In control scheme, it may be noted that the magnetizing current component $I_{m1}$ (and $I_{m2}$) is related to field current $I_{frk}$ associated with winding sets *a b c* and *a′b′c′*. Field current commands $I^{*'}_{fr1}$ and $I^{*'}_{fr2}$ are synthesized using Equation (23) in feedback loop. In field control loop, the field current error is fed to the field controller (PI controller) to establish the required field excitation. It may be noted here that the magnitude of field current magnitude associated with each winding set *a b c* and *a′b′c′* is assumed to be same $I'_{fr} (= 0.5 I_{fr})$. Now, the flux component of stator current is generated by

$$\left.\begin{aligned} I^*_{M1} &= I^*_{m1} - I'_{fr}\cos\delta_1 \\ I^*_{M2} &= I^*_{m2} - I'_{fr}\cos\delta_2 \end{aligned}\right\} \tag{24}$$

Above relation will yield a finite value of $I^*_{M1}$and $I^*_{M2}$. But at steady state, it becomes zero (i.e. $I^*_{M1} = I^*_{M2} = 0$) and Equation (23) will be satisfied. As soon as stator current component $I^*_{Tk}$ and $I^*_{Mk}$ are synthesized for winding sets *a b c* ($k = 1$) and *a′b′c′* ($k = 2$), the reference value of current in stationary reference frame is generated. For this purpose, following two-step transformation is carried out.

(1) Current component $I^*_{Tk}$ and $I^*_{Mk}$ are transformed to *d*–*q* component in stationary reference frame, using angle $\alpha_k$ in transformation in relation (15).

(2) Above obtained stationary *d*–*q* component of current is transformed into its equivalent three-phase current (Krause et al., 2004).

The reference current generated in above steps in stationary reference frame are then compared with actual phase current of stator windings which results in current error. This current error is fed to the hysteresis current controller to regulate switching of inverter circuit feeding the motor.

## 5. Simulation results

The developed system of vector-controlled six-phase synchronous motor drive was implemented in Matlab/Simulink environment. For this purpose, a 3.7-kW motor (parameters are given in Appendix 1) was operated in four quadrant. Initially, speed command was given at time $t = 0.1$ s. in ramp way, following to which motor starts to run at synchronous speed after time $t = 0.65$ s, showing its operation in first quadrant. A load of 50% of base torque was applied at time $t = 1.5$ s which results a small dip in speed by 0.03 rad/s, but regains its original speed (i.e. synchronous speed) after time $t = 0.5$ s, as shown in Figure 3. In order to examine the motor operation in second quadrant, the direction of load torque was reversed at time $t = 3$ s, resulting in a small increase in rotor speed by 0.05 rad/s. A small change in rotor speed due to sudden change in load torque signifies the disturbance rejection property of the drive system. A zoomed view of speed variation has been shown in Figure 3(a). Following to the change in load torque, not only the variation in $q$-component of stator current but also resulted a small variation in $d$-component of current, as shown in Figure 5. Change in stator current is also reflected in $T$-axis component of motor current flowing in winding sets $a\,b\,c$ and $a'b'c'$, in Figures 4 and 6. In order to shift the motor operation to third quadrant, a speed reversal command was initiated at time $t = 4$ s in ramp way. The motor is then finally operates in reverse motoring mode after time $t = 5$ s at synchronous speed. Further, at time $t = 6$ s the load torque is reversed to operate the drive in fourth quadrant. It is important to note here that the switching of motor operation to different quadrant results in a large variation in $d$-component stator current whose effect is compensated by rotor field current, as shown in Figure 7(b). But the field circuit has a larger time constant, making the response slow (Das & Chattopadhyay, 1997; Jain & Ranganathan, 2011; Krause et al., 2004). This sluggish response of field circuit is substantially improved due to the magnetizing current injection along flux direction, as depicted by reference value of current in Figure 4(b and c) as well as in actual current in Figure 6(c and d). Input phase voltage fed to winding sets $a\,b\,c$ and $a'b'c'$, is shown in Figure 8(a) and (b), respectively.

## 6. Conclusion

The scheme of vector control applicable for asymmetrical six-phase synchronous motor has been developed and extensively investigated in four quadrant operation. In this control technique, stator current is maintained for motor operation at unity power factor, which minimizes the stator current

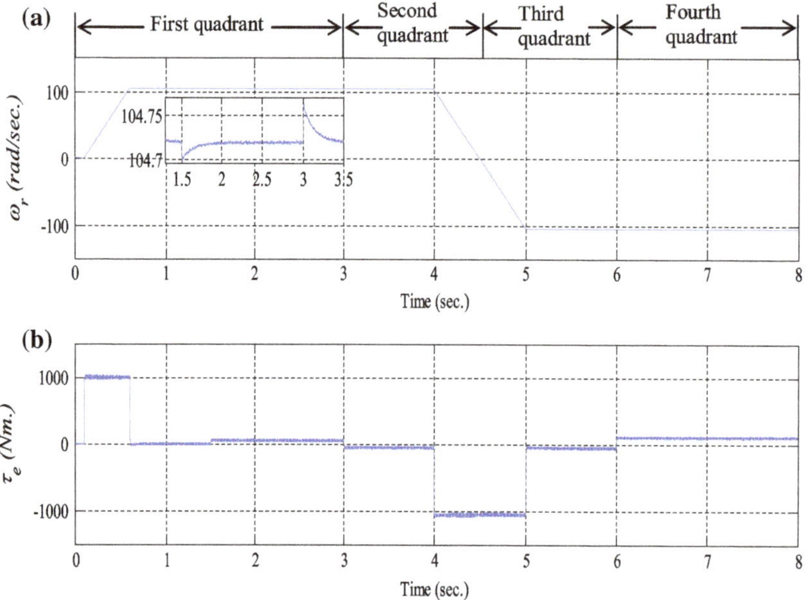


**Figure 3. Motor response (a) rotor speed $\omega_r$ and (b) torque $T_e$.**

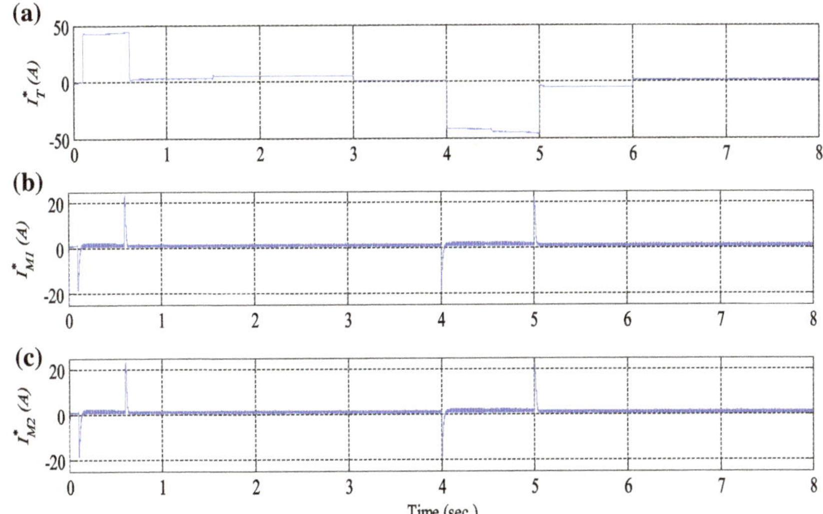


**Figure 4. Reference current in *M–T* coordinate (a) $I_{T\ 1}^{*}$, (b) $I^{*}_{M1}$ and (c) $I^{*}_{M2}$.**

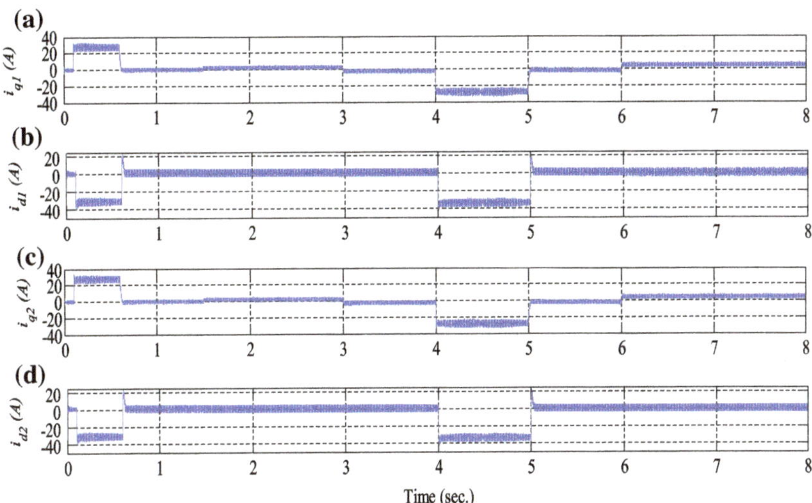


**Figure 5. Stator current in *d–q* coordinate (a) $i_{q1}$ (b) $i_{d1}$ (c) $i_{q2}$ (d) $i_{d2}$.**

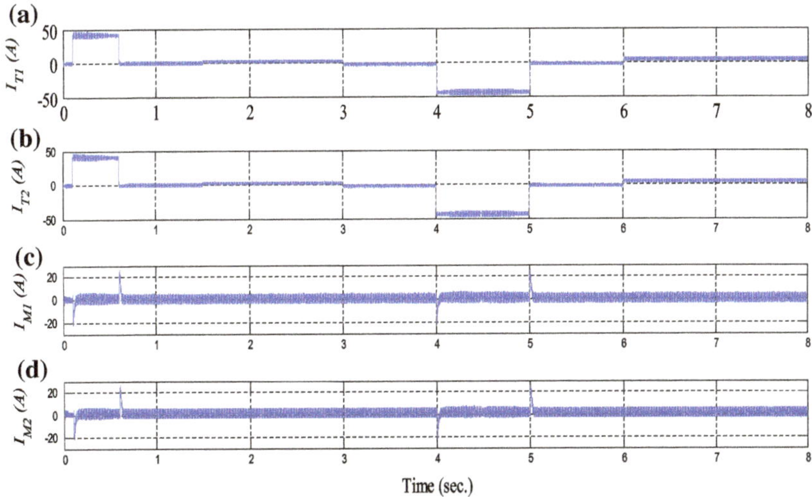


**Figure 6. Actual current in *M–T* coordinate (a) $I_{T1}$, (b) $I_{T2}$ (c) $I_{M1}$ and (d) $I_{M2}$.**

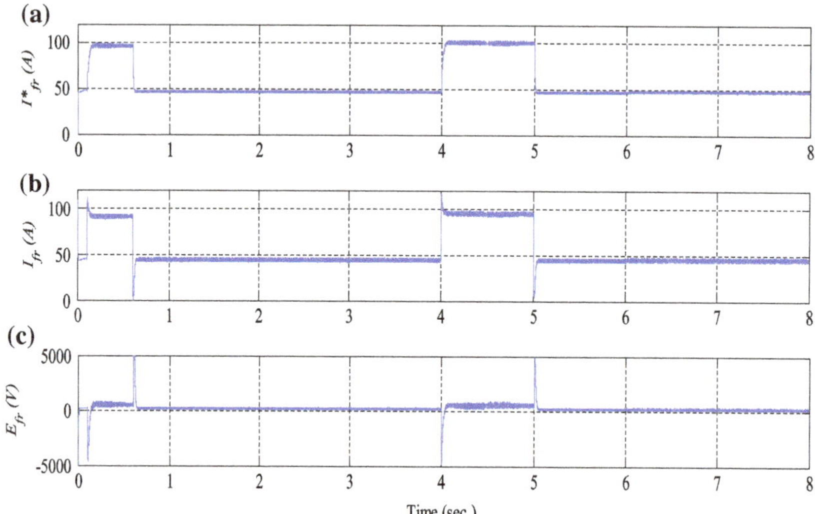


Figure 7. Voltage–current in field circuit (a) reference $I^*_{fr}$, (b) actual $I_{fr}$ (c) voltage $E_{fr}$.

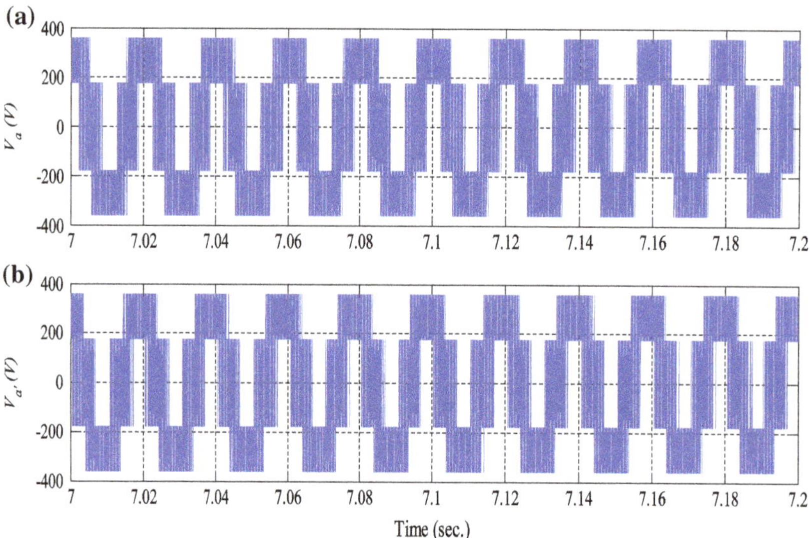


Figure 8. Inverter output voltage (a) phase *a* (b) phase *a'*.

and hence less losses. The dynamic behavior of six-phase synchronous motor in different quadrant operation was found to be substantially improved because of the decoupled/independent control of flux as well as torque component of current along *M–T* axis, respectively. Furthermore, sluggish response of the rotor field circuit was also noted to be improved because of the magnetizing current injection from armature side.

The present work may be further extended and investigated under different practical application with their experimental validation.

**Funding**
The authors received no direct funding for this research.

**Author details**
Arif Iqbal[1]
E-mail: arif.iqbal.in@gmail.com, arif0548@gmail.com
ORCID ID: http://orcid.org/0000-0002-7113-6007
G.K. Singh[1]
E-mail: gksngfee@gmail.com
Vinay Pant[1]
E-mail: vpantfee@gmail.com
[1] Department of Electrical Engineering, Indian Institute of Technology Roorkee, Roorkee 247667, Uttarakhand, India.

**References**

Aghamohammadi, M. R., & Pourgholi, M. (2008). Experience with SSSFR test for synchronous generator model identification using Hook–Jeeves optimization method. *International Journal of System Applications, Engineering and Development, 2*, 122–127.

Alger, P. L. (1970). *Induction machine*. New York, NY: Gorden and Breach.

Bojoi, R., Lazzari, M., Profumo, F., & Tenconi, A. (2003). Digital field-oriented control for dual three-phase induction motor drives. *IEEE Transactions on Industry Applications, 39*, 752–760. http://dx.doi.org/10.1109/TIA.2003.811790

Bose, B. K. (2002). *Mordern power electronics and AC drives*. Upper Saddle River, NJ: Prentice Hall.

Das, S. P., & Chattopadhyay, A. K. (1997). Observer based stator flux oriented vector control of cycloconverter-fed synchronous motor drive. *IEEE Transactions on Industry Applications, 33*, 943–955. http://dx.doi.org/10.1109/28.605736

Iqbal, A., Singh, G. K., & Pant, V. (2014). Steady-state modeling and analysis of six-phase synchronous motor. *System Science & Control Engineering, 2*, 236–249.

Iqbal, A., Singh, G. K., & Pant, V. (in press). Stability analysis of asymmetrical six-phase synchronous motor. *Turkish Journal of Electrical Engineering & Computer Sciences.*

Jain, A. K., & Ranganathan, V. T. (2011). Modeling and field oriented control of salient pole wound field synchronous machine in stator flux coordinates. *IEEE Transactions on Industrial Electronics, 58*, 960–970. http://dx.doi.org/10.1109/TIE.2010.2048295

Jones, C. V. (1967). *The unified theory of electric machine*. London: Butterworths.

Krause, P. C., Wasynczuk, O., & Sudhoff, S. D. (2004). *Analysis of electrical machinery and drive Systems*. Piscataway, NJ: IEEE Press and Wiley.

Levi, E. (2008). Multiphase electric machines for variable-speed applications. *IEEE Transactions on Industrial Applications, 38*, 1893–1909.

Schiferl, R. F., & Ong, C. M. (1983). Six phase synchronous machine with AC and DC stator connections, part I: Equivalent circuit representation and steady-state analysis. *IEEE Transactions on Power Apparatus and Systems, 102*, 2685–2693. http://dx.doi.org/10.1109/TPAS.1983.317674

Singh, G. K. (2002). Multi-phase induction machine drive research—A survey. *Electric Power Systems Research, 61*, 139–147. http://dx.doi.org/10.1016/S0378-7796(02)00007-X

Singh, G. K. (2011). Modeling and analysis of six-phase synchronous generator for stand-alone renewable energy generation. *Energy, 36*, 5621–5631. http://dx.doi.org/10.1016/j.energy.2011.07.005

Singh, G. K., Nam, K., & Lim, S. K. (2005). A simple indirect field-oriented control scheme for multiphase induction machine. *IEEE Transactions on Industrial Electronics, 52*, 1177–1184. http://dx.doi.org/10.1109/TIE.2005.851593

## Appendix 1

### Motor parameter of 3.7 kW, 6 poles, 36 slots are

| | | |
|---|---|---|
| $r_1 = 0.210\ \Omega$ | $x_{mq} = 3.9112\ \Omega$ | $r_{fr} = 0.056\ \Omega$ |
| $r_2 = 0.210\ \Omega$ | $x_{md} = 6.1732\ \Omega$ | $x_{ldq} = 0$ |
| $r_{Kq} = 2.535\ \Omega$ | $x_{lKq} = 0.66097\ \Omega$ | $x_{lm} = 0.001652\ \Omega$ |
| $r_{Kd} = 140.0\ \Omega$ | $x_{lKd} = 1.550\ \Omega$ | $x_{lfr} = 0.2402\ \Omega$ |
| $x_{l1} = x_{l2} = 0.1758\ \Omega$ | | |

$$x = \begin{bmatrix} (x_{l1}+x_{lm}+x_{md}) & 0 & (x_{lm}+x_{md}) & x_{ldq} & x_{md} & 0 & x_{md} \\ 0 & (x_{l1}+x_{lm}+x_{md}) & -x_{ldq} & (x_{lm}+x_{mq}) & 0 & x_{mq} & 0 \\ (x_{lm}+x_{md}) & -x_{ldq} & (x_{l2}+x_{lm}+x_{md}) & 0 & x_{md} & 0 & x_{md} \\ x_{ldq} & (x_{lm}+x_{mq}) & 0 & (x_{l2}+x_{lm}+x_{mq}) & 0 & x_{mq} & 0 \\ & x_{md} & 0 & x_{md} & 0 & (x_{lKd}+x_{md}) & \\ 0 & & & & & & \\ & x_{md}0 & x_{mq} & 0x_{mq} & 0 & (x_{lKq}+x_{mq}) & 0 \\ x_{md} & 0 & x_{md} & 0 & x_{md}0 & (x_{lfr}+x_{md}) & \end{bmatrix}$$

# 13

# Reliable terminal sliding mode control for uncertain high-order MIMO systems with actuator faults

Jiaye Fu[1]* and Yugang Niu[1]

*Corresponding author: Jiaye Fu, Key Laboratory of Advanced Control and Optimization for Chemical Process (East China University of Science & Technology), Ministry of Education, Shanghai 200237, China E-mail: 940324037@qq.com

Reviewing editor: James Lam, University of Hong Kong, Hong Kong

**Abstract:** This paper considers the problem of terminal sliding mode control (TSMC) for uncertain MIMO systems in which the actuator faults may happen in any channel of actuators. By means of state transformation and exponent-logarithmic sliding surface method, a reliable TSMC is proposed such that both the finite-time stability of sliding mode dynamics and the reachability of sliding surface are ensured, despite actuator faults and parameter uncertainties. Finally, simulation results are provided to illustrate the effectiveness of the proposed controller.

**Subjects: Automation Control; Control Engineering; Engineering & Technology; Systems & Control Engineering; Systems Engineering**

**Keywords: MIMO systems; actuator faults; terminal sliding mode control; exponent-logarithmic terminal sliding mode**

## 1. Introduction

Sliding mode control (SMC) is a useful and effective scheme to deal with parameter uncertainties and external disturbances for both linear and nonlinear systems (Seshagiri & Khalil, 2002; Vicente & Gerd, 2001; Xu, Lee, & Pan, 2003). In general, the conventional sliding surface is the linear function of system states. Although the parameters in linear sliding surface can be adjusted to get faster convergence rate, the system states cannot usually reach the equilibrium points in finite time. To overcome this drawback, the terminal sliding mode control (TSMC) (Yu, Yu, & Man, 2000) was proposed and studied by many researchers, e.g. (Lin, 2006; Liu & Li, 2009; Tao, Taur, & Chan, 2004; Yu & Man, 2002; Zou, Kumar, Hou, & Liu, 2011) and the references therein.

### ABOUT THE AUTHOR

Miss Jiaye Fu received the BE degree in electrical engineering and automation form East China University of Science and Technology , Shanghai, China, in 2013. From 2013, she has been pursuing her Master degree in control science and engineering in East China University of Science and Technology, Shanghai, China. Her research interests include sliding mode control, terminal sliding mode control, fault-tolerant control and their applications.

### PUBLIC INTEREST STATEMENT

Sliding mode control (SMC) is a useful and effective scheme to deal with parameter uncertainties and external disturbances for both linear and nonlinear systems. To tackle the problems of globally asymptotic stabilization, terminal sliding mode control scheme has been developed to achieve finite-time stabilization. It provides some superior properties such as fast, finite time convergence, and high static tracking precision.

In safety-critical systems, a minor fault of a single component can result in severe performance deterioration or may even produce catastrophic effects. To overcome this problem and realize finite time convergence, this paper introduces an exponent-logarithmic terminal sliding surface to improve the speed of response, and designs a reliable SMC approach to guarantee the stability of the overall closed-loop systems and finite time convergence of system states despite the presence of actuator faults.

Compared with the linear hyperplane-based SMC, TSMC provides some superior properties, such as fast, finite-time convergence, and high static tracking precision. For example, for second-order systems, a fast terminal sliding mode (FTSM) was proposed in Yu and Man (2002), which achieved better dynamic property by employing the FTSM concept in both reaching phase and sliding phase. In He, Liu, Liu, Liu, and Liu (2008), an exponential TSMC scheme was introduced into the learning algorithm to improve approximation ability for an unstable nonlinear system. Feng, Han, and Wang (2007) further considered the problem of high-order MIMO systems and proposed a hierarchical TSMC method. However, the TSMC method in Feng et al. (2007) only ensured that partial system states could be driven to zero in finite time and the rest converged to zero asymptotically. Wang, Yang, and Zhang (2007) then considered a new TSMC scheme which could drive all system states to equilibrium points in finite time, but its convergence performance was not so better compared to the one in second-order systems. Li, Ma, Zheng, and Geng (2014) presented a fast nonsingular integral TSMC which could avoid the singularity problem without any constraint, and provided faster responses by tuning the parameters. More recently, Hu, Jiang, Chen, and Liu (2014) developed an exponent-logarithmic terminal sliding surface for a special of *SISO strict-feedback* nonlinear systems, which could attain higher convergence performances. However, it is worthy of noting that the aforementioned works were made under the assumption that the actuator/sensor worked normally, i.e. there didn't exist the faults of actuator/sensor.

As is well known, in safety-critical systems, a minor fault of a single component can result in severe performance deterioration or may even produce catastrophic effects. An effective way to maintain an acceptable stability/performance against undesired actuator/sensor failures is to utilize fault-tolerant control (FTC) strategies for the controlled systems (Bateman, Noura, & Ouladsine, 2011; Hu, 2010; Huo, Li, & Tong, 2012; Liu, Cao, & Shi, 2013). In the past decades, many reliable control methodologies have been applied, e.g. the coprime factorization approach (Vidyasagar & Viswanadham, 1985), the Hamilton–Jacobi based approach (Yang, Wang, & Soh, 2000), the linear matrix inequality based approach (Liao, Wang, & Yang, 2002), and so on. Moreover, the reliable SMC approach has been proposed (Chen, Niu, Zou, & Jia, 2013; Liang, Liaw, & Lee, 2000; Niu & Wang, 2009). Among them, Liang et al. (2000) first addressed the FTC design based on SMC, which did not require the solution of any Hamilton–Jacobi equation and could retain the advantages of conventional SMC designs Niu and Wang (2009) solved the problem of SMC with partial actuator degradation through adopting a model of actuator faults such that both normal operation and partial actuator degradation were covered. The results in Niu and Wang (2009) were further extended to the Markovian jumping systems in Chen et al. (2013). More recently, some interesting researches on FTC based on TSMC were made. For example, Xu and Liu (2014) studied the TSMC design based on the T–S fuzzy system models and presented both of the active and passive FTC schemes. Qu, Gao, Huang, Mei, and Zhai (2014) proposed a finite-time FTC scheme for the faulty UAV attitude control systems by utilizing a fault detection strategy, which ensured that the dynamic systems converge to a stable state in a finite-time in the case of actuator faulty. However, to the author's best knowledge, the TSMC for high-order MIMO systems with actuator actuators has been not well addressed and remains still open. Moreover, the characteristic of TSMC structure also makes those existing works not to be simply extended to the present case.

Motivated by the above discussion, this paper considers the problem of TSMC for a class of uncertain high-order MIMO systems subject to actuator faults. Firstly, the model of actuator faults is presented, in which only the bounds of actuator faults are known. It should be pointed out that in the design of TSMC for uncertain high-order MIMO systems with actuator faults, one has to consider the realization of finite-time convergence for whole system states and the characteristic of actuator faults, which result in the aforementioned work that cannot be directly utilized in this paper. Hence, a state transformation is utilized and a high-order exponent-logarithmic sliding surface is designed, which can achieve the finite-time stability of sliding mode dynamics with shorter convergent time. And then, a reliable sliding mode controller is designed such that the system states can be driven onto the specified sliding surface in finite time. It is shown that the effect of actuator faults can be coped with by the present reliable TSMC method.

Notations: Throughout this work, $\|\cdot\|$ denotes the Euclidean norm of a vector or the spectral of a matrix. $\lambda_{max}(\cdot)$ denotes the maximum eigenvalue of a matrix. $I$ represents an identity matrix of appropriate dimensions. ln $(\cdot)$ is a logarithmic function (base e). $\mathrm{diag}(\cdot)$ is a diagonal matrix. $\mathrm{sgn}(\cdot)$ is a sign function. Matrices, if not explicitly stated otherwise, are assumed to have compatible dimensions.

## 2. Problem statement

Consider a class of uncertain high-order MIMO systems

$$\dot{X}(t) = [A + \Delta A(t)]X(t) + BU(t) + F(t) \tag{1}$$

where $X(t) \in R^n$ are the system states, $U(t) \in R^m$ are the control inputs, $A$ and $B$ are the known constant matrixes. Without loss of generality, it is assumed that the pair $(A, B)$ is controllable and the matrix $B$ has full column rank. $\Delta A(t)$ and $F(t)$ represent the unknown parameter uncertainties and external disturbances, respectively, satisfying:

$$\Delta A(t) = BN_1(t), \quad \|N_1(t)\| \le l_1 \tag{2}$$

$$F(t) = BN_2(t), \quad \|N_2(t)\| \le l_2 \tag{3}$$

where $l_1$ and $l_2$ are known constants.

As discussed in the introduction, the actuator fault is usually inevitable in actual application. Hence, in this work, it is assumed that the actuator faults may happen and the control signal received by the system is $U^F(t)$ satisfying:

$$U^F(t) = E(t)U(t) \tag{4}$$

where $E(t) = \mathrm{diag}[e_1(t), e_2(t), \dots, e_m(t)]$ with faults factor $e_i(t)$ $(i = 1, 2, \dots, m)$ satisfying:

$$0 < \underline{e}_i \le e_i(t) \le \bar{e}_i, \ \ \underline{e}_i \le 1, \bar{e}_i \ge 1\ , \ \ i = 1, 2, \dots, m \tag{5}$$

In this work, it is assumed that both the lower and upper bounds of $e_i(t)$ are known.

Define

$$e_{0i} = \frac{1}{2}(\underline{e}_i + \bar{e}_i), \quad r_i = \frac{\bar{e}_i + \underline{e}_i}{\bar{e}_i - \underline{e}_i}, \quad \delta_i(t) = \frac{e_i(t) - e_{0i}}{e_{0i}} \tag{6}$$

and $E_0 = \mathrm{diag}[e_{01}, e_{02}, \dots, e_{0m}]$, $R = \mathrm{diag}[r_1, r_2, \dots, r_m]$, $\delta(t) = \mathrm{diag}[\delta_1(t), \delta_2(t), \dots, \delta_m(t)]$. It is easily shown that $\|\delta_i(t)\| \le r_i < 1 \ \ (i = 1, 2, \dots, m)$.

Then, the actuator faults model (4) can be written as:

$$U^F(t) = E_0[I + \delta(t)]U(t), \quad \|\delta(t)\| \le \|R\| < 1 \tag{7}$$

*Remark* 1 It can be seen that the actuator faults model in (4)–(5) covers the normal operation case (as $\underline{e}_i = \bar{e} = 1$) and partial faults case (as $0 < \underline{e}_i < \bar{e}_i$), and each actuator may be a failure independently.

In the sequel, a lemma is given, which is useful for the development of the main results.

Lemma 1 (Yu, Yu, Shirinzadeh, & Man, 2005) Assume that a continuous, positive-definite function $V(t)$ satisfies the following differential inequality:

$$\dot{V}(t) \le -\varphi V^{\gamma}(t) \qquad \forall t > 0 \tag{8}$$

where $\varphi > 0,\ 0 < \gamma < 1$ are constants. Then it can be found that $V(t)$ which starts from $V(0)$ can reach $V = 0$ in finite time. Moreover, the reaching time $t_r$ is given by:

$$t_r = \frac{V^{1-\gamma}(0)}{\varphi(1-\gamma)} \tag{9}$$

## 3. High-order exponent-logarithmic TSM design

### 3.1. State transformation

Firstly, a state transformation is made for the system model (1). Due to $rank(B) = m$, there exists an invertible matrix $T$ satisfying $TB = \begin{bmatrix} 0 & B_2 \end{bmatrix}^T$, where $B_2 \in R^{m\times m}$ is full rank. Moreover, we have

$$\begin{aligned} T(A+\Delta A(t))T^{-1} &= TAT^{-1} + TBN_1T^{-1} = \begin{bmatrix} A_{11} & A_{12} \\ A_{21} & A_{22} \end{bmatrix} + \begin{bmatrix} 0 \\ B_2N_1(t)T^{-1} \end{bmatrix} \\ TF(t) &= TBN_2(t) = \begin{bmatrix} 0 \\ B_2N_2(t) \end{bmatrix} \end{aligned} \tag{10}$$

where $A_{11} \in R^{(n-m)\times(n-m)}$, $A_{12} \in R^{(n-m)\times m}$,$A_{21} \in R^{m\times(n-m)}$, $A_{11} \in R^{(n-m)\times(n-m)}$, $A_{22} \in R^{m\times m}$, $B_2 \in R^{m\times m}$.

By means of the following state transformation:

$$Z(t) = TX(t) \tag{11}$$

the system (1) can be transformed to the following form:

$$\begin{cases} \dot{Z}_1(t) = A_{11}Z_1(t) + A_{12}Z_2(t) & (12a) \\ \dot{Z}_2(t) = A_{21}Z_1(t) + A_{22}Z_2(t) + B_2U(t) + B_2N_1(t)\bar{T}_1Z_1(t) + B_2N_1(t)\bar{T}_2Z_2(t) + B_2N_2(t) & (12b) \end{cases}$$

where $Z_1(t) = [z_1, \cdots z_{n-m}]^T \in R^{n-m}$, $Z_2(t) = [z_{n-m+1}, \cdots z_n]^T \in R^m$, $T^{-1} \triangleq [\bar{T}_1 \ \ \bar{T}_2]$, $\bar{T}_1 \in R^{n\times(n-m)}$, $\bar{T}_2 \in R^{n\times m}$.

Note that $Z(t) = TX(t)$, when $Z(t) \to 0$, $X(t) \to 0$. So the studied problem is equally transformed to the design of TSM surface and robust controller for system (12).

### 3.2. An exponent-logarithmic TSM

It is well known that the TSM has some superior properties, e.g. higher precision, faster convergence, etc. Hence, in this work, a new high-order exponent-logarithmic TSM is proposed as follows:

$$S(t) = C_1Z_1(t) + C_2Z_2(t) + C_3G(t) + C_4H(t) = 0 \tag{13}$$

where the vector functions $G(t)$ and $H(t)$ are given as:

$$\begin{cases} G(t) = \left[(2e^{|z_1|}-1)\ln(2-e^{-|z_1|})\mathrm{sgn}(z_1), \ldots, (2e^{|z_{n-m}|}-1)\ln(2-e^{-|z_{n-m}|})\mathrm{sgn}(z_{n-m})\right]^T \\ H(t) = \left[(2e^{|z_1|}-1)\ln^{q_1/p_1}(2-e^{-|z_1|})\mathrm{sgn}(z_1), \ldots, (2e^{|z_{n-m}|}-1)\ln^{q_{n-m}/p_{n-m}}(2-e^{-|z_{n-m}|})\mathrm{sgn}(z_{n-m})\right]^T \end{cases} \tag{14}$$

where $p_i,\ q_i,\ (i = 1, \ldots, n-m)$ are positive odd integers satisfying $q_i < p_i < 2q_i$.

In (13), the matrices $C_1 \in R^{m\times(n-m)}$, $C_2 \in R^{m\times m}$, $C_3 \in R^{m\times(n-m)}$ and $C_4 \in R^{m\times(n-m)}$ are chosen by the following expressions:

$$\begin{cases} A_{11} - A_{12}C_2^{-1}C_1 = 0 \\ A_{12}C_2^{-1}C_3 = \mathrm{diag}(\alpha_1, \dots, \alpha_{n-m}) \\ A_{12}C_2^{-1}C_4 = \mathrm{diag}(\beta_1, \dots, \beta_{n-m}) \end{cases} \tag{15}$$

where $\alpha_i > 0$, $\beta_i > 0$, $(i = 1, \dots, n - m)$. The invertible matrix $C_2$ is usually chosen as an identity matrix for convenience. And then, the matrices $C_1$, $C_3$ and $C_4$ can be obtained as follows:

$$\begin{cases} C_1 = C_2A_{12}^{+}A_{11} \\ C_3 = C_2A_{12}^{+}\mathrm{diag}(\alpha_1, \dots, \alpha_{n-m}) \\ C_4 = C_2A_{12}^{+}\mathrm{diag}(\beta_1, \dots, \beta_{n-m}) \end{cases} \tag{16}$$

where $A_{12}^{+}$ is the right inverse matrix of $A_{12}$ given as follows:

$$A_{12}^{+} = A_{12}^{T}(A_{12}A_{12}^{T})^{-1} \tag{17}$$

Apparently, the matrix $A_{12}$ should be full row rank for attaining (17).

According to the SMC theory, when the system states reach the sliding surface $S = 0$ and move along it, the sliding mode dynamics in $S = 0$ is a $n-m$ order system. For $S = 0$, we obtain from (12a):

$$Z_2 = -C_2^{-1}(C_1Z_1 + C_3G + C_4H) \tag{18}$$

which substituted into system (12b) yields the reduced-order system as follows:

$$\dot{Z}_1 = (A_{11} - A_{12}C_2^{-1}C_1)Z_1 - A_{12}C_2^{-1}C_3G - A_{12}C_2^{-1}C_4H \tag{19}$$

In the following, we shall analyze the finite-time stability of sliding mode dynamics (19).

THEOREM 1 *Consider the uncertain high-order MIMO system (12).The exponent-logarithmic TSM is designed as (13) with the parameters satisfying (14 and 15). Then the sliding mode dynamics (19) on $S = 0$ is finite-time stable. That is, the state $Z_1$ of sliding mode dynamics will converge to the equilibrium points in finite time. Moreover, the states $Z_2$ will also converge to equilibrium points in finite time.*

*Proof* By means of the condition (15), it follows from (19) that

$$\dot{Z}_1 = \begin{bmatrix} -\alpha_1(2e^{|z_1|} - 1)\ln(2 - e^{-|z_1|})\mathrm{sgn}(z_1) - \beta_1(2e^{|z_1|} - 1)\ln^{q_1/p_1}(2 - e^{-|z_1|})\mathrm{sgn}(z_1) \\ \vdots \\ -\alpha_{n-m}(2e^{|z_{n-m}|} - 1)\ln(2 - e^{-|z_{n-m}|})\mathrm{sgn}(z_{n-m}) \\ -\beta_{n-m}(2e^{|z_{n-m}|} - 1)\ln^{q_{n-m}/p_{n-m}}(2 - e^{-|z_{n-m}|})\mathrm{sgn}(z_{n-m}) \end{bmatrix} \tag{20}$$

which shows that all elements of $Z_1$ is independent. Thus, we can analyze every state variables $z_i$ $(i = 1, 2, \dots, n - m)$, respectively. To this end, choose the candidate Lyapunov function $V_1$ as follows

$$V_1 = \frac{1}{2}z_i^2 \tag{21}$$

we have

$$\dot{V}_1 = z_i\dot{z}_i = -\alpha_i(2e^{|z_i|} - 1)\ln(2 - e^{-|z_i|})|z_i| - \beta_i(2e^{|z_i|} - 1)\ln^{q_i/p_i}(2 - e^{-|z_i|})|z_i| \le 0$$

which means that the state $z_i$ is stable.

In the following, we shall analyze the finite-time convergence performance of state $z_i$. Firstly, it follows from the $i$th term in (20) that

$$-\beta_i = f(z_i)^{-q_i/p_i}\frac{df(z_i)}{dt} + \alpha_i f(z_i)^{1-q_i/p_i} \tag{22}$$

where $f(z_i) = \ln(2 - e^{-|z_i|})\text{sgn}(z_i)$. Define $y = f(z_i)^{1-q/p}$. We have:

$$\frac{dy}{dt} = (1 - q_i/p_i)f(z_i)^{-q_i/p_i}\frac{df(z_i)}{dt} \tag{23}$$

And then, from (22) to (23), we can get:

$$-\beta_i\frac{p_i - q_i}{p_i} = \frac{dy}{dt} + \alpha_i\frac{p_i - q_i}{p_i}y \tag{24}$$

By solving first-order linear differential equation (24), one obtains:

$$y = -\frac{\beta_i}{\alpha_i} + \frac{\beta_i}{\alpha_i}e^{-\frac{p_i-q_i}{p_i}\alpha_i t} + y(t_r)e^{-\frac{p_i-q_i}{p_i}\alpha_i t} \tag{25}$$

where $t_r$ denotes the instant when the system states arrive at the sliding surface $S(t)$=0 from any initial state. Define $t_{si}$ as the instant when the state $z_i$ converges to equilibrium point 0. Due to $f(z_i, t_{si}) = 0$ and $y(t_{si}) = 0$, we have from (25):

$$t_{si} = \frac{p_i}{\alpha_i(p_i - q_i)}\ln\frac{\alpha_i[\ln(2 - e^{-|z_i(t_r)|})\text{sgn}(z_i(t_r))]^{(p_i-q_i)/p_i} + \beta_i}{\beta_i} \tag{26}$$

From the above analysis, the state $Z_1$ is finite-time stable. That is, the state $Z_1$ of sliding mode dynamics will converge to the equilibrium points within a finite time $t_s = \max\limits_{1\leq i\leq n-m}(t_{si})$.

Furthermore, it is seen from (14) and (18) that the state vector $Z_2$ is fully determined by $Z_1$. Hence, when the states $Z_1$ reach equilibrium points in finite-time, the states $Z_2$ will also converge to equilibrium points in finite time. □

### 3.3. Reliable TSMC design

In the sequel, we shall design a reliable sliding mode controller $U(t)$ to guarantee that the system states from any initial states $Z(0) \neq 0$ are driven onto the sliding mode surface $S = 0$ in finite time.

In this work, the reliable SMC law is designed as:

$$U(t) = \begin{cases} U_l + U_m, & S \neq 0 \\ 0, & S = 0 \end{cases} \tag{27}$$

with

$$\begin{cases} U_l = -(C_2B_2E_0)^{-1}[(C_1A_{11} + C_2A_{21} + C_3\tilde{G}A_{11} + C_4\tilde{H}A_{11})Z_1 \\ \qquad +(C_1A_{12} + C_2A_{22} + C_3\tilde{G}A_{12} + C_4\tilde{H}A_{12})Z_2] \\ U_m = -\dfrac{(C_2B_2E_0)^T S}{(1 - \lambda_{\max}(R))\left\|S^T C_2B_2E_0\right\|}[l_1\|E_0\|\left\|\bar{T}_1\right\|\|Z_1\| + l_1\|E_0\|\left\|\bar{T}_2\right\|\|Z_2\| \\ \qquad +\lambda_{\max}(R)\|U_l\| + l_2\|E_0\| + \eta] \end{cases} \tag{28}$$

where $\eta > 0$ is a positive constant, $\tilde{G} = \mathrm{diag}(\tilde{g}_1, \ldots, \tilde{g}_{n-m})$ and $\tilde{H} = \mathrm{diag}(\tilde{h}_1, \ldots, \tilde{h}_{n-m})$ with

$$\begin{cases} \tilde{g}_i = \left[2e^{|z_i|}\ln(2-e^{-|z_i|})+1\right]\mathrm{sgn}(z_i), i=1,\ldots,n-m \\ \tilde{h}_i = \left[2e^{|z_i|}\ln^{q_i/p_i}(2-e^{-|z_i|})+\frac{q_i}{p_i}\ln^{q_i/p_i-1}(2-e^{-|z_i|})\right]\mathrm{sgn}(z_i), i=1,\ldots,n-m \end{cases} \tag{29}$$

*Remark 2* The TSMC law (28)–(29) includes the matrix $E_0$ and the scalar $\lambda_{max}(R)$, which depend on $\bar{e}_i$ and $\underline{e}_i$, reflecting the effect of the faulty actuator. Hence, the present TSMC method can effectively deal with the effect of actuator faults.

THEOREM 2 *Consider the uncertain high-order MIMO system described by (12) with actuator faults as in (7). The reliable SMC law (28)–(29) can ensure that the system states are driven onto the sliding surface S = 0 in finite time.*

*Proof* Consider the following Lyapunov function candidate $V_2$:

$$V_2 = \frac{1}{2}S^T S \tag{30}$$

Taking the time derivative of $V_2$ yields:

$$\begin{aligned} \dot{V}_2 &= S^T\dot{S} = S^T(C_1\dot{Z}_1 + C_2\dot{Z}_2 + C_3\tilde{G}\dot{Z}_1 + C_4\tilde{H}\dot{Z}_1) \\ &= S^T[C_1A_{11}Z_1 + C_1A_{12}Z_2 + C_2A_{21}Z_1 + C_2A_{22}Z_2 + C_2B_2U^F + C_2B_2N_1\bar{T}_1Z_1 \\ &\quad + C_2B_2N_1\bar{T}_2Z_2 + C_2B_2N_2 + C_3\tilde{G}A_{11}Z_1 + C_3\tilde{G}A_{12}Z_2 + C_4\tilde{H}A_{11}Z_1 + C_4\tilde{H}A_{12}Z_2] \\ &= S^T[(C_1A_{11} + C_2A_{21} + C_3\tilde{G}A_{11} + C_4\tilde{H}A_{11})Z_1 + (C_1A_{12} + C_2A_{22} + C_3\tilde{G}A_{12} + C_4\tilde{H}A_{12})Z_2 \\ &\quad + C_2B_2E_0U + C_2B_2E_0\delta U + C_2B_2N_1\bar{T}_1Z_1 + C_2B_2N_1\bar{T}_2Z_2 + C_2B_2N_2] \end{aligned} \tag{31}$$

By substituting the control law (28 and 29) into (31), one has:

$$\begin{aligned} \dot{V}_2 &= S^TC_2B_2[E_0U_m + E_0\delta(U_l + U_m) + N_1\bar{T}_1Z_1 + N_1\bar{T}_2Z_2 + N_2] \\ &= S^TC_2B_2E_0U_m + S^TC_2B_2E_0\delta U_m \\ &\quad + S^TC_2B_2E_0[\delta U_l + E_0^{-1}N_1\bar{T}_1Z_1 + E_0^{-1}N_1\bar{T}_2Z_2 + E_0^{-1}N_2] \\ &\le S^TC_2B_2E_0U_m + S^TC_2B_2E_0\delta U_m \\ &\quad + \left\|S^TC_2B_2E_0\right\|[\|\delta\|\|U_l\| + l_1\left\|E_0^{-1}\right\|\left\|\bar{T}_1\right\|\|Z_1\| + l_1\left\|E_0^{-1}\right\|\bar{T}_2\|Z_2\| + l_2\left\|E_0^{-1}\right\|] \\ &\le S^TC_2B_2E_0U_m + S^TC_2B_2E_0\delta U_m \\ &\quad + \left\|S^TC_2B_2E_0\right\|[\|\lambda_{max}(R)\|\|U_l\| + l_1\left\|E_0^{-1}\right\|\left\|\bar{T}_1\right\|\|Z_1\| + l_1\left\|E_0^{-1}\right\|\left\|\bar{T}_2\right\|\|Z_2\| + l_2\left\|E_0^{-1}\right\|] \end{aligned} \tag{32}$$

By means of $U_m$ in (28), we further get:

$$\begin{aligned} \dot{V}_2 &\le -\left\|S^TC_2B_2E_0\right\|\frac{1}{1-\lambda_{max}(R)} \\ &\quad \times[l_1\left\|E_0^{-1}\right\|\left\|\bar{T}_1\right\|\|Z_1\| + l_1\left\|E_0^{-1}\right\|\left\|\bar{T}_2\right\|\|Z_2\| + \|\lambda_{max}(R)\|\|U_l\| + l_2\left\|E_0^{-1}\right\| + \eta] \\ &\quad + \frac{\left\|S^TC_2B_2E_0\delta(C_2B_2E_0)^TS\right\|}{\left\|S^TC_2B_2E_0\right\|}\frac{1}{1-\lambda_{max}(R)} \\ &\quad \times[l_1\left\|E_0^{-1}\right\|\left\|\bar{T}_1\right\|\|Z_1\| + l_1\left\|E_0^{-1}\right\|\left\|\bar{T}\right\|\|Z_2\| + \|\lambda_{max}(R)\|\|U_l\| + l_2\left\|E_0^{-1}\right\| + \eta] \\ &\quad + \left\|S^TC_2B_2E_0\right\|[\|\lambda_{max}(R)\|\|U_l\| + l_1\left\|E_0^{-1}\right\|\left\|\bar{T}_1\right\|\|Z_1\| + l_1\left\|E_0^{-1}\right\|\left\|\bar{T}_2\right\|\|Z_2\| + l_2\left\|E_0^{-1}\right\|] \end{aligned} \tag{33}$$

Note

$$\begin{aligned}\left\|S^TC_2B_2E_0\delta(C_2B_2E_0)^TS\right\| &\le \|\lambda_{max}(\delta)\|\left\|S^TC_2B_2E_0(C_2B_2E_0)^TS\right\| \\ &\le \lambda_{max}(R)\left\|S^TC_2B_2E_0(C_2B_2E_0)^TS\right\|\end{aligned} \tag{34}$$

So, we have

$$\begin{aligned}\dot V_2 \le &-\left\|S^TC_2B_2E_0\right\|\frac{1}{1-\lambda_{max}(R)} \\ &\times[l_1\left\|E_0^{-1}\right\|\left\|T_1^{-1}\right\|\|Z_1\| + l_1\left\|E_0^{-1}\right\|\left\|T_2^{-1}\right\|\|Z_2\| + \|\lambda_{max}(R)\|\|U_l\| + l_2\left\|E_0^{-1}\right\| + \eta] \\ &+\lambda_{max}(R)\left\|S^TC_2B_2E_0\right\|\frac{1}{1-\lambda_{max}(R)} \\ &\times[l_1\left\|E_0^{-1}\right\|\left\|\bar T_1\right\|\|Z_1\| + l_1\left\|E_0^{-1}\right\|\left\|\bar T_2\right\|\|Z_2\| + \|\lambda_{max}(R)\|\|U_l\| + l_2\left\|E_0^{-1}\right\| + \eta] \\ &+\left\|S^TC_2B_2E_0\right\|[\|\lambda_{max}(R)\|\|U_l\| + l_1\left\|E_0^{-1}\right\|\left\|\bar T_1\right\|\|Z_1\| + l_1\left\|E_0^{-1}\right\|\left\|\bar T_2\right\|\|Z_2\| + l_2\left\|E_0^{-1}\right\|]\end{aligned}$$

$$\begin{aligned}&= -\left\|S^TC_2B_2E_0\right\|[l_1\left\|E_0^{-1}\right\|\left\|\bar T_1\right\|\|Z_1\| + l_1\left\|E_0^{-1}\right\|\left\|\bar T_2\right\|\|Z_2\| + \|\lambda_{max}(R)\|\|U_l\| + l_2\left\|E_0^{-1}\right\| + \eta] \\ &\quad+\left\|S^TC_2B_2E_0\right\|[\|\lambda_{max}(R)\|\|U_l\| + l_1\left\|E_0^{-1}\right\|\left\|\bar T_1\right\|\|Z_1\| + l_1\left\|E_0^{-1}\right\|\left\|\bar T_2\right\|\|Z_2\| + l_2\left\|E_0^{-1}\right\|] \\ &= -\eta\left\|S^TC_2B_2E_0\right\|\end{aligned} \tag{35}$$

Furthermore, the expression (35) can be written as:

$$\dot V_2 \le -\eta\left\|S^TC_2B_2E_0\right\| \le -\eta\|C_2B_2E_0\|\|S\| = -\eta\|C_2B_2E_0\|(2V)^{1/2} = -\varphi V^\gamma \tag{36}$$

with $\varphi = 2^{1/2}\eta\|C_2B_2E_0\| > 0, \gamma = 1/2 < 1$.

Thus, according to Lemma 1, it follows from (36) that the system (12) will reach the sliding mode $S = 0$ within a finite time $t_r$ given by:

$$t_r = \frac{V^{1-\gamma}(0)}{\varphi(1-\gamma)} \tag{37}$$

□.

*Remark 3* It is easily obtained from Theorem 1 and Theorem 2 that the system state $Z_1(t)$ and $Z_2(t)$ reach the equilibrium points in finite time $t = t_s + t_r$ from any initial states. That implies that the system states $X(t)$ will converge to the equilibrium points within finite time.

*Remark 4* It can be seen from (13 to 15) and (27 to 29) that the proposed SMC method in this work involves some exponent and logarithmic computations. Besides, the state transformation (10) and the right inverse matrix (17) are required. Apparently, these computations don't require any complex computing techniques and are simpler compared with some existing methods, e.g. linear matrix inequality.

*Remark 5* It should be pointed out that the transformation matrix $T$ needs to ensure that the block matrix $A_{12}$ is full row rank, which may bring some difficulty for the choice of matrix $T$.

*Remark 6* It should be pointed out that although both this work and Wang (2014) are concerned with the uncertain MIMO-controlled systems, the former focuses on the problem of system states' finite-time convergence with actuator faults, and the latter considered the adaptive tracking control problem under the assumption that the actuators work normally. This also results in the method in Wang (2014) that cannot be utilized to deal with the problem in the present work.

## 4. Simulation example

Consider the high-order MIMO system (1) with parameters as

$$A = \begin{bmatrix} 0 & 0 & 0 & 1 \\ 5 & -5 & -4 & 1 \\ -1 & 1 & 0 & 0 \\ -4 & 0 & 0 & -5 \end{bmatrix}, B = \begin{bmatrix} 0 & 0 \\ 4 & 0 \\ 0 & 0 \\ 0 & 3 \end{bmatrix}, F = \begin{bmatrix} 0 \\ 4.8\sin t \\ 0 \\ 3.6\cos t \end{bmatrix}$$

$$\Delta A = \begin{bmatrix} 0 & 0 & 0 & 0 \\ 0.8\sin t & 1.6\cos t & 4\sin 2t & 1.6\cos\frac{t}{2} \\ 0 & 0 & 0 & 0 \\ 1.2\sin t & 1.8\cos 2t & 0.6\sin t & 0.6\cos\frac{t}{2} \end{bmatrix}$$

And the boundaries of uncertainties and disturbances are given as $l_1 = \sqrt{1.96}$ and $l_2 = \sqrt{2.88}$. It is assumed that the actuator fault parameters $\underline{e}_1 = 0.8, \bar{e}_1 = 1.2, \underline{e}_2 = 0.6, \bar{e}_2 = 1.4$.

Define the state transformation as:

$$Z = TX = \begin{bmatrix} 0 & 0 & 1 & 0 \\ 1 & 0 & 0 & 0 \\ -1 & 1 & 0 & 0 \\ 0 & 0 & 0 & 1 \end{bmatrix} X$$

we have

$$A_{11} = \begin{bmatrix} 0 & 0 \\ 0 & 0 \end{bmatrix}, A_{12} = \begin{bmatrix} 1 & 0 \\ 0 & 1 \end{bmatrix}, A_{21} = \begin{bmatrix} -4 & 0 \\ 0 & -4 \end{bmatrix}, A_{22} = \begin{bmatrix} -5 & 0 \\ 0 & -5 \end{bmatrix}, B_2 = \begin{bmatrix} 4 & 0 \\ 0 & 3 \end{bmatrix},$$

$$\bar{T}_1 = \begin{bmatrix} 0 & 1 \\ 0 & 1 \\ 1 & 0 \\ 0 & 0 \end{bmatrix}, \bar{T}_2 = \begin{bmatrix} 0 & 0 \\ 1 & 0 \\ 0 & 0 \\ 0 & 1 \end{bmatrix}, N_1 = \begin{bmatrix} 0.2\sin t & 0.4\cos t & \sin 2t & 0.4\cos\frac{t}{2} \\ 0.4\sin t & 0.6\cos 2t & 0.2\sin t & 0.2\cos\frac{t}{2} \end{bmatrix},$$

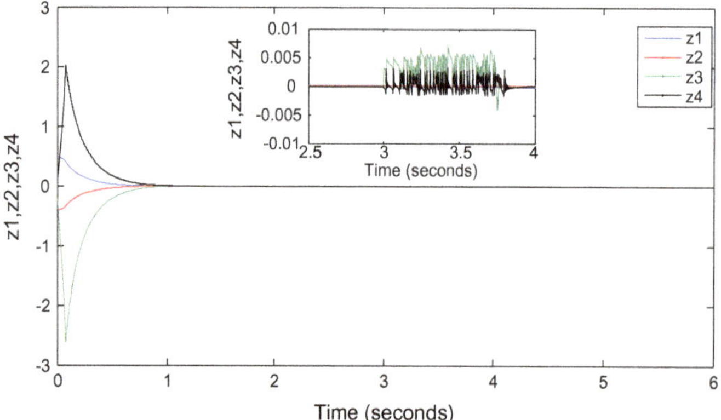


**Figure 1. The trajectories of system states $z_1, z_2, z_3, z_4$.**

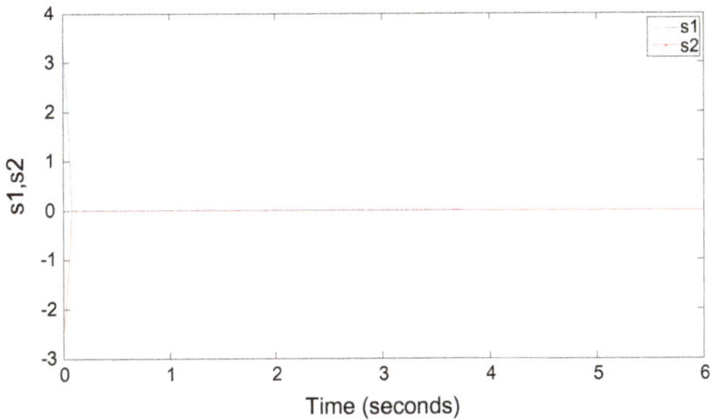


**Figure 2. The trajectories of sliding variables $s_1$, $s_2$.**

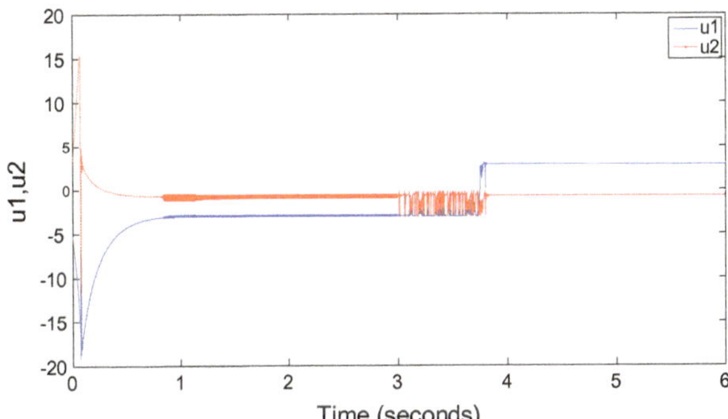


**Figure 3. The control signals $u_1$, $u_2$.**

$$N_2 = \begin{bmatrix} 1.2\sin t \\ 1.2\cos t \end{bmatrix}$$

According to Theorem 1, the terminal sliding mode parameters are designed as follows:

$$C_1 = \begin{bmatrix} 0 & 0 \\ 0 & 0 \end{bmatrix}, C_2 = \begin{bmatrix} 1 & 0 \\ 0 & 1 \end{bmatrix}, C_3 = \begin{bmatrix} 2 & 0 \\ 0 & 2 \end{bmatrix}, C_4 = \begin{bmatrix} 2 & 0 \\ 0 & 2 \end{bmatrix}, p_1 = p_2 = 11, q_1 = q_2 = 9.$$

In order to illustrate the performance of the proposed reliable SMC in this work, it is assumed that the actuator faults happen from 0.3 s with fault factors $e_1(t) = 0.9$ and $e_2(t) = 0.8$.

Simulation results are shown in Figures 1–3. It can be seen from Figure 1 that from 0 to 3 s (as actuator in fault free case), system states converge to the equilibrium points within finite time; after 3 s (as actuator in fault case), system states have a large-range change firstly, and then converge to the equilibrium points in finite time under the designed robust control.

## 5. Conclusion

In this work, the problem of TSMC for uncertain high-order MIMO systems subject to actuator faults has been considered. By means of exponent-logarithmic TSM technique, a reliable SMC law has been designed to attain the finite-time convergence of the closed-loop system, despite actuator faults. However, it should be pointed out that the underlying systems in this work only involve actuator faults. In practical applications, especially, in the network-based environment, there may exist time delay, packet dropouts, and quantization, which need to be further considered in future research, and some novel techniques may be required as in (Dong, Wang, Ding, & Gao, 2014, 2015; Dong, Wang, & Gao, 2013; Ding, Wang, Shen, & Dong, 2015).

**Funding**
This work was supported by NNSF from China [grant number 61273073].

**Author details**
Jiaye Fu[1]
E-mail: 940324037@qq.com
ORCID ID: http://orcid.org/0000-0002-7038-3233
Yugang Niu[1]
E-mail: acniuyg@ecust.edu.cn
[1] Key Laboratory of Advanced Control and Optimization for Chemical Process (East China University of Science & Technology), Ministry of Education, Shanghai 200237, China.

## References

Bateman, F., Noura, H., & Ouladsine, M. (2011). Fault diagnosis and fault-tolerant control strategy for the aerosonde UAV. *IEEE Transactions on Aerospace and Electronic Systems, 47*, 2119–2137. doi:10.1109/TAES.2011.5937287

Chen, B., Niu, Y., Zou, Y., & Jia, T. (2013). Reliable sliding-mode control for markovian jumping systems subject to partial actuator degradation. *Circuits Systems and Signal Processing, 32*, 601–614. doi:10.1007/s00034-012-9468-0

Ding, D., Wang, Z., Shen, B., & Dong, H. (2015). Envelope-constrained H∞ filtering with fading measurements and randomly occurring nonlinearities: The finite horizon case. *Automatica, 55*, 37–45. doi:10.1016/j.automatica.2015.02.024

Dong, H., Wang, Z., Ding, S., & Gao, H. (2014). Finite-horizon estimation of randomly occurring faults for a class of nonlinear time-varying systems. *Automatica, 52*, 355–362. doi:10.1016/j.automatica.2014.10.026

Dong, H., Wang, Z., Ding, S., & Gao, H. (2015). Finite-horizon reliable control with randomly occurring uncertainties and nonlinearities subject to output quantization. *Automatica, 50*, 3182–3189. doi:10.1016/j.automatica.2014.11.020

Dong, H., Wang, Z., & Gao, H. (2013). Distributed H∞filtering for a class of Markovian jump nonlinear time-delay systems over lossy sensor networks. *IEEE Transactions on Industrial Electronics, 60*, 4665–4672. doi:10.1109/TIE.2012.2213553

Feng, Y., Han, X., & Wang, Y. (2007). Second order terminal sliding mode control of uncertain multivariable systems. *International Journal of Control, 80*, 856–862. doi:10.1080/00207170601185046

He, M., Liu, Y., Liu, L., Liu, G., & Liu, H. (2008). A novel fuzzy neural network approximator with exponential fast terminal sliding mode. *Word Congress on Intelligent Control and Automation*, 4736–4740. doi:10.1109/WCICA.2008.4593689

Hu, Q. (2010). Robust adaptive sliding-mode fault-tolerant controlwith$L_2$-gain performance for flexible spacecraft using redundant reaction wheels. *IET Control Theory & Applications, 4*, 1055–1070. doi:10.1049/iet-cta.2009.0140

Hu, Z., Jiang, B., Chen, Y., & Liu X. (2014, August 9–11). An exponent-logarithmic terminal sliding mode control. In *The 25th Chinese Process Control Conference*. Dalian.

Huo, B., Li, Y., & Tong, S. (2012). Fuzzy adaptive fault-tolerant output feedback control of multi-input and multi-output non-linear systems instrict-feedback form. *IET Control Theory & Applications, 6*, 2704–2715. doi:10.1049/iet-cta.2012.0435

Li, P., Ma, J., Zheng, Z., & Geng, L. (2014). Fast nonsingular integral terminal sliding mode control for nonlinear dynamical systems. *Decision and Control*, 4739–4746. doi:10.1109/CDC.2014.7040128

Liang, Y. W., Liaw, D. C., & Lee, T. C. (2000). Reliable control of nonlinear systems. *IEEE Transactions on Automatic Control, 45*, 706–710. doi:10.1109/9.847106

Liao, F., Wang, J. L., & Yang, G. H. (2002). Reliable robust flight tracking control: An LMI approach. *IEEE Transactions on Control Systems Technology, 10*, 76–89. doi:10.1109/87.974340

Lin, C. (2006). Nonsingular terminal sliding mode control of robot manipulators using fuzzy wavelet networks. *IEEE Transactions on Fuzzy Systems, 14*, 849–859. doi:10.1109/TFUZZ.2006.879982

Liu, H., & Li, J. (2009). Terminal sliding mode control for spacecraft formation flying. *IEEE Transactions on Aerospace and Electronic Systems, 45*, 835–846. doi:10.1109/TAES.2009.5259168

Liu, M., Cao, X., & Shi, P. (2013). Fault estimation and tolerant control for fuzzy stochastic systems. *IEEE Transactions on Fuzzy Systems, 21*, 221–229. doi:10.1109/TFUZZ.2012.2209432

Niu, Y., & Wang, X. (2009). Sliding mode control design for uncertain delay systems with partial actuator degradation. *International Journal of Systems Science, 40*, 403–409. doi:10.1080/00207720802436265

Qu, Q., Gao, S., Huang, D., Mei, J., & Zhai, B. (2014). Fault-tolerant control for UAV with finite-time convergence. *Chinese Control and Decision Conference*, 2857–2862. doi:10.1109/CCDC.2014.6852660

Seshagiri, S., & Khalil, H. K. (2002). On introducing integral action in sliding mode control. *Decision and Control, 2*, 1473–1478. doi:10.1109/CDC.2002.1184727

Tao, C., Taur, J. S., & Chan, M. L. (2004). Adaptive fuzzy terminal sliding mode controller for linear systems with mismatched time-varying uncertainties. *IEEE Transactions on Systems, Man and Cybernetics, Part B (Cybernetics), 34*, 255–262. doi:10.1109/TSMCB.2003.811127

Vicente, P. V., & Gerd, H. (2001). Chattering free sliding mode control for a class of nonlinear mechanical systems. *International Journal of Robust and Nonlinear Control, 11*, 1161–1178. doi:10.1002/rnc.598

Vidyasagar, M., & Viswanadham, N. (1985). Reliable stabilization using a multi-controller configuration. *Automatica, 21*, 599–602. doi:10.1109/CDC.1983.269643

Wang, C. (2014). Adaptive tracking control of uncertain MIMO switched nonlinear systems. *International Journal of Innovative Computing Information and Control, 10*, 1149–1159. Retrieved from http://www.ijicic.org/ijicic-13-06007.pdf

Wang, C., Yang, C., & Zhang, K. (2007). Terminal sliding mode control for singular systems with unmatched uncertainties. *IEEE International Conference on Control and Automation, 2929–2933*. doi:10.1109/ICCA.2007.4376898

Xu, J., Lee, T., & Pan, Y. (2003). On the sliding mode control for DC servo mechanisms in the presence of unmodeled dynamics. *Mechatronics, 13*, 755–770. doi:10.1016/S0957-4158(02)00062-4

Xu, S. S., & Liu, Y. (2014). Study of Takagi-Sugeno fuzzy-based terminal-sliding mode fault-tolerant control. *IET Control Theory & Applications, 8*, 667–674. doi:10.1049/iet-cta.2013.0535

Yang, G., Wang, J., & Soh, Y. C. (2000). Reliable guaranteed cost control for uncertain nonlinear systems. *IEEE Transactions on Automatic Control, 45*, 2188–2192. doi:10.1109/9.887682

Yu, X., & Man, Z. (2002). Fast terminal sliding-mode control design for nonlinear dynamical systems. *IEEE Transactions on Circuits and Systems, 49*, 261–264. doi:10.1109/81.983876

Yu, S., Yu, X., & Man, Z. (2000). Robust global terminal sliding mode control of SISO nonlinear uncertain systems. *Conference on Decision and Control, 3*, 2198–2203. doi:10.1109/CDC.2000.914122

Yu, S., Yu, X., Shirinzadeh, B., & Man, Z. (2005). Continuous finite-time control for robotic manipulators with terminal sliding mode. *Automatica, 41*, 1957–1964. doi:10.1109/ICIF.2003.177408

Zou, A. M., Kumar, K. D., Hou, Z. G., & Liu, X. (2011). Finite-time attitude tracking control for spacecraft using terminal sliding mode and neural network. *IEEE Transactions on Systems, Man, and Cybernetics, 41*, 950–963. doi:10.1109/TSMCB.2010.2101592

# 14

# QNP_SHELL: A computerized tool for improving decision-making skills for nuclear power plant operators

Hassan Qudrat-Ullah[1]*

*Corresponding author: Hassan Qudrat-Ullah, School of Administrative Studies, York University, 4700 Keele Street, Toronto, Ontario, Canada M9V 3K7.
E-mail: hassnq@yorku.ca

Reviewing editor: Zude Zhou, Wuhan University of Technology, China

**Abstract:** Decision-making in complex systems such as nuclear power plants (NPPs) is a difficult task at best. The safety and integrity of many such high-capital cost-intensive installations depend on the operator's capability to correctly diagnose and take appropriate measures to avoid any abnormal operations of an NPP. Therefore, the role of the expert systems in the offline training programs for the operators is ever increasing. In this paper, we describe the development of an expert system, "QNP_SHELL," to assist, offline QNPP operators and plant personnel in a better familiarization to infer the anticipated and foreseen malfunctions from the observed symptoms. QNP_SHELL's inferencing mechanism is of the "Rule-based" type and to search the knowledge base it adopts the "Depth First" technique. The diagnostic performance of the trainee operators using QNP_SHELL on various accidents at QNPP has been found, through both the qualitative and quantitative evaluations, satisfactory.

**Subjects: Information / Knowledge Management; Management Education; Operations Management**

**Keywords: nuclear power plants; operators; training; expert system; fault diagnosis**

## ABOUT THE AUTHOR

Hassan Qudrat-Ullah is an associate professor of management science, at the School of Administrative Studies, York University, Canada. He has over 18 years of university teaching, research, and consulting experience in the US, Canada, Singapore, Norway, UK, Korea, China, Saudi Arabia, Switzerland, and Pakistan. His research focuses on: How people can make better decision in complex tasks? He has published 17 journal articles, 3 books, 8 book chapters, and over 40 papers in refereed conference proceedings. He is an editor-in-chief of *International Journal of Complexity in Applied Science and Technology*. His latest book, *Better Decision Making in Complex, Dynamic Tasks* (Springer: 2014), focuses on the design, development, and applications of computer simulation-based decision support systems. The research reported here, relating to his research on the thematic area of decision-making in complex tasks, describes the development and utility of an expert system for the training of nuclear plant operators.

## PUBLIC INTEREST STATEMENT

Decision-making in complex systems such as nuclear power plants is a difficult task at best. The safety and integrity of many such high-capital cost-intensive installations depend on the operator's capability to correctly diagnose and take appropriate measures to avoid any abnormal operations of a nuclear power plant. Therefore, training of nuclear power plant operator takes the center stage among the possible initiatives for the safe and successful plant operations. It is imperative that training for this kind of complex tasks needs to be specific and focused to the needs of the particular power plant. How should we train these operators? Various tools and programs are used for this purpose. Use of computerized decision support is common. Among such tools, the knowledge-based expert systems are particular useful for the fault diagnosis training tasks. In this paper, we describe the design, development, and use of an expert system, "QNP_SHELL."

## 1. Introduction

Fault diagnosis is a critical skill required for operators of high-stakes complex systems such as nuclear power plants (NPPs). To prevent accidents in NPPs, the operators are required to detect early signs of potential abnormal operations. It requires not only better understanding of operations of the NPP, but also agile diagnostic skills and judgmental decision-making abilities. To develop such skills, expert systems are extensively used in their training programs and protocols (Abu-Khader, 2009; Heo, Chang, Choi, Choi, & Jee, 2005; Lee, 2002; Moshkbar-Bakhshayesh & Ghofrani, 2013; Najdawi, Chung, & Salaheldin, 2008; Naser, 1990; Santhosh et al., 2011). On the development and utility of expert systems, the interested readers are referred to Buchanan's comprehensive review (Buchanan, 1986; Ma & Jiang, 2011). The management of QNPP,[1] a 300 MWe power plant in the category of SMRs (Locatelli, Bingham, & Mancini, 2014), in an effort to develop indigenous capabilities, commissioned this project to develop an expert system for the training of its operators for fault diagnosis of QNPP.

There are many methods available for fault detection and diagnosis in NPPs (Liu, Xie, Peng, & Ling, 2014). We can categorize them into two major types of methods that are applied for transient identification in NPP. They are: (i) comprehensive data-driven techniques [e.g. artificial neural networks, fuzzy logic-based systems, and heuristic techniques (Yong-kuo, Min-jun, Chun-li, & Ya-xin, 2013)] and (ii) model-based methods [e.g. hard computing intensive mathematical models (Angeli, 2010)]. While a fairly large number of applications of data-driven systems have been successfully applied, the application of model-driven systems is very limited [an excellent comparative review is presented in Ma and Jiang (2011) and Moshkbar-Bakhshayesh and Ghofrani (2013)]. In particular, the ability of heuristic techniques not to suffer from the local minima problem makes them suitable for the development of fault diagnosis system. Therefore, consistent with our objective "to develop an expert system for the off-line training of QNPP's operators on fault diagnosis," we adopted the rule-based approach. Figure 1 presents the basic structure of a rule-based expert system, the production system model for problem-solving/decision-making (Newell & Simon, 1972). Compared with traditional mathematical and numerical approaches for fault diagnosis, a rule-based expert system:

(i) processes knowledge expressed in the form of rules and use heuristics to arrive at the conclusion,
(ii) clearly separates knowledge from the processing program (i.e. inference engine),
(iii) traces fully the chain of reasoning behind any conclusion reached and data-set used,
(iv) can deal with incomplete, uncertain, and fuzzy data, and
(v) provides flexibility to change or add new knowledge (Sydenham & Thorn, 2005). In-house availability of knowledge engineers provided additional incentive for the adoption of the rule-based approach.

In this paper, we contribute with the design, development, and assessment of such a training tool, QNP_SHELL. The main objective of the development of QNP_SHELL is to augment QNPP operators' and plant personnel's training by making expert systems available to them to assist them offline, to

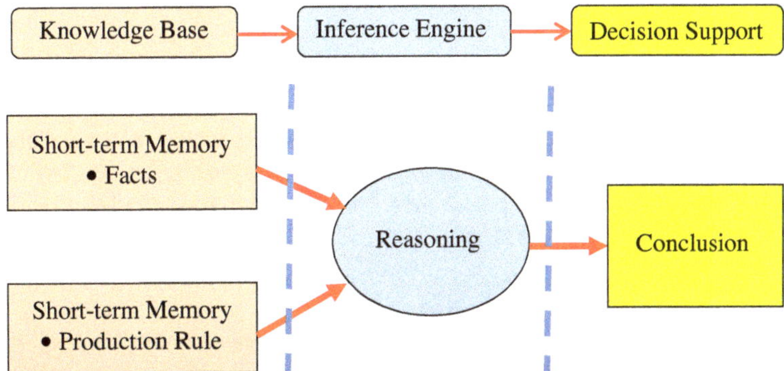


**Figure 1. The production rule-oriented problem-solving model.**

become more familiar with anticipated and unforeseen transients of the plant. The principal benefit attained will be the improved productivity of fault diagnostic types of expert systems. Instead of buying an off-the-shelf expensive solution (e.g. VP-expert) or developing various expert systems to assist on offline basis in diagnosing various types of faults, a general inferencing mechanism, independent of any domain knowledge (Francis Cheong Yiu Fung, 1989; Vinod, Babar, Kushwaha, & Raj, 2002), we have designed, developed, and tested an expert system shell, QNP_SHELL.

To utilize this problem-independent expert system shell, the knowledge about various faults to be diagnosed is to be organized and coded into various files. The QNP_SHELL can then be directed during a consultation session, to access knowledge from the specified files. Therefore, the same QNP_SHELL can be used to study different types of diagnostics by directing it to consult different knowledge base files. This use of QNP_SHELL facilitates the diagnostic study and, as a result, various types of fault diagnosis are made readily accessible to the plant personnel to become better familiar with the plant transient and fault diagnoses. The utilization of the developed QNP_SHELL will help in reducing the cost and man hours on the design and development of expert systems to assist plant personnel in offline fault diagnoses (Naser, 1990; Negnevitsky, 2005). The efficiency is achieved by cutting the job down to the only requirement of coding and organizing the knowledge of specific domains into files.

In Section 2, we describe the methodology including (i) program structure and (ii) system description. The knowledge base development scheme is presented in Section 3. Section 4 presents the testing and evaluation of the developed QNP_SHELL. Finally, Section 5 concludes this paper with the highlights on potential future research.

## 2. Methodology

### *2.1. A brief account of relevant theoretical and practical perspectives*

Fault diagnosis in technical systems such as NPPs in general and the design, development, and use of expert systems in particular have received considerable theoretical and practical attention over the past several decades (Angeli, 2010; Li, Upadhyaya, & Perillo, 2012; Locatelli et al., 2014). In expert systems research, regardless of the purpose of an expert system (e.g. in our case the purpose is the fault diagnosis), the fundamental organization principle is the separation of the knowledge base and the program structure (i.e. inference engine) (Buchanan, 1986). In the design process of an expert system, this fundamental principle not only results in a flexible and modular expert system, but also enhances the productivity of designers/builders of expert systems—they don't have to worry about the development of different data structures and algorithms.

In fact, rule-based expert systems have been successfully applied in fault detection, monitoring, and diagnosis of complex systems such as NPPs (Abraham, 2005; Abu-Khader, 2009; Angeli, 2010). These expert systems can explain the reasoning process (e.g. what chain of symptoms leads to a particular fault of a NPP) and deal with uncertainty, which traditional computational methods do not handle (Abraham, 2005). There are three core components of a rule-based expert system—program structure, knowledge base, and user interface (with a dedicated explanation facility).

The program structure of an expert system acts like a brain—handles users' queries and provides casual explanation of events in an effective and efficient way. Two types of inferencing mechanisms are used. Forward chaining process begins with the facts and works forward to arrive at a conclusion. Backward chaining is the process that starts with conclusions [e.g. a specific accident like loss of coolant accident (LOCA) has happened] and works backward to identify the relevant chain of facts (Abraham, 2005; Angeli, 2010; Yong-kuo et al., 2013). QNP_SHELL is based on both the forward and backward chaining processes.

The development of the knowledge base is another critical task in the development of an expert system. The most common data structures generally used for symbolic representations are production rules, semantic networks, frames, predicate logic, and hybrid (Angeli, 2010). With the objective of achieving a relatively higher level of effectiveness and efficiency in the knowledge base of the QNP_SHELL, we have adopted a hybrid approach where production rules and predicate logic are our main data structures. The selection of knowledge acquisition and representation methods primarily depends on the purpose of the expert system (Naser, 1990; Liu et al., 2014). The knowledge acquisition process includes various techniques including interviews, protocol analysis, repertory grid analysis, and automatic induction techniques (Angeli, 2010). However, when it comes to elicitation of expert knowledge, interviews, although time-consuming, are considered as the most popular and widely used form of expert knowledge acquisition (Jiang, 2008; Milton, 2007). Given the specialized and experiential nature of the knowledge of the experts, a knowledge engineering team at QNNP employed interviews as well as utilized the existing archival data to acquire knowledge for the knowledge base of the QNP_SHELL. For knowledge representation, an exhaustive but reliable set of production rules is developed.

Finally, besides the program structure and the knowledge base, a user interface that allows the operators or users of any expert system to interact with its inference engine is an equally critical component of an expert system. The ecological interface design (Vicente, 2002) and HCI design principles (Howie, Sy, Ford, & Vicente, 2000) provide sound theoretical basis for the development of the user interface of any decision support system albeit an expert system. Drawing on HCI design principles, the developed user interface of QNP_SHELL:

- allows user to interact with the QNP_SHELL in a systematic and flexible manner,
- and provides the rich causal explanation of the user's actions.

In the following subsections, both the development and the operational details of the program structure (i.e. inference engine), knowledge base, explanation facility, and user interface of the QNP_SHELL are presented.

### 2.2. Program structure

A modular approach has been adopted during QNP_SHELL program development (Angeli, 2010; Waterman, 1986). The modules developed and incorporated in the design and development of QNP_SHELL can, according to their functionality, be classified as modules that make up the inference engine, the user interface, and the explanation facility. The QNP_SHELL is problem independent and can be utilized for performing various types of diagnoses. The only restriction is that the information should be coded and organized according to a specific format. The programming language used is PROLOG (Townsend, 1989). Figure 1 presents the functional schema of QNP_SHELL.

#### *2.2.1. The inference engine*

Figure 2 presents the schematic of the inference engine of QNP_SHELL. The major modules that constitute the inference engine are (i) a top most repetitive loop, which controls the program execution, (ii) the main diagnostic module that decides the control strategy, invokes the various file manipulation modules and the inferencing modules, and (iii) the inferencing modules that process various inference conditions associated with a fault. To process each different type of inference condition, a different inferencing module is called upon. The invoking of the various inferencing modules is to be determined by the main diagnostic module and the file manipulation modules. These inferencing modules scan the information stored in different files and extract the necessary information as and when required by the main diagnostic module and the information-processing modules including the conflict resolution module. Moreover, these inferencing modules acknowledge the user's response to various queries as well as to the selection of an option from the multi-option list. Additionally, these modules update the dynamic database.

The inference engine accesses the knowledge base in a particular sequence by entering into a dialog with the user (Figure 3 presents its schematic diagram). It attempts to infer new information based on the user's response. It then uses these facts and attempts to match them with the information stored in the knowledge base files. If all the inferences about the existence of a fault stand

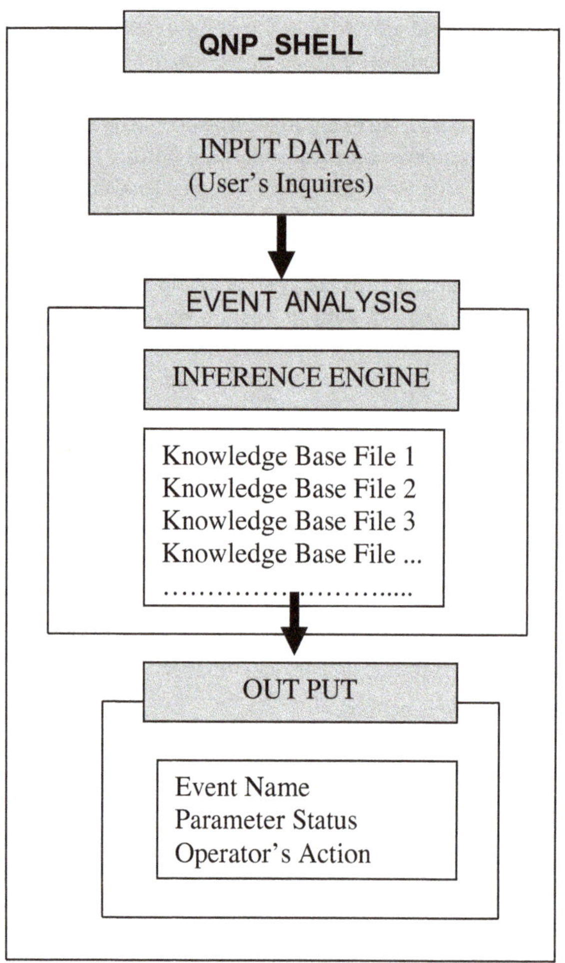


**Figure 2. Functional block diagram of QNP_SHELL.**

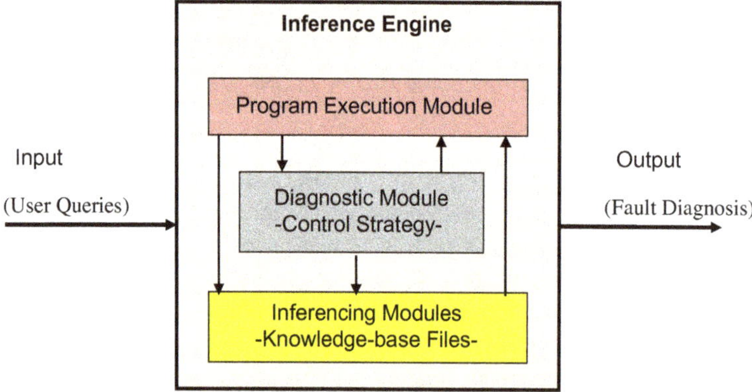


**Figure 3. Schematic of the inference engine of QNP_SHELL.**

true, then the concluding module decides whether the existence of the fault under consideration can be claimed with certainty or its existence can only be suspected. If the inference test to a fault fails then the next fault is selected for analysis.

### *2.2.2. User interface*

To enhance the user friendliness of the system, a minimum of typing has been demanded of the user. Wherever possible, the available choices have been displayed as multiple options through pop-up windows.

Independent modules have been employed to control the prompt of various queries on the monitor screen and to acknowledge the user's response. Whenever the user's response is an affirmation to a query, the string-processing modules are automatically invoked. These modules transfer the query into a grammatically correct sentence. The latter is then displayed on the monitor screen in the form of symptoms. Another independent module caters for alarm generation and their display on the monitor screen.

#### *2.2.3. Explanation facility*

At the user's demand for an explanation to the current state of diagnosis, an independent module is invoked. This module displays the required information (e.g. an accurate causal chain from systems to the respective fault) to the user—*a true learning and knowledge building facility.*

### ***2.3. System description***

#### *2.3.1. The execution of the QNP_SHELL*

On the execution of the QNP_SHELL, a configuration setup module is invoked. The module either by reading a configuration file or through an interactive session with the user, selects:

(i) The knowledge base files to be consulted during the problem diagnosis.
(ii) The dir wherefrom the knowledge base files are to be accessed.

#### *2.3.2. The configuration setup*

Once the user has agreed to the files and to the disk/dir, the configuration setup module disengages itself and a repetitive loop is invoked. This repetitive loop performs the diagnosis by executing the remainder of the program for any number of consultations. The repetitive loop remains intact as long as the user's response is in the affirmative to the query that is put to him at the end of each consultation.

#### *2.3.3. Initialization and fault diagnosis process*

For every execution of the repetitive loop, the necessary variables are initialized and the main diagnostic module is invoked. The main diagnostic module in turn, and at first, invokes the problem set consulting module. The latter makes its access to the file that contains the problem set and reads a term in serial order. The fault as well as the associated list of inferences and confirmations is passed on to the main diagnostic module. The latter sequentially accesses the various condition sets that constitute the list of inferences and passes them to the inferencing mechanism for further analysis. The inferencing mechanism automatically caters for invoking the various inferencing modules depending upon the type of condition set under process. To analyze each different type of condition set, a separate module is triggered by the inferencing mechanism. Based upon a user's responses to the queries contained in a condition set, the inferencing mechanism determines the diagnostic path to be followed.

#### *2.3.4. Backtracking feature*

If during the analysis the inferencing mechanism negates the possibility of existence of a fault, then backtracking occurs. It automatically picks up the term that lies next in serial order and the above analyses are repeated.

#### *2.3.5. The confirmatory test*

If the inferences associated with a fault are true, then the fault under consideration is considered to have been diagnosed. At this stage, the main diagnostic module finally invokes the confirmatory test module. This module attempts to match the facts with the information in the knowledge base files. If an exact pattern matching occurs, the diagnosed fault is displayed on the monitor screen along with a message of certainty to its existence. If, however, an exact pattern matching is not observed, then the display of the fault is accompanied by a message of suspicion about its existence.

## 3. Knowledge base development

### *3.1. The overall functional architecture*

A maximum of five files are required to be developed for use by the QNP_SHELL. A configuration file through which the QNP_SHELL reads the file names, where the user has stored the knowledge base. Four other files are required to be developed. These files contain the knowledge base according to the format as specified in the following sections. Depending on the nature of the domain on which the knowledge base is required, the users of the QNP_SHELL will have to develop all of the four or at least the first two of the following files.

(1) A file containing the PROBLEM SET. Each term of this file contains a problem to be diagnosed and the associated dependencies that lead to affirmations or negation to the existence of the problem. This file is the heart of the knowledge base. Its development cannot be skipped, no matter on which domain the knowledge base is developed.

(2) A file containing the queries to be put to the user during problem analysis. Each term of the file contains a query along with two switches, (i) option/no_option and (ii) alarm/no-alarm. These switches indicate or deny the presence of multiple options and alarms/warnings associated with each query. The associated multiple options will pop up on the monitor screen if the user has affirmed in response to the query. The user will then have to select one of the options. Likewise, the associated alarms will be displayed on the monitor screen, if the user has affirmed to the query. This file is one of the two files of primary importance and cannot be skipped.

(3) A file containing multiple options for all such queries for which the OPTION switch has been set ON by the knowledge engineer.

(4) A file containing alarms/warnings associated to all those queries as well as to the multiple options for which the ALARM switch has been set ON by the knowledge engineer.

### *3.2. Operational description*

During its operation, the QNP_SHELL first makes its access to the file containing the problem set. It then picks up a problem, in serial order along with the condition-dependent sets of query numbers. The program then performs analysis on the condition set in a sequential order. The success or failure of a condition set depends on the nature of the set as well as the user's response to queries, whose numbers are contained in the set. If a condition set is at success, the system will take up the next condition set. If all the condition sets associated with a problem are at success, the problem is considered to be diagnosed. However, the system proceeds further to decide about the confirmation or suspicion to the existence of the problem. If a condition set fails, then the system may take up the next set or it may discard the possibility of the existence of the problem under consideration. While processing a condition set, the program sequentially picks up a query number from the list contained in the condition set and reads the relevant query from the file containing the primary queries. The query is then displayed and the user is supposed to respond accordingly. If the user negates the query, then depending on the type of the condition set, any one of the following situations may occur:

(1) A query with the next query number from the list may be displayed for the user's response.

(2) The program may move over to the next condition set for the same problem.

The program may discard the current problem and may proceed to the next problem in the problem set. If, however, the user affirms to the query, then depending on the ON/OFF conditions of the alarms and multi-option switches, the alarms and/or options may or may not be read from appropriate files.

### *3.3. File format for knowledge base development*

#### *3.3.1. Format for "Diagnstc.kba" containing the problem set*

The file, "Diagnostics.kba" lists all the problem sets in the following format:

> ***faults (Sno,[A list containing a problem to be diagnosed], [A complex list structure containing the primary condition sets], [A complex list structure containing the confirmation sets]).***

As an example, we present the coding scheme for the diagnostic of a problem: Is charging tank lever is decreasing? The command syntax is:

> **faults(2,["Is Charging Tank Level Decreasing"], [and_cond([1,5]), or_cond[([2,6]), and_cond([3,10])be_or([1,2,3])], [must_be([1.2,3.2,4.0]), must_not([2.3,2.0])]).**

The description of this command's variables is as follows:

**(i) Sno**

A unique integral number associated with each of the problems diagnosed (e.g. 2 in this example). The problems in the "Diagnstc.kba" need not be arranged in an ascending/descending order of Sno.

**(ii) A list containing a fault or a group of faults to be diagnosed**

The problem to be diagnosed may contain a single statement describing the problem or it may be a multi statements problem. Each single statement is to be considered as an element of the list. The only limitation is that the number of characters in each statement should not be more than 76.

**(iii) A complex list structure containing the condition sets**

The third variable is a complex list structure. Any one, or all, or any combination of the following conditions can be used. Each of the following condition sets should contain a list of query number as its first value. An empty list is also a valid entry. The following are some examples of the permissible condition sets along with their description:

**(1) AND condition**

Its format is: and_cond ([A list of integral numbers]). For example,

**and_cond([1,l5,10,15]).**

If any query is associated with the query number contained in the list and the user has responded with negation, the "and_cond" will fail. This means that the problem to which the "and_cond" is associated has been discarded from analysis. For any of the "and_cond" set to be true, all the queries in the list must be affirmed.

**(2) OR condition**

The format of this condition is: or_cond ([A list of integral numbers]). For example,

**or_cond([1,2,6,9]).**

It means, no matter what the user has replied, the system will proceed to process the next query number contained in the "or_cond" set. The decision will be taken after analyzing the complete list contained in the "or_cond" set. If the user's response to all the question numbers contained in the list was in negative, then the "or_cond" will fail, which means that the problem to which the "or_cond" is associated has been discarded from analysis. For an "or_cond" to be true, at least one of the queries associated with the corresponding query numbers in the list must be affirmed.

**(iv) A complex list structure containing the confirmation sets**

The fourth variable of the command is also a complex list structure. An empty list is a valid entry.

### *3.3.2. Format for "PRIM.KBA" containing the primary queries*

The following is the general format for of this file:

**prim (Qno,[List containing a query], Option_choice, Alarm_choice)**

As an example, we present the command syntax of a query, "Is there an increase in temperature?":

*prim(1,["Check the Channel Temperature recorder", "Is there an increase in temperatures"], "no_option", "alarm")*

The description of this command's variables is given below.

**(i) Qno**

A unique integral number associated with each of the queries to be put to the user.

**(ii) List containing a query**

A string list containing a query along with its description, if required, is to be inserted for the user's information. The following conditions must be met during file development:

- The query to be put to the user must be the last element of the list.
- An individual element of the list must not contain more than 77 characters.

**(iii) Option choice**

Option_choice = "option"/"no_option"

A flag, associated with each query which indicates or denies the presence of further options corresponding to the query. In the "PRIM.KBA" file, if the flag "option" has been associated with none of the queries, then the file "SEC.PRO" containing the secondary options need not be developed.

**(iv) Alarm choice**

Alarm_choice = "alarm"/"no_alarm"

A flag associated with each query which indicates or denies the availability of alarms associated with the primary query, irrespective of the fact whether secondary options are associated with the query or not. If none of the queries in "PRIM.PRO" has been associated and none of the secondary options is associated, the file "ALARM.PRO" containing the alarms need not be developed.

### *3.3.3. Format for SEC.KBA containing the secondary options*

The file, "SEC.KBA" deals with the secondary options. The format of this file is as:

**sec (Qno, Information-description, [A list of options], alarms_for ([A list containing the option numbers to which alarms are associated])**

Here is an example of this command syntax.

sec(2, "The charging tank level decrease is", ["fast", "gradual", "slow"], alarms_for([1,2]))

The variables of this command syntax are described below.

**(i) Qno**

The same unique query number (an integer) which is associated with the corresponding query in "PRIM.KBA".

**(ii) Information description**

This is a string providing explanation of the options which are to be selected by the user from the following defined list.

**(iii) A list of options**

A list of options that will be displayed to the user and the user will have to select one. The number of characters contained in the option list must be such that the total length of the string should not exceed 76 characters.

**(iv) Alarms for ([A list containing the option numbers to which alarms are associated])**

The term "alarms_for ([...])" contains a list of integers corresponding to the nth option contained in the option list to which certain alarm/warning is associated. If the alarm is not associated with any of the secondary options, the knowledge engineer will have to enter an empty list "[]" within the term "alarms_for(...)".

## 4. Testing and preliminary evaluation of QNP_SHELL

For the utility and performance assessment of QNP_SHELL, we applied a multi-method approach—in addition to the application of the Turing test, the performance of the trainee operators was assessed through statistical validation procedures.

### 4.1. Applying the Turing test

#### 4.1.1. Selection of the testing event

LOCA is the worst kind of accident scenario that one can expect in a NPP (Apostolakis, Kafka, & Mancini, 1988). LOCA accident occurs where there is a leak or break in the primary coolant loop of a PWR reactor such as our QNPP, a 300 MWe plant. If LOCA accident is not identified and mitigation actions are not taken in time, catastrophic consequences including core melt and release of radioactivity to the atmosphere (if containment systems also fails) will happen. Therefore, we decided to test QNP_SHELL on LOCA. The information on LOCA diagnostics, from the written form of the standard procedures, as made available by QNPP management, was coded and organized into four different files. These files were then utilized to have a demonstration of the working QNP_SHELL.

#### 4.1.2. Performance evaluation of QNP_SHELL on LOCA at QNPP

The purpose of this expert system is to advise you, the user, to identify LOCA types in a PWR, when high-pressure injection signal alarm appears in the control room of QNPP.

##### 4.1.2.1. Event definitions based on manual procedures of QNPP.

3.1 LOCA-1 is medium LOCA in the main loop (QMP 14.21.2)

3.2 LOCA-2 is leakage due to pressurizer safety/relief valve open (inadvertently) (QMP 14.23.1)

3.3 LOCA-3 is leakage due to steam generator (SG) tube rupture (primary leak into secondary side) (QMP 14.31)

3.4 LOCA-4 is leakage due to steam line break outside the containment building (QMP 14.11)

3.5 LOCA-5 is leakage due to steam line or feedwater line (inside the containment building) without SG tube rupture (QMP 14.6-14.10)

3.6 LOCA-6 is leakage due to steam line or feedwater line (inside the containment building) with SG tube rupture (QMP 14.21.4)

3.7 LOCASG is an additional logic signal to confirm one of the above break types.

Please note that corresponding to each of above-defined LOCA types, there are separate procedures for safety actions (called "QMP" Manual Procedures in this PWR), which the operator is required to follow after confirming the leakage type. For the benefit of the user, the QMP procedure numbers are also given in the expert system with each break definition.

*4.1.2.2. User's queries within QNP_SHELL expert system.*

| | | |
|---|---|---|
| *ASK* | *HPINSG:* | *"Is there high pressure injection signal?"* |
| CHOICES | HPINSG: | yes, no |
| ASK | PRLVL: | "Is pressurizer level <2.85 m?" |
| CHOICES | PRLVL: | yes, no |
| ASK | PPCL: | "Is pressure in primary coolant loop <110 bar?" |
| CHOICES | PPCL: | yes, no |
| ASK | PDECA: | "Is diff. pressure of equipment compartment w.r.t. atmosphere >30 mbar?" |
| CHOICES | PDECA: | yes, no |
| ASK | NASG: | "Is N16 activity behind SG 1 or 2 becomes >limit?" |
| CHOICES | NASG: | yes, no |
| ASK | NBSG: | "Is N16 activity behind SG 3 or 4 becomes >limit?" |
| CHOICES | NBSG: | yes, no |
| ASK | DIRIC: | "Is direct ionizing radiation level (inside containment > limit?" |
| CHOICES | DIRIC: | yes, no |
| ASK | PRLRT: | "Is pressure level in pressurizer relief tank >12 bar?" |
| CHOICES | PRLRT: | yes, no |
| ASK | PRSVON: | "Is pressure safety/relief valve open?" |
| CHOICES | PRSVON: | yes, no |
| ASK | LOCASG: | "Is corresponding LOCASG logic alarm signal available?" |
| CHOICES | LOCASG: | yes, no |

*4.1.2.3. Rules formulations within the QNP_SHELL expert system (only few rules are presented here as an example).*

| | | |
|---|---|---|
| Rule 1 | IF | HPINSG = yes AND |
| | | PDECA = yes AND |
| | | NBSG = yes AND |
| | | PRLRT = yes |
| | THEN | LOCA type is LOCA-1 (Ref. Definition 3.1, above) |
| Rule 3 | IF | HPINSG = yes AND |
| | | PDECA = yes AND |
| | | NBSG = yes AND |
| | | PRLRT = yes AND |
| | | PRSVON = yes |
| | THEN | LOCA type is LOCA-2 (Ref. Definition 3.2, above) |
| Rule 11 | IF | HPINSG = yes AND |
| | | PDECA = yes AND |
| | | DIRIC = yes AND |
| | | LOCASG = yes |
| | THEN | LOCA type is LOCA-6 (Ref. Definition 3.6, above) |

In a computer-simulated LOCA accident condition, operators using QNP_SHELL expert system were able to identify the type of LOCA correctly. Not only were these operators able to run down the plant to "cold shutdown and depressurized" condition, a required condition as specified by the emergency procedures of QNPP, their dynamic behavior during the control of the accident was judged as "excellent" by the group of experienced plant operators. Therefore, the performance of the QNP_SHELL has been found satisfactory by the experienced plant operators at QNPP, through a procedure known as the Turing test (Spring, 1993).

### *4.2. Statistical validation of QNP_SHELL*

The utility and effectiveness of QNP_SHELL on the diagnostics performance of its users were also assessed on a range of events through a quasi-experimental manner. The management of QNPP has an intensive training program for the operators. In a six-week modular program, it includes classroom instructions based on QNPP's operating manuals (three weeks), exercises and written tests (one week), and simulator-based fault diagnostic practice and learning (two weeks). Compared with the group of trainee operators who did not have access to QNP_SHELL (i.e. prior to the induction of QNP_SHELL), this new group with access to QNP_SHELL had only one week of classroom instructions. The performance of both groups was measured on their ability to correctly identify the faults. First, they had to do a written test consisting of multiple-choice questions on the symptoms of the fault and then their performance was tested on QNPP's simulator that fully replicates QNPP. All the participants were tested on seven events, as listed in Table 1. The order of the events was random. The diagnostic accuracy of the trainee operators, assessed through the written tests, is the same for both groups. Although the total number of errors (i.e. how many faults were not correctly identified by the trainee operator) by Group 1 (those without QNP_SHELL) was higher (i.e. eight versus four) than Group 2 (with QNP_SHELL), the overall diagnostics performance on the seven events did not differ statistically ($F = 1.61$; $p = 0.21$). We ran several *t*-tests for all the events separately and found no statistical difference between both groups (please see the *t*-test results in Tables 2 and 3). Since all the trainee operators correctly identified LOCA, the statistical testing for LOCA was not required.

**Table 1. Trainee participants and list of diagnostic events**

| | Group 1 (without QNP_SHELL) | Group 2 (with QNP_SHELL) | Diagnostic performance events |
|---|---|---|---|
| *N* | 15 | 13 | LOCA: loss of coolant accident |
| Age (average) | 24 years | 25 years | |
| Gender | 12 males; 3 females | 11 males; 2 females | FLBA: feed line break of loop A<br>FLBB: feed line break of loop B |
| Instruction | 3 weeks | 1 week | SGTRA: steam generator A's tube rupture |
| Education | Postgraduates | Postgraduates | SGTRB: steam generator B's tube rupture<br>SLBA: steam line break of loop A<br>SLBB: steam line break of loop B |

**Table 2. Diagnostic performance on FLBA, FLBB, and SGTRA events**

| Event | FLBA | | FLBB | | SGTRA | |
|---|---|---|---|---|---|---|
| Group | Group 1 | Group 2 | Group 1 | Group 2 | Group 1 | Group 2 |
| Mean | 0.13 | 0.08 | 0.13 | 0.076923 | 0.07 | 0 |
| Variance | 0.12 | 0.08 | 0.12 | 0.076923 | 0.07 | 0 |
| Observations | 15 | 13 | 15 | 13 | 15 | 13 |
| Pooled variance | 0.10 | | 0.10 | | 0.04 | |
| *df* | 26 | | 26 | | 26 | |
| *t* Stat | 0.465731 | | 0.465731 | | 0.928571 | |
| $P(T \leq t)$ one-tail | 0.322643 | | 0.322643 | | 0.180826 | |
| *t* Critical one-tail | 1.705618 | | 1.705618 | | 1.705618 | |
| $P(T \leq t)$ two-tail | 0.645286 | | 0.645286 | | 0.361652 | |
| *t* Critical two-tail | 2.055529 | | 2.055529 | | 2.055529 | |

Table 3. Diagnostic performance on SGTRB, SLBA, and SLBB events

| Events | SGTRB | | SLBA | | SLBB | |
|---|---|---|---|---|---|---|
| **Group** | **Group 1** | **Group 2** | **Group 1** | **Group 2** | **Group 1** | **Group 2** |
| Mean | 0.07 | 0 | 0.07 | 0.08 | 0.07 | 0.08 |
| Variance | 0.07 | 0 | 0.07 | 0.08 | 0.07 | 0.08 |
| Observations | 15 | 13 | 15 | 13 | 15 | 13 |
| Pooled variance | 0.04 | | 0.07 | | 0.07 | |
| *df* | 26 | | 26 | | 26 | |
| *t* Stat | 0.928571 | | −0.10129 | | −0.10129 | |
| $P(T \leq t)$ one-tail | 0.180826 | | 0.460047 | | 0.460047 | |
| *t* Critical one-tail | 1.705618 | | 1.705618 | | 1.705618 | |
| $P(T \leq t)$ two-tail | 0.361652 | | 0.920094 | | 0.920094 | |
| *t* Critical two-tail | 2.055529 | | 2.055529 | | 2.055529 | |

These diagnostics performance assessment results were shared with the trainee operators and feedback was provided in two debriefing sessions, one for each group. Two expert operators delivered these debriefing sessions. All the participants were instructed to review QNPP's manuals and prepare for the diagnostics assessment through the QNPP's simulator. The expert operators emphasized the recognition and memorization of each symptom related to the propagation path of each event.

Again, in the simulator-based diagnostic assessment of the trainee operators, the same measure, *diagnostic accuracy*, was used. Indeed, one can argue that the timing measure (i.e. how fast an operator can identify a fault) should have been assessed as well. In fact, when we looked at the computer logs of all the participants, they all had completed each event's diagnosis in less than 80s. The expert operators at QNPP termed such a timing performance quite satisfactory. How about their statistical performance on simulator-based diagnostic testing? Because participants prepared well and were eager to perform—soon they have to apply for licensing, both the groups performed successfully—all of trainee operators identified each of the events correctly.

We recognize the limitations of these evaluations (e.g. small sample size and only one performance measure) of QNP_SHELL. Nevertheless, with the reported results of our qualitative (i.e. Turing testing) and quantitative (i.e. statistical validation) assessments of the diagnostic performance of the trainee operators, we are confident about the continued utility and success of QNP_SHELL at QNPP and elsewhere (e.g. other NPPs of our client). In fact, as the quality of training is the leading factor that shapes the actual performance of the NPP operators (Alvarenga, Frutuoso e Melo, & Fonseca, 2014), we can expect improved decision-making skills by the QNP_SHELL trained operators.

## 5. Concluding remarks

The role of human operators in increasing the efficiency and availability of complex systems such NPPs is critical. Therefore, the operator's training on the accurate identification of the root causes of faults is indispensable. The developed framework system allows the builders of expert systems to focus on the development of knowledge bases without having to develop a new program structure that otherwise is needed to process them. Utilizing this framework, an expert system for fault diagnosis training of the power plant operators at QNPP has been developed as a part of indigenous capability development program of our client. The developed, QNP_SHELL-based, expert system was tested on a range of events (e.g. LOCA, SGTR, FLB, and SLB) that can occur in a PWR reactor. Using the analysis and advice of QNP_SHELL expert system, the operators were able to (i) identify the correct type of event, and (ii) execute the relevant emergency procedures in time—a *raison d'etre* of any diagnostic system. Therefore, the developed expert system can be applied in diagnosing an accident situation like LOCA and act as an additional confirmatory aid to plant operators.

Additionally, the modular architecture of QNP_SHELL allows users to study any kind of fault diagnosis over any domain. The only limitation is that the symptoms associated with the faults to be diagnosed can be broken down into a set of questions.

The QNP_SHELL can diagnose a fault or a group of faults having the same unique symptoms. Once the diagnosis has been performed, it can further analyze the confirmation or suspicion to the existence of the diagnosed results. Provision has been kept to further enhance the capabilities of the QNP_SHELL to accommodate the analysis of the group of faults that may have the same symptoms but different confirmation requirements. Furthermore, the capabilities of the QNP_SEHLL can be enhanced to base the predictions about the existence of fault on probabilistic evaluations. In the next phase of the research project, we intend to develop a web-enabled version of QNP_SHEL enabling simultaneous multi-site training sessions for the NPP operators.

**Acknowledgments**
The author would like to thank the two respected anonymous reviewers for their useful comments and critique. Also, an earlier version of this paper (a short paper of five pages) was accepted for presentation and publication in the proceedings of the conference, ISMS2012, Malaysia.

**Funding**
The author received no direct funding for this research.

**Author details**
Hassan Qudrat-Ullah[1]
E-mail: hassnq@yorku.ca
ORCID ID: http://orcid.org/ 0000-0002-8018-2363
[1] School of Administrative Studies, York University, 4700 Keele Street, Toronto, Ontario, Canada M9V 3K7.

**Note**
1. At the request of our client, we have used anonymous names, QNP_SHELL and QNPP, as referenced in this paper.

**References**

Abraham, A. (2005). Rule-based expert systems. In H. Sydenham & R. Thorn (Eds.), *Handbook of measuring system design* (pp. 909–919). New York, NY: Wiley.

Abu-Khader, M. (2009). Recent advances in nuclear power: A review. *Progress in Nuclear Energy, 51*, 225–235. http://dx.doi.org/10.1016/j.pnucene.2008.05.001

Alvarenga, M. A. B., Frutuoso e Melo, P. F., & Fonseca, R. A. (2014). A critical review of methods and models for evaluating organizational factors in human reliability analysis. *Progress in Nuclear Energy, 75*, 25–41. http://dx.doi.org/10.1016/j.pnucene.2014.04.004

Angeli, C. (2010). Diagnostic expert systems: From expert's knowledge to real-time systems. In P. Sajja & R. Akerkar (Eds.), *Advanced knowledge based systems: Model, applications & research* (Vol. 1, pp. 50–73). Kolhapur: Technomathematics Research Foundation.

Apostolakis, A., Kafka, P., & Mancini, G. (1988). *Accidence sequence modeling*. London: Elsevier Applied Science.

Buchanan, B. G. (1986). Expert systems: Working systems and the research literature. *Expert Systems, 3*, 32–50. http://dx.doi.org/10.1111/exsy.1986.3.issue-1

Francis Cheong Yiu Fung, F. (1989). Framework for building rule-based machine diagnostic expert systems. *Knowledge-Based Systems, 2*, 228–238. http://dx.doi.org/10.1016/0950-7051(89)90067-1

Heo, G., Chang, S., Choi, S., Choi, G., & Jee, M. (2005). Advisory system for the diagnosis of lost electric output in nuclear power plants. *Expert Systems with Applications, 29*, 747–756. http://dx.doi.org/10.1016/j.eswa.2005.06.002

Howie, E., Sy, S., Ford, L., & Vicente, K. J. (2000). Human-computer interface design can reduce misperceptions of feedback. *System Dynamics Review, 16*, 151–171. http://dx.doi.org/10.1002/(ISSN)1099-1727

Jiang, H. (2008). Study on knowledge acquisition techniques. In *The Proceedings of 2nd International Symposium on Intelligent Information Technology Applications* (pp. 181–185). Washington, DC.

Lee, M. (2002). Expert system for nuclear power plant accident diagnosis using a fuzzy inference method. *Expert Systems, 19*, 201–207. http://dx.doi.org/10.1111/exsy.2002.19.issue-4

Li, F., Upadhyaya, B. R., & Perillo, S. R. P. (2012). Fault diagnosis of helical coil steam generator systems of an integral pressurized water reactor using optimal sensor selection. *IEEE Transactions on Nuclear Science, 59*, 403–410. http://dx.doi.org/10.1109/TNS.2012.2185509

Liu, Y., Xie, C., Peng, M., & Ling, S. (2014). Improvement of fault diagnosis efficiency in nuclear power plants using hybrid intelligence approach. *Progress in Nuclear Energy, 76*, 122–136. http://dx.doi.org/10.1016/j.pnucene.2014.05.001

Locatelli, G., Bingham, C., & Mancini, M. (2014). Small modular reactors: A comprehensive overview of their economics and strategic aspects. *Progress in Nuclear Energy, 73*, 75–85. http://dx.doi.org/10.1016/j.pnucene.2014.01.010

Ma, J., & Jiang, J. (2011). Applications of fault detection and diagnosis methods in nuclear power plants: A review. *Progress in Nuclear Energy, 53*, 255–266. http://dx.doi.org/10.1016/j.pnucene.2010.12.001

Milton, N. (2007). *Knowledge acquisition in practice: A step by step guide*. London: Springer Verlag.

Moshkbar-Bakhshayesh, K., & Ghofrani, M. B. (2013). Transient identification in nuclear power plants: A review. *Progress in Nuclear Energy, 67*, 23–32. http://dx.doi.org/10.1016/j.pnucene.2013.03.017

Najdawi, M., Chung, Q. B., & Salaheldin, S. (2008). Expert systems for strategic planning in operations management: A framework for executive decisions. *International Journal of Management and Decision Making, 9*, 310–327. http://dx.doi.org/10.1504/IJMDM.2008.017412

Naser, J. (1990). *Expert systems applications for the electric power industry*. Singapore: Taylor & Francis.

Negnevitsky, M. (2005). *Artificial intelligence: A guide to intelligent systems* (2nd ed.). Englewood Cliffs, NJ: Addison Wesley.

Newell, A., & Simon, H. (1972). *Human problem solving.* The University of Michigan: Prentice-Hall.

Santhosh, T. V., Kumar, M., Thangamani, I., Srivastava, A., Dutta, A., Verma, V., ..., Ghosh, A. K. (2011). A diagnostic system for identifying accident conditions in a nuclear reactor. *Nuclear Engineering and Design, 241,* 177–184. http://dx.doi.org/10.1016/j.nucengdes.2010.10.024

Spring, G. (1993). Validating expert system prototypes using the Turing test. *Transportation Research Part C: Emerging Technologies, 1,* 293–301. http://dx.doi.org/10.1016/0968-090X(93)90003-X

Sydenham, H., & Thorn, R. (2005). *Handbook of measuring system design.* New York, NY: Wiley. http://dx.doi.org/10.1002/0471497398

Townsend, C. (1989). *Introduction to turbo prolog.* San Francisco, CA: Sybex.

Vicente, K. J. (2002). Ecological Interface design: Progress and challenges. *Human Factors: The Journal of the Human Factors and Ergonomics Society, 44,* 62–78. http://dx.doi.org/10.1518/0018720024494829

Vinod, S., Babar, A., Kushwaha, H., & Raj, V. (2002). Symptoms based diagnostic system for nuclear power plant operations using artificial neural networks. *Reliability Engineering & System Safety, 82,* 33–40.

Waterman, D. (1986). *A guide to expert systems.* Boston, MA: Addison-Wesley.

Yong-kuo, L., Min-jun, P., Chun-li, X., & Ya-xin, D. (2013). Research and design of distributed fault diagnosis system in nuclear power plant. *Progress in Nuclear Energy, 68,* 97–110. http://dx.doi.org/10.1016/j.pnucene.2013.06.002

# 15

# Soil salinity and moisture measurement system for grapes field by wireless sensor network

M.K. Bhanarkar[1]* and P.M. Korake[1]

*Corresponding author: M.K. Bhanarkar, Communication Electronics Research Laboratory, Department of Electronics, Shivaji University, Kolhapur 416004, India

E-mail: mkb_eln@unishivaji.ac.in

Reviewing editor: Shashi Dubey, Hindustan College of Engineering, India

**Abstract:** Soil moisture and salinity measurement are the essential factors for crop irrigation as well as to increase the yield. Grapes eminence depends on the water volume contents in soil and soil nutrients. Based on these conditions, we determined water demand for best quality of grapes by wireless sensor network (WSN). Using lot of chemical fertilizers increases soil salinity but reduces soil fertility, soil salinity defines electrical conductivity or salty soil. Precise agriculture systems are integrated with multiple sensors to monitor and control the incident. Integrated WSN is designed and developed to measure soil moisture and salinity. ATmega328 microcontroller, XBee and Soil sensors are integrated across the system. This system is more competent, it can be helpful to automatic irrigation system and soil salinity monitoring.

**Subjects: Engineering & Technology; Environment & Agriculture; Food Science & Technology; Physical Sciences**

**Keywords: wireless sensor network; micro-controller ATmega328; XBEE; radio communication; communication protocol; signals processing; environmental sustainability engineering**

## 1. Introduction

Soil content measurement is the major part in agriculture research. In rainy season, all nutrients of soil are carried away because of water flow and soil erosion in summer season, it is harmful for the growth of crop. Soil moisture and salinity-related crops yield (Katerji, Van Hoornb, Hamdyc, & Mastrorillid, 2003),

## ABOUT THE AUTHORS

M.K. Bhanarkar received his MSc in Physics with Electronics specialization in 2001 and PhD in Physics with molecular study by time domain and frequency domain techniques in 2007 from Dr. Babasaheb Ambedkar Marathwada University, Aurangabad, INDIA. His research areas are antenna design, Wireless sensor network, smart antenna system for bio-medical applications. He has published 55 + papers in journals and conferences. He is an editorial member and reviewer of various journals. He is working as an Assistant Professor, Department of Electronics, Shivaji University, Kolhapur, India.

At my group, P.M. Korake is a PhD research scholar who is working on wireless sensor network for agricultural interest. He received his BSc and MSc in Electronics from Department of Electronics Solapur University, Solapur, Maharashtra, India. Currently he is pursuing PhD at Department of Electronics, Shivaji University, Kolhapur.

## PUBLIC INTEREST STATEMENT

Suggested wireless sensor system will be useful for farmer, having low cost and easily can be operated. It is mostly useful for grapes forming where there is lack of water. It will be more beneficial to handle and carry because of compact size and low weight.

water management as well as soil nutrients are essential for growth and quality of grapes. Grapes are grown in the different area across India, but each area has different soil conditions and quality as well. Some important factors are affecting the quality of grapes and those are also based on the soil.

To know, some important principles for the study of grapes are most essential i.e. symptoms on the vine of water anxiety and physiological stages i.e. size and quality of the grapes, depth, and size of the root, because of lack of water; all these factors depend on water storage capacity of the soil (Coggan, 2002). Mostly, water management depends on types of soil such as alluvial sandy soil with different layers of sand and silt. If clay content having less than 5% and water retention capacity is less than 50 $mm^3/mm^3$, then it has to irrigate every 7 days. If well-drained red soil of high physical potential, but with low pH and clay is equal to 15–20%, then it has to irrigate every 20 days. The quality of grapes depends on the type of water supply; water supply can be from different resources such as gravitational water, free drainage, rainfall, and available water (Holzman, Rivas, & Piccolo, 2014).

## 2. Literature

### *2.1. Physiological stages*

After harvesting of grapes, building of reserves for next season period of active root growth maintain adequate moisture levels, but do not encourage any active growth that can utilize reserves (very dangerous for freeze damage), leaves should stay active only, manage the ripening of the shoots for protection against winter chilling.

### *2.2. Practical methods for evaluating soil moisture*

Various systems can be used to evaluate the soil moisture, important facts that to know the rooting depth (make profile pit) by Tensiometers, Neutron Probe, and Resistance Blocks.

There are many automatic irrigation systems such as full surface wetting flood systems; it covers total soil area, overhead sprinkler systems, and micro jets are concentrated wetting such as drip system. Water penetration time depends on the type of irrigation system as well as rate of water penetration depends on soil type and volume of water applied. Water is irrigated which depends upon the physiological stages of soil type (water holding capacity), climate (temperature, relative humidity, wind), type of irrigation (flood, sprinkler, drip), and water quality. Most important thing is availability of water (how much? and when available?). Extraction of water is 40% from the top and 25% of rooting zone for planting (Coggan, 2002).

### *2.3. Practical methods to decide timing and size of irrigation*

There are certain methods in estimation of soil water holding capacity, plant factors, depletion level of soil water, effectiveness of irrigation system, and water volume.

#### *2.3.1. Water quality*

Grape vines have moderate to good resistance with salinity. Water for irrigation containing saline salts or elements has direct and indirect effects on the plant. In field, direct effect is uptake nutrients and indirect effect is the physical effect on the soil.

#### *2.3.2. Soil sensors*

Soil sensors that measure necessary soil properties. These sensors can be used to control conflicting rate application equipment in real-time soil parameter measurement. There are many types of sensors like electromagnetic, optical, mechanical, electrochemical, airflow, and acoustic for different applications such as industrial, defense, agriculture, medical, etc.

#### *2.3.3. Measuring soil properties*

In ideal precision agriculture system, sensor is connected directly with a "black box" which can read the data, processes it and controls output changes immediately. The system contains real-time signal detected by the sensor.

(1) Some sensors and controllers systems require a certain time for measurement, integration, and amendment, which decreases the operation speed.

(2) Design of treatment algorithms.

### 3. Sensor data-based systems

Today, various systems that are vehicle-based soil sensors are available; only electromagnetic sensors are commercially available and commonly used. Ideally, developer would like to operate sensors that provide inputs for existing treatment algorithms. Instead, commercially available sensors provide measurements such as electrical conductivity that cannot be used directly since the absolute value depends on a number of physical and chemical soil properties such as texture, organic matter, salinity, moisture content. On the other hand, electromagnetic sensors provide precious information about soil differences and similarities, which makes it possible to divide the field into smaller and relatively reliable areas referred to as management zones (De Benedetto, Castrignanò, Quarto, 2013).

As new active soil sensors available, different real-time and map-based system treatments may be economically applied for much smaller field areas, reducing the effect of soil contradiction within each management zone.

#### *3.1. Wireless sensor network (WSN)*

Wireless sensor network (WSN) have many applications such as industrial automation, automated and smart homes video surveillance, traffic monitoring, medical device monitoring, monitoring of weather conditions, air traffic control, robot control, personal body area network (Majone et al., 2013).

Different types of communication topologies are available for WSN i.e. mesh, star, etc. But in distributed sensor network, data sharing is not centralized because such WSN networks consist if central node is collapsed then entire network will be collapsed. It will help to better collection of data, to provide nodes with backup in case the failure of the central node.

*Characteristics of WSN* (Shruthi, Shaila, Venugopal, & Patnaik, 2015):

(1) Power consumption constrains for node using battery.

(2) Mobility of nodes.

(3) Scalability to large scale of exploitation.

(4) Capability to stand with inclement environment conditions.

(5) Ease to use.

(6) Cross-layer design.

We have developed such wireless sensor node for soil moisture monitoring of crops. This system is energy efficient, low cost, portable, and small size. From soil, sensor read real-time moisture level is displayed in analog and digital form. Same sensor is also used for measurement of soil conductivity (salinity).

### 4. WSN architecture

It is wireless automatic distributed sensor network. It has multiple sense stations called sensor nodes. It is operated on IEEE 802.15.4 protocol and 2.4 GHz frequency. We can also use 6LoWPAN which stands for IP^6 protocol for Internet of things (IoT). Sensed data send wirelessly to router and then to coordinator. Every sensor node is equipped with sensor, microcontroller, XBee, and power supply.

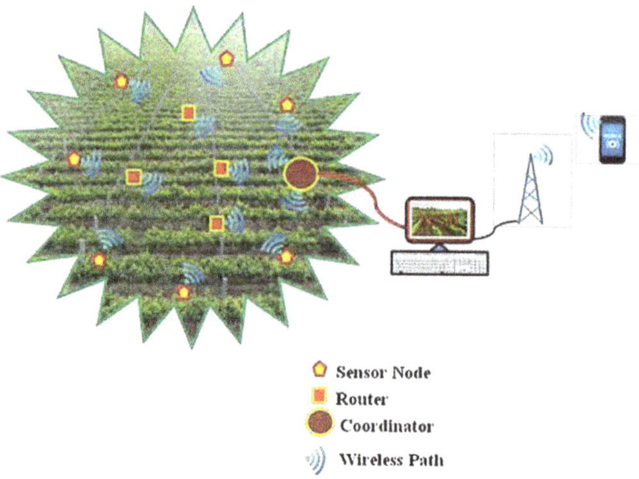


**Figure 1. Architecture of WSN.**

For grapes quality monitoring, we have developed Soil moisture or conductivity sensing WSN system. Figure 1 shows that architecture of WSN for grapes quality monitoring. Also end device (WSN node or mote), routers (mediator), and coordinator (gateway) are shown in Figure 1 with wireless communication mode.

### *4.1. WSN node structure*

#### *4.1.1. Soil moisture sensor with interface module*

A simple water sensor can be used to detect soil moisture. When the soil moisture module detects low moisture, outputs will be high level. The sensing level is adjustable using a small potentiometer. It has both analog and digital outputs using a LM393 comparator chip. It requires the operating voltage 3.3–5 V and some extra features power indicator (red) and digital output indicator (green). The module is basically testing resistance between the input terminals. It can be used for sensing soil moisture, or can be a "raindrop" detector if you make many close-together wires on an insulator. Long-term stability as soil sensor will be better with two stainless steel probes such as cooking skewers, etc.

#### *4.1.2. ATmega 328 microcontroller*

It is 8-bit high-performance low power microcontroller. It is advanced RISC architecture, high endurance non-volatile memory segments; it has many peripheral features and operating voltage 1.8–5 V. It has three power consumption modes i.e. active mode, power mode, and power save mode. Operating temperature range is −40°C to 85°C. It has 32 KB flash, 1 KB EEPROM, and 2 KB RAM.

#### *4.1.3. XBee*

XBee is the brand name of Digi International for a family of the form factor compatible radio module. Such communication is wireless IEEE 802.15.4 based protocol. All XBee's are available in 20 pin packages and two form factors surface mount (SMT) and through the hole. Typical antenna modules have certain antenna options.

XBee operates still in a transparent data mode or in a packet-based application programming interface (API) mode. In the transparent mode, data coming into the Data IN (DIN) pin are directly transmitted over-the-air to the intended receiving radios without any modification. Incoming packets can either be directly addressed to one target (point-to-point) or broadcast to multiple targets (star). This mode is primarily used in instances where an existing protocol cannot tolerate changes to the data format. AT commands are used to control the radio's settings. In API mode, the data are wrapped in a packet structure that allows for addressing, parameter setting, and packet delivery feedback, including remote sensing and control of digital I/O and analog input pins. XBee programming uses software tool XTU. It is a special tool for programming of ZigBee different modes.

## 5. Simulation survey of WSN

WSN simulation is a significant role for WSN development. Protocols, schemes, even new ideas can be evaluated in a very large range in simulation study. WSN simulators permit users to isolate different factors by tuning configurable parameters. This section illustrates using several simulation tools like NS-2, NS-3, TOSSIM, EmStar, OMNeT++, J-Sim, ATEMU, and Avrora; and can analyze the advantage and disadvantage of each simulation tool (Yu, 2011; Yu, Wu, Han, & Zhang, 2013). Today's most of WSN simulation is done in NS-3. It is network simulator series 3 which is discrete event computer network simulator and WSN simulator.

## 6. Development of WSN system

Soil moisture sensor reads the analog data and provides to the on-chip ADC of ATmega 328. Those data are converted and it transmits serially through XBee. The entire system contains WSN node as shown Figure 1, it shows the architecture of WSN. Figure 2 indicates wireless node architecture. Figure 3 shows WSN node circuit diagram. Those data are sensed by sensor node and collecting at the coordinator.

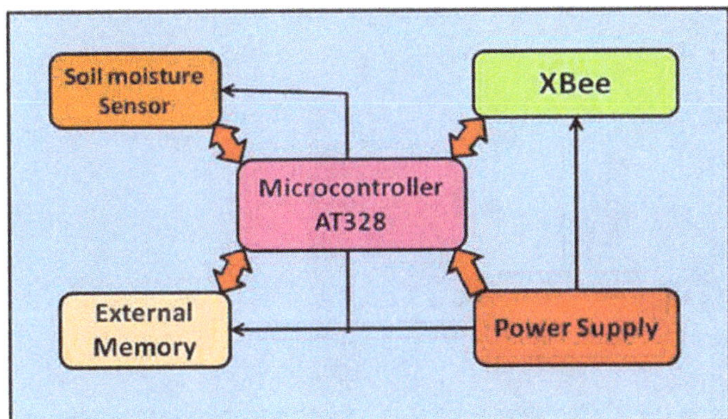


**Figure 2. WSN node architecture.**

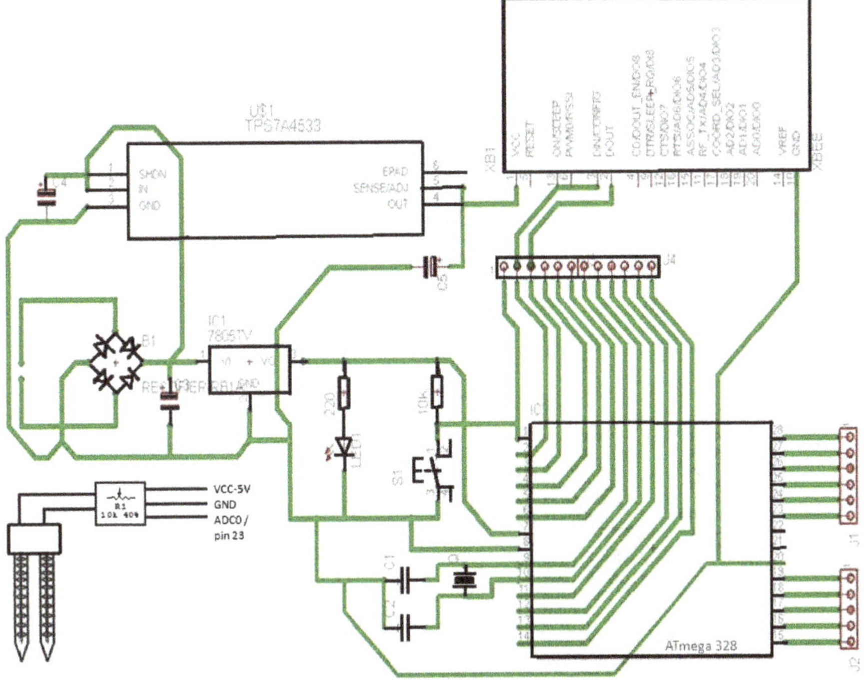


**Figure 3. Circuit schematic of WSN node.**

The circuit schematic is shown in Figure 3, it shows that supply section that contains the regulator 7805 is 5 V for microcontroller and TPS7A4533 is 3.3 V voltage regulator for Zig-Bee. Soil moisture sensor is an interface to ATmega 328 pin 23. The circuit schematic and circuit layout is designed in eagle PCB designing software to routing tracks automatically within given area. We have also interfaced Zig-Bee to the microcontroller for wireless communication. This circuit diagram shows many sections that are integrated on a single board mean we reduced the size. Microcontroller ATmega 328 is low cost, low power, and having many features on a single chip. Figures 4 and 5 show water and humidity sensor nodes. Figure 6 indicate, the coordinator window which displayed real time soil moisture analog data of WSN nodes.

**Figure 4. Sensor node 1.**

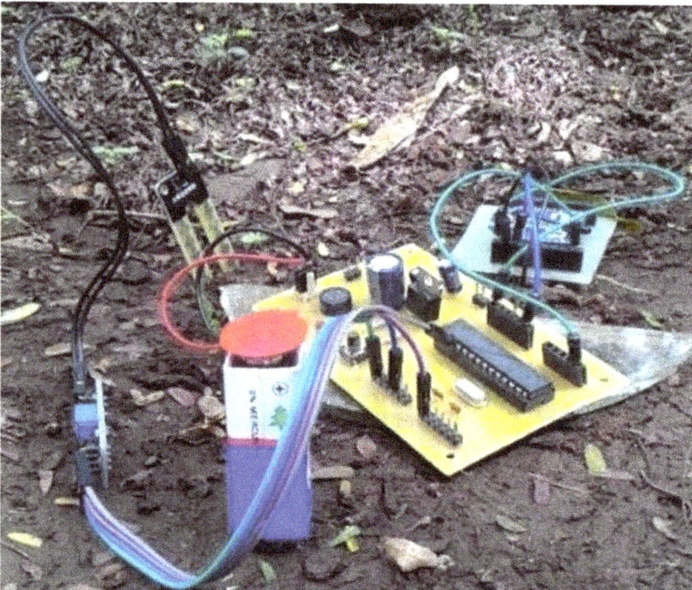

**Figure 5. Sensor node 2.**

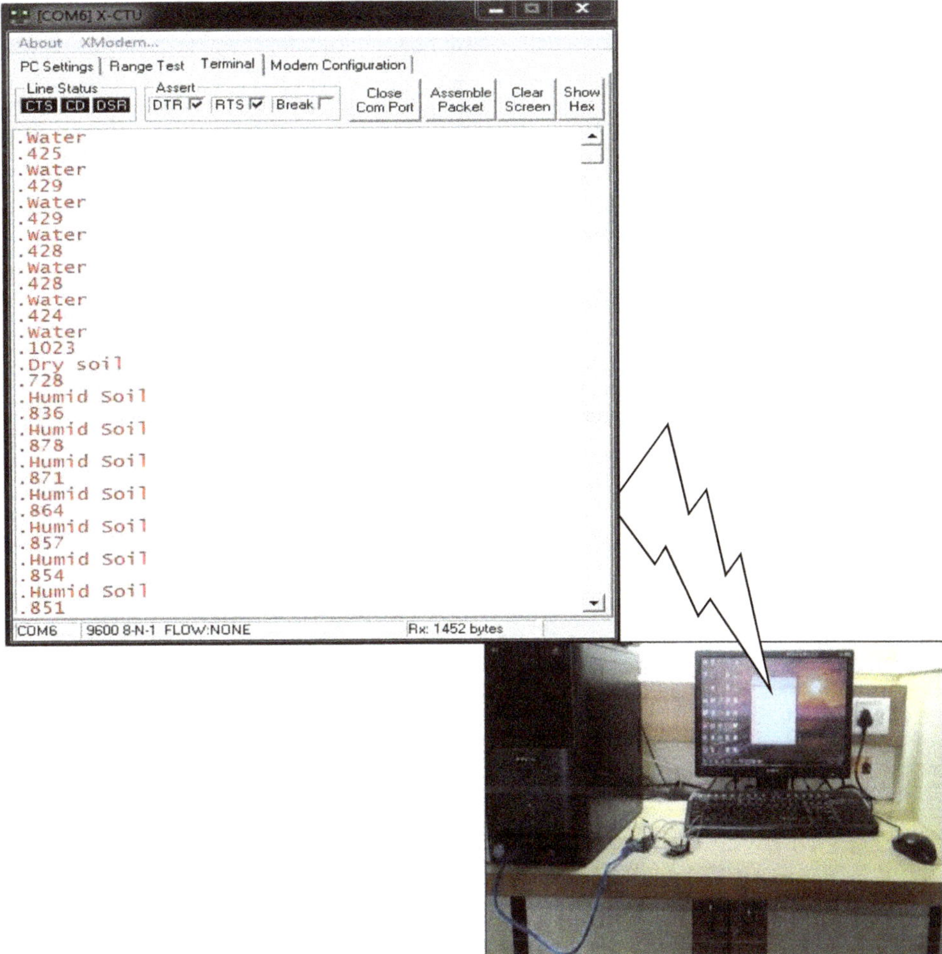


**Figure 6. Result display on coordinator window.**

## 7. Results

Analog data are displayed on the coordinator window read by the system and this received data could be shared with mobile and the internet. As per given conditions, programmed WSN system read data i.e. if analog value is greater than 900, then it is said to be dry soil, if a value greater than

**Table 1. Results from sensor node 1 and 2**

| Sensor node 1 | Sensor node 2 |
|---|---|
| **425 (Water)** | **836 (Humid soil)** |
| 429 | 878 |
| 424 | 871 |
| 428 | 864 |
| 430 | 857 |
| 422 | 860 |
| 431 | 861 |
| 430 | 866 |
| 426 | 864 |
| 425 | 862 |
| 423 | 863 |
| 424 | 861 |

500 and less than 900, then it is humid soil and if a value greater than 250 and less than 500, then it could be water. As per these conditions, both sensor nodes data from the coordinator window is shown in Table 1.

## 8. Conclusion

The proposed system is low cost and energy efficient to measure the soil salinity and moisture for grapes quality monitoring wirelessly. System performance is exceedingly superior for multiple sensor nodes across single coordinator and multiple routers. Such system we can use for several crop monitoring and their environment study as well as precision agriculture systems. The said system can build up within Indian currency Rs 2,500/node.

**Acknowledgment**
Authors are thankful to the head of the department, Electronics, Shivaji University, Kolhapur, India to provide research facility.

**Funding**
The authors received no direct funding for this research.

**Author details**
M.K. Bhanarkar[1]
E-mail: mkb_eln@unishivaji.ac.in
ORCID ID: http://orcid.org/0000-0002-1252-6806
P.M. Korake[1]
E-mail: korake.prakash@gmail.com
ORCID ID: http://orcid.org/0000-0002-3595-8265
[1] Communication Electronics Research Laboratory, Department of Electronics, Shivaji University, Kolhapur 416004, India.

### References

Coggan (2002). Water measurement in soil and vines. *Vineyard and Winery Management*, 43–53.

De Benedetto, D., Castrignanò, A., & Quarto, R. (2013). A geostatistical approach to estimate soil moisture as a function of geophysical data and soil attributes. *Procedia Environmental Sciences, 19*, 436–445. http://dx.doi.org/10.1016/j.proenv.2013.06.050

Holzman, M. E., Rivas, R., & Piccolo, M. C. (2014). Estimating soil moisture and relationship with crop yield using surface temperature and vegetation index. *International journal of Applied Earth Observation and Geo-information, 28*, 181–192.

Katerji, N., Van Hoornb, J. W., Hamdyc, A., & Mastrorillid, M. (2003). Salinity effect on crop development and yield, analysis of salt tolerance according to several classification methods. *Agricultural Water Management, 62*, 37–66.

Majone, B., Viani, F., Filippi, E., Bellin, A., Massa, A., Toller, ... Salucci, M. (2013). Wireless sensor network deployment for monitoring soil moisture dynamics at the field scale. *Procedia Environmental Sciences, 19*, 426–435. http://dx.doi.org/10.1016/j.proenv.2013.06.049

Shruthi, B. S., Shaila, K., Venugopal, K. R., & Patnaik, L. M. (2015). The paradigm of major issues and challenges in wireless sensor networks. *International Journal of Advanced Research in Computer Science and Software Engineering, 5*, 601–609 (ISSN: 2277 128X).

Yu, F. (2011). A survey of wireless sensor network simulation tools (pp. 1–10). Retrieved from http://www1.cse.wustl.edu/~jain/cse567-11/ftp/sensor/index.html

Yu, X., Wu, P., Han, W., & Zhang, Z. (2013). A survey on wireless sensor network infrastructure for agriculture. *Computer Standards & Interfaces, 35*, 59–64.

# 16

# Adaptive sliding mode control and its application in chaos control

Liqun Shen[1*], Chengwei Li[1] and Mao Wang[2]

*Corresponding author: Liqun Shen, School of Electrical Engineering and Automation, Harbin Institute of Technology, Harbin 150001, P.R. China
E-mail: liqunshen@gmail.com
Reviewing editor: James Lam, The University of Hong Kong, Hong Kong

**Abstract:** The sliding motion of sliding mode control system is studied in this paper. Using the measure concept, two new quantities about the sliding motion are introduced, and a new relationship about the sliding motion is derived with the new quantities. According to this relationship, an adaptive law of the magnitude of the controller's switching part is proposed, which can minimize the chattering phenomenon according to the predefined robust margin. To verify the effectiveness of the proposed control scheme, it is applied to Rössler system with uncertain disturbances. Simulation results show that the proposed control method can stabilize Rössler system with the magnitude of the controller's switching part adjusted adaptively with the disturbances.

**Subjects: Automation Control, Control Engineering, Dynamical Control Systems, Process Control - Chemical Engineering, Systems & Controls**

**Keywords: sliding mode control, adaptive control, measure concept, chaos control**

## 1. Introduction

Variable structure control with sliding mode was first proposed and studied in the early 1950s in the Soviet Union and it has been developed greatly since its introduction to the world by several research papers as Utkin (1977, 1978). It is a robust control method with relatively simple controller design

## ABOUT THE AUTHOR

Liqun Shen received the BE degree in electrical engineering and automation, the ME degree in control theory and control engineering, and the PhD degree in control science and engineering from Harbin Institute of Technology, Harbin, China, in 2002, 2004, and 2008, respectively.

From 2004 to 2008, he was with the School of Automation, Harbin University of Science and Technology. From 2008 to 2011, he was a postdoctoral researcher in the School of Electrical Engineering and Automation, Harbin Institute of Technology. In 2008, he joined the School of Electrical Engineering and Automation, Harbin Institute of Technology. He is currently a lecturer in the School of Electrical Engineering and Automation, Harbin Institute of Technology, Harbin, China. His research starts with chaos control and chaos synchronization. Later, he made some attempts in chaos-based detection and the control of power electronic systems. His research interests include control techniques, control theory, signal detection, and their applications.

## PUBLIC INTEREST STATEMENT

Sliding mode control is a robust control technique with relatively simple design procedures. It can be applied in various control systems and is especially suitable for the systems where the control input is inherently switching, such as the power electronic systems.

Chattering phenomenon is a main drawback of this method, that is to say, the control input will switch at a high frequency. To alleviate chattering, this paper finds a method to reduce the amplitude of the switching part of the controller. The amplitude of the switching part can be adjusted according to the disturbances and a predefined robust factor. So, we can adjust the chattering according to a predefined robust level.

procedures. There are generally two steps to design a sliding mode controller (Edwards & Spurgeon, 1998; Hung, Gao, & Hung, 1993; Utkin, 1977, 1978; Young, Utkin, & Ozguner, 1999). In the first step, a sliding surface that determines the system performance should be designed. In the second step, a nonlinear switching controller that satisfies the reachability condition should be designed. When the sliding motion is induced, a robust control system with reduced order is obtained (Edwards & Spurgeon, 1998).

Because the sliding mode control method is a robust control method, it can be applied in nonlinear or noisy environments (Chen, Liu, Ma, & Zhang, 2012; Chen, Zhang, Ma, & Liu, 2012; Chen, Zhang, Sprott, Chen, & Ma, 2012; Chen, Zhao, Ma, & Zhang, 2011; Dadras & Momeni, 2009; Fang, Li, Li, & Li, 2013; Herterng, Chen, & Chen, 2000; Li & Chang, 2009; Roopaei, Sahraei, & Lin, 2010; Yau, 2004). However, a main drawback of sliding mode control is the chattering phenomenon, which will not only cause the wear and tear of the actuator but also deteriorate the system performance seriously. To alleviate the chattering phenomenon, there are many methods proposed, such as the boundary layer methods (Slotine, 1984; Slotine & Sastry, 1983) and the second-order sliding mode control methods (Bartolini, 1989; Bartolini, Ferrara, & Usani, 1998; Bartolini & Pydynowski, 1993). These methods utilize the spirit of sliding mode control with the control input filtered before its sending out. The main purpose of these methods is to manage the frequency of the control input and slow down the variation of it. This paper does not eliminate chattering, but deals with chattering on the other direction. If we try to reduce the magnitude of the controller's switching part, then the chattering phenomenon will also be alleviated. In other words, the magnitude of the controller's switching part should not be too large. If an adaptive law can be designed to make the magnitude of the controller's switching part adapt to the disturbances of the system, the chattering phenomenon will also be alleviated. The adaptive law of the magnitude of the controller's switching part was studied in an early time (Yoo & Chung, 1992). And till now, researchers are still doing efforts in this direction (Dadras & Momeni, 2009; Fang et al., 2013; Li & Chang, 2009; Roopaei et al., 2010; Yau, 2004). But in these adaptive schemes, there are some parameters which always have the positive derivatives. The magnitude of such parameters cannot be reduced even when the disturbances get small. Moreover, when disturbances and calculation errors are taken into account, these parameters will go to infinity in real-time algorithms. To overcome this problem, the general sliding motion is analyzed, and more information about the sliding motion is presented.

In this paper, the sliding motion of sliding mode control system is studied in detail. Two new quantities about the sliding motion are introduced using the measure concept, and a new relationship about the sliding motion is derived. According to this relationship, an adaptive law of the magnitude of the controller's switching part is proposed, which can minimize the chattering phenomenon according to the predefined robust margin. This method can provide a new tool to minimize the chattering phenomenon.

This paper is organized as follows. In Section 2, the sliding motion is analyzed and a new relationship is presented using the measure concept (Dudley, 2002; Royden, 1988; Rudin, 1987). And the result is consistent with Filippov's equivalent control theory (Filippov, 1960). Then, an adaptive law of the magnitude of the controller's switching part is proposed in Section 3. Section 4 presents an application of the proposed control scheme in chaos control. Section 5 makes the conclusion that summarizes the contribution of this paper.

## 2. Sliding motion analysis

### *2.1. General sliding mode control system*

Consider a general sliding mode control system as follows:

$$\dot{x} = f(t, x) + bu \tag{1}$$

where $x \in R^n$ is the state vector, $f(t, x) \in R^n$ represents a bounded continuous function, $b=[0,\ldots, 0, 1]^T$, $u=\begin{cases} u^+, S=C^T x>0 \\ u^-, S=C^T x<0 \end{cases}$ is the scalar control input, where $C \in R^n$, $S=C^T x=0$ is the sliding surface, $u^+$ and $u^-$ represent two bounded continuous control inputs in the right-hand side of (1) while $S>0$ and $S<0$, respectively.

ASSUMPTION 1. *To focus on the sliding motion analysis, we suppose system (1) satisfies the reachability condition.*

ASSUMPTION 2. *To simplify the analysis, we assume that for all the time t during the sliding motion, the control input must be $u(t)=u^+$ or $u(t)=u^-$.*

By Assumption 1, the trajectory of system (1) will converge to the sliding surface, and satisfies

$$S\dot{S}<0 \tag{2}$$

On the sliding surface, sliding motion will take place and we have (Edwards & Spurgeon, 1998)

$$S=0 \tag{3}$$

and

$$\dot{S}=0 \tag{4}$$

Equations 3 and 4 are two basic equations when analyzing the sliding motion.

### 2.2. Sliding motion analysis of the discretized system

If the control input of system (1) is discretized as a real-time sliding mode control system, we can get the following system

$$\dot{x}(t)=f(t,x)+bu(k\Delta t) \tag{5}$$

where $u(k\Delta t)=\begin{cases} u^+, S(k\Delta t)=C^T x(k\Delta t)>0 \\ u^-, S(k\Delta t)=C^T x(k\Delta t)<0 \end{cases}, t \in [k\Delta t, (k+1)\Delta t), k=0, 1, 2, \ldots$. While $\Delta T$ approaches zero, system (5) approaches system (1). If Assumption 1 holds, then system (5) satisfies Equation 2 when $t=k\Delta T$, $k=0, 1, 2, \ldots$.

To focus on the sliding motion analysis, we suppose system (5) has already been in sliding motion while $t=0$. In the following, the discrete time version of Equations 3 and 4 will be derived. Observe the dynamic of system (5) in the interval $[0, T)$, where $T$ is a positive real number. Let $\Delta T=\frac{T}{n}$ and $\Delta t=\frac{\Delta T}{m}$, where $m$ and $n$ are two positive integers. Then the interval $[0, T)$ can be divided into $n$ intervals as $[0, \Delta T)$, $[\Delta T, 2\Delta T)$, ..., $[(n-1)\Delta T, T)$, and each $\Delta T$ interval has $m$ $\Delta T$ intervals. On a single point of a time interval, system (5) and system (1) are generally different. But on each $\Delta T$ interval, system (5) and system (1) will behave similarly while $\Delta t$ approaches zero. From the definition of the sliding surface, in each $\Delta t$ interval we have

$$\dot{S}(t)=C^T f(t,x)+C^T bu(k\Delta t) \tag{6}$$

where $t \in [k\Delta t, (k+1)\Delta t)$, $k=0, 1, 2, \ldots, m \cdot n-1$, as well as the reachability condition is satisfied. During the sliding motion, it is obvious that

$$|S(t)| \leq M \cdot \Delta t \tag{7}$$

where $M = \max_{t\in[0,T)} \left\{ \left| C^T f(t,x) + C^T bu(t) \right| \right\}$.

*Remark 1.* During the sliding motion, the system trajectory will be chattering around the sliding surface. Because the reachability condition is assumed to be satisfied, the system trajectory cannot go away from the sliding surface in the successive two steps. Then, $S(t)$ can be bounded by the maximum one step out of the sliding surface, which is represented by Equation 7.

Because $f(t, x)$ and $u(t)$ are bounded, the maximum one step out of the sliding surface will approach zero while $\Delta t$ approaches zero, which is shown in Equation 7. To build a bridge between system (1) and system (5), we observe the sliding motion in small $\Delta T$ intervals. And we use the average value in $\Delta T$ interval to represent the corresponding quantities in system (1). According to Equation 7, it is obvious that

$$\begin{aligned} \lim_{\Delta t\to 0} \left|\bar{S}_i\right| &= \lim_{\Delta t\to 0} \left| \frac{1}{\Delta T} \int_{i\Delta T}^{(i+1)\Delta T} S(t)dt \right| \\ &\le \lim_{\Delta t\to 0} M \cdot \Delta t = 0 \end{aligned} \tag{8}$$

where $\bar{S}_i$ represents the average value of $S(t)$ on the $i$th $\Delta T$ interval, $i=0, 1, \dots, n-1$. Then, we have

$$\lim_{\Delta t\to 0} \bar{S}_i = 0 \tag{9}$$

$i = 0, 1, \dots, n-1$. This is the discrete time version of Equation 3. Also according to Equation 7, we can get the following relationship

$$\begin{aligned} \lim_{\Delta t\to 0} \left|\bar{\dot{S}}_i\right| &= \lim_{\Delta t\to 0} \left| \frac{1}{\Delta T} \int_{i\Delta T}^{(i+1)\Delta T} \dot{S}(t)dt \right| \\ &= \lim_{\Delta t\to 0} \left| \frac{1}{\Delta T} (S_{i+1} - S_i) \right| \\ &\le \lim_{\Delta t\to 0} \frac{\Delta t}{\Delta T} M \\ &= 0 \end{aligned} \tag{10}$$

where $\Delta t \ll \Delta T$ needs to be satisfied, $\bar{\dot{S}}$ represents the average value of $\dot{S}(t)$ on the $i$th $\Delta T$ interval, $S_i = S(i\Delta T)$, $i=0, 1, \cdots, n-1$. Then, we have

$$\lim_{\Delta t\to 0} \bar{\dot{S}}_i = 0 \tag{11}$$

This is the discrete time version of Equation 4. Because $f(t, x)$ is bounded and continuous, we have

$$\lim_{\Delta t\to 0} (S_{i+1} - S_i) = \lim_{\Delta t\to 0} \sum_{j=0}^{m-1} \dot{S}_{ij} \cdot \Delta t \tag{12}$$

where $\dot{S}_{ij} = S(i\Delta T + j\Delta t)$, $i=0, 1, \dots, n-1, j=0, 1, \dots, m-1$. Substitute Equation 12 into Equation 10, a new relationship can be found as

$$\lim_{\Delta t\to 0} \frac{\Delta t}{\Delta T} \sum_{j=0}^{m-1} \dot{S}_{ij} = 0 \tag{13}$$

Equations 9 and 11 are the discrete time correspondences of Equations 3 and 4. Equation 13 is a new relationship that provides additional information about the sliding motion. According to Equations 10 and 12, Equations 11 and 13 can be combined as

$$\lim_{\Delta t \to 0} \bar{\dot{S}}_i = \lim_{\Delta t \to 0} \frac{\Delta t}{\Delta T} \sum_{j=0}^{m-1} \dot{S}_{ij} = 0 \tag{14}$$

And Equation 14 can represent the sliding motion more exactly. This is also a relationship between the average value $\bar{\dot{S}}_i$ and the instant value $\dot{S}_{ij}$. In the following, the continuous time version of Equation 14 will be derived using the measure concept.

### 2.3. A new relationship about the sliding motion

Concerning about system (5), let $U^+(t)=\{t' : t' \in [0, t), u(t')=u^+\}$ and $U^-(t)=\{t' : t' \in [0, t), u(t')=u^-\}$ be two sets of time to be studied.

And let $m(U^+(t))$ and $m(U^-(t))$ represent the measure of $U^+(t)$ and $U^-(t)$, respectively. According to Assumption 2, we have

$$U^+(t) \cup U^-(t) = [0, t) \tag{15}$$

and

$$U^+(t) \cap U^-(t) = \phi \tag{16}$$

Then, the following relationship holds

$$m(U^+(t)) + m(U^-(t)) = t \tag{17}$$

Concerning about system (5), for each $\Delta t$ interval which belongs to $[0, T)$, there must be $u^+$ or $u^-$ in active. So, each $\Delta t$ interval must belong to $U^+(T)$ or $U^-(T)$. Because $f(t, x)$, $u^+$, and $u^-$ are all continuous, while $\Delta T$ approaches zero, Equation 14 can be reformulated as

$$\begin{aligned}\lim_{\substack{\Delta t \to 0 \\ \Delta T \to 0}} \bar{\dot{S}}_i = & \, m(U^+((i+1)\Delta T) \backslash U^+(i\Delta T)) \\ & \times C^T[f(i\Delta T, x(i\Delta T)) + bu^+(i\Delta T)]/\Delta T + m(U^-((i+1)\Delta T) \backslash U^-(i\Delta T)) \\ & \times C^T[f(i\Delta T, x(i\Delta T)) + bu^-(i\Delta T)]/\Delta T = 0\end{aligned} \tag{18}$$

where $U^+((i+1)\Delta T) \backslash U^+(i\Delta T)$ and $U^-((i+1)\Delta T) \backslash U^-(i\Delta T)$ represent the relative complement operation. While $\Delta t$ approaches zero, Equation 5 will approach Equation 1. Now consider the general sliding mode control system as Equation 1. While $\Delta T$ approaches zero, Equation 18 can be rewritten in the differential form as

$$\dot{S}(t) = \mu^+ C^T f^+ + \mu^- C^T f^- = 0 \tag{19}$$

where $\mu^+ = \frac{dm(U^+(t))}{dt}$, $\mu^- = \frac{dm(U^-(t))}{dt}$, $f^+ = f(t, x) + bu^+(t)$, $f^- = f(t, x) + bu^-(t)$. And this is the continuous time version of Equation 14. According to Equation 17, we can get the following relationship

$$\mu^+ + \mu^- = 1 \tag{20}$$

And the important concept of equivalent control can be derived as

$$f_{eq} = \mu^+ f^+ + \mu^- f^- \tag{21}$$

and

$$u_{eq} = \mu^+ u^+ + \mu^- u^- \tag{22}$$

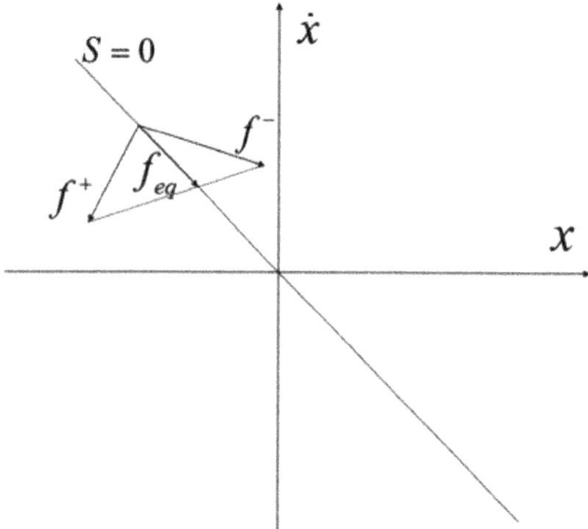


**Figure 1. The result of the equivalent system.**

where $f_{eq}$ is the equivalent right-hand side of Equation 1, and $u_{eq}$ represents the equivalent control. Filippov's equivalent control theory is an important result in sliding mode control theory. And the result of Equation 21 is consistent with Filippov's equivalent control theory, which is shown in Figure 1.

From Equations 19 to 22, we can see that two important quantities $\mu^+$ and $\mu^-$ are introduced. And more information can be obtained about the sliding motion. As a sum up, the following theorem can be concluded.

THEOREM 1. *By introducing two quantities as $\mu^+$ and $\mu^-$, a new relationship about the sliding motion of system* (1) *can be derived as*

$$\dot{S}(t)=\mu^{+}C^{T}f^{+}+\mu^{-}C^{T}f^{-}=0$$

*And the status of system* (1) *can be revealed by $\mu^+$ and $\mu^-$, and the following relationship holds*

$$\begin{cases} \mu^{+}=1, & S<0 \\ \mu^{-}=1, & S<0 \\ \mu^{+}C^{T}f^{+}+\mu^{-}C^{T}f^{-}=0, & S=0\, and\, \dot{S}=0 \end{cases}$$

*Proof.* See the deriving process of (19). □

By observing $\mu^+$ and $\mu^-$, the adaptive law of the magnitude of the controller's switching part can be designed.

## 3. Adaptive sliding mode controller design

Consider a sliding mode control system with time-varying disturbances as follows:

$$\dot{x}=Ax+bn(t,x)+bu \tag{23}$$

where $x\in R^n$ is the state vector, $A\in R^{n\times n}$ and $b\in R^n$ are constant matrices, $u$ is the control input, and $n(t, x)$ represents the time-varying disturbances.

To design a sliding mode controller, a sliding surface should be designed first as follows:

$$S=C^{T}x=0 \tag{24}$$

where $C \in R^n$ is a constant vector, which determines the system performance. Then, a controller that satisfies the reachability condition should be designed. Suppose $C^T b$ is invertible, a classical sliding mode controller can be designed as

$$u = -(C^T b)^{-1} \left( C^T A x + \hat{K} \cdot sgn(S) \right) \tag{25}$$

where $sgn(S) = \begin{cases} 1, & S > 0 \\ -1, & S < 0 \end{cases}$ is the signal function.

To satisfy the reachability condition, we can choose a Lyapunov function as

$$V = \frac{1}{2} S^2 \tag{26}$$

and make sure that its derivative along system (23) is negative definite. According to Equations 25 and 26, we have

$$\begin{aligned} \dot{V} &= S^T \dot{S} \\ &= x^T C C^T [Ax + bn(t,x) + bu] \\ &= x^T C C^T A x + x^T C C^T bn(t,x) - x^T C (C^T A x + \hat{K} \cdot sgn(S)) \\ &= C^T bn(t,x) \cdot S - \hat{K} |S| \\ &\le \left( \left| C^T bn(t,x) \right| - \hat{K} \right) |S| \end{aligned} \tag{27}$$

Obviously, if $\hat{K} > \left| C^T bn(t,x) \right|$, the reachability condition will be satisfied. But to alleviate the chattering phenomenon, the magnitude of the controller's switching part $\hat{K}$ should not be too large. We can use the following theorem as the updating law of $\hat{K}$.

THEOREM 2. *Considering system* (23) *with the sliding mode controller as* Equation 25, *an adaptive law of* $\hat{K}$ *can be designed as follows:*

$$\begin{cases} \hat{K} = \frac{\gamma}{1 - |\mu^+ - \mu^-|}, & S = 0 \text{ and } \dot{S} = 0 \\ \dot{\hat{K}} = |S|, & \text{else} \end{cases}$$

*Then, the reachability condition will be satisfied and* $\hat{K}$ *will approach the magnitude of the disturbances with the predetermined margin* $\gamma$.

*Proof.* While the trajectory of system (23) is out of the sliding surface, the control scheme should make sure the reachability condition be satisfied. Let $K_{max} = \max |C^T bn(t, x)|$, we can choose a Lyapunov function as

$$V = \frac{1}{2} S^2 + \frac{1}{2} \left( \hat{K} - K_{max} \right)^2 \tag{28}$$

According to Equations (23), (25), (28), and Theorem 2, we have

$$\begin{aligned} \dot{V} &= S^T \dot{S} + (\hat{K} - K_{max}) \dot{\hat{K}} \\ &= S^T C^T [Ax + bn(t,x) + bu] + (\hat{K} - K_{max}) |S| \\ &= S^T C^T A x + S^T C^T bn(t,x) - S^T C^T A x - \hat{K} |S| + (\hat{K} - K_{max}) |S| \\ &= C^T bn(t,x) \cdot S - K_{max} \cdot |S| \\ &\le 0 \end{aligned} \tag{29}$$

We can see that the reachability condition is satisfied.

Next, the robust margin $\gamma$ should be studied. According to Theorem 1 and the definition of the controller, a relationship between the system behavior and the two introduced quantities $\mu^+$ and $\mu^-$ can be built as

$$\begin{cases} \mu^+ = 1, & S > 0 \\ \mu^- = 1, & S < 0 \\ \hat{K}(\mu^+ - \mu^-) = C^T bn(t,x), & S = 0 \text{ and } \dot{S} = 0 \end{cases} \tag{30}$$

According to Equation 30, we have

$$\hat{K}\,|(\mu^+ - \mu^-)| = \left|C^T bn(t,x)\right| \tag{31}$$

during the sliding motion. As stated in Theorem 2, $\hat{K} = \frac{\gamma}{1-|\mu^+ - \mu^-|}$ during the sliding motion. Eliminating $|\mu^+ - \mu^-|$ in Equation 31, we have

$$\hat{K} = \left|C^T bn(t,x)\right| + \gamma \tag{32}$$

where $\gamma$ is the predefined robust margin. □

*Remark 2.* In real-time algorithms, $\mu^+$ and $\mu^-$ can be approximated as in Equation 18 by setting $\Delta T$ a fixed small value and $\Delta t$ the sampling interval.

## 4. Adaptive sliding mode control of Rössler system

To show the effectiveness of the proposed control scheme, Rössler system is used as an example in the simulation. The state equation of Rössler system can be described as

$$\dot{x} = Ax + bf(x) + bd(t) + bu \tag{33}$$

where $x = [x_1\ x_2\ x_3]^T$ is the state vector, $A = \begin{bmatrix} 0 & -1 & -1 \\ 1 & 0.2 & 0 \\ 0 & 0 & -5.7 \end{bmatrix}$ and $f(x) = x_1 x_3 + 0.2$ represent the dynamic structure of Rössler system, $d(t)$ stands for the external disturbances, $b = [0\,0\,1]^T$, $u$ is the control input. It has been proved that system (33) presents chaos when $d(t) = 0$ and $u = 0$.

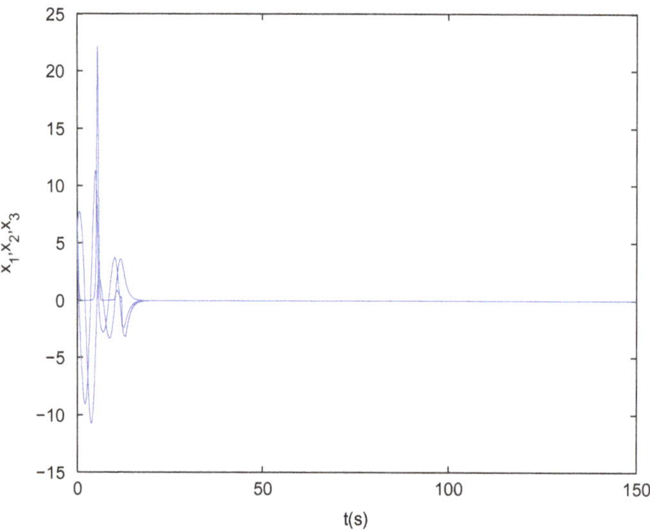


**Figure 2. The state response of Rössler system.**

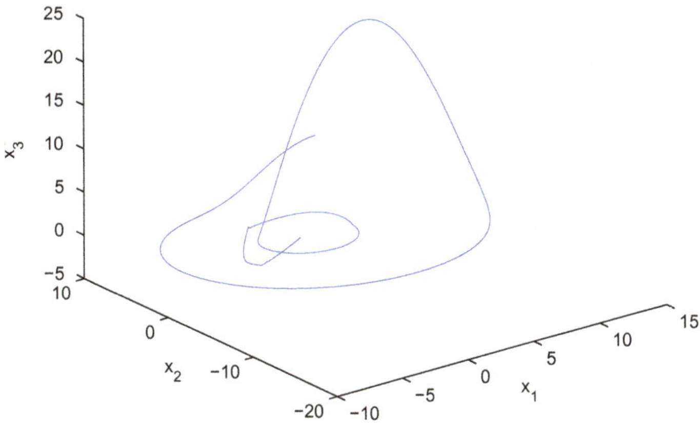


Figure 3. The trajectory of Rössler system.

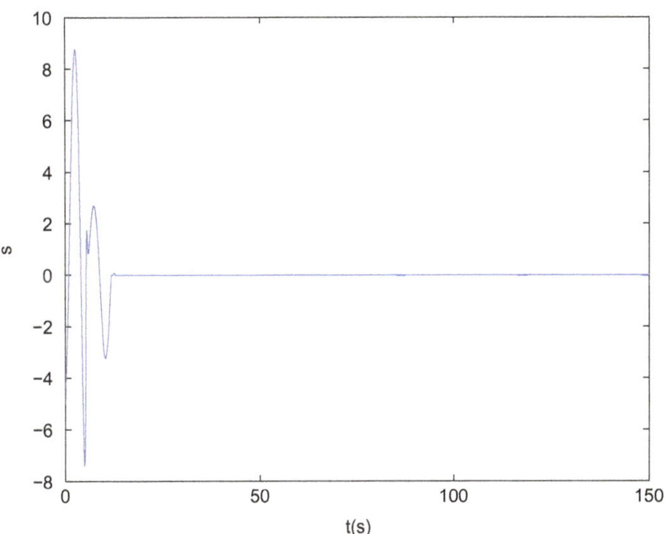


Figure 4. The response of the sliding surface.

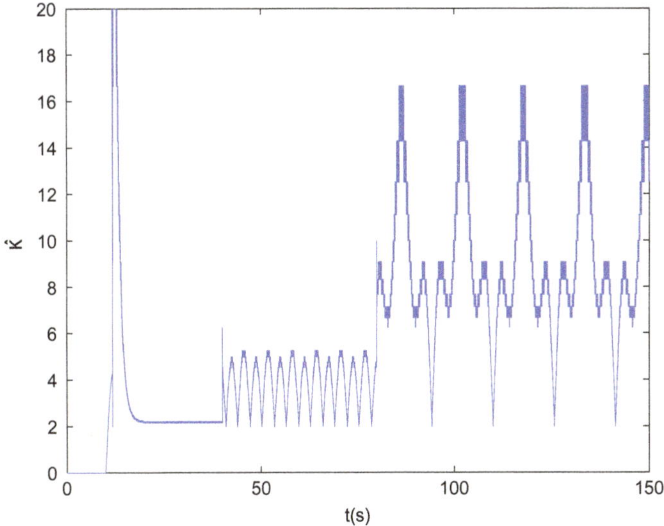


Figure 5. The magnitude of the controller's switching part.

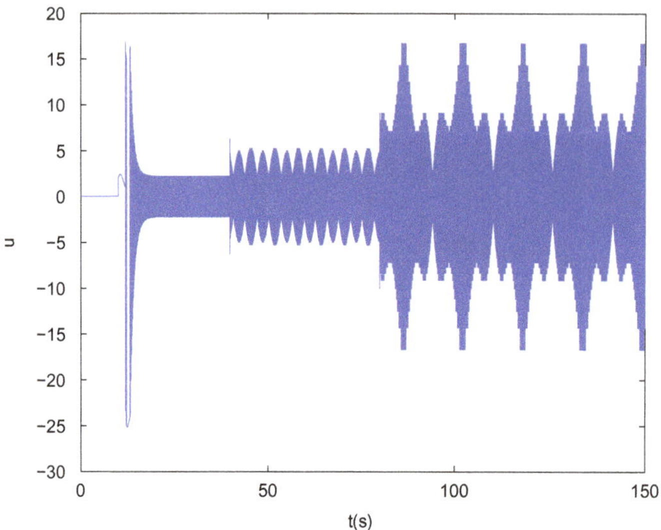


**Figure 6. The control input.**

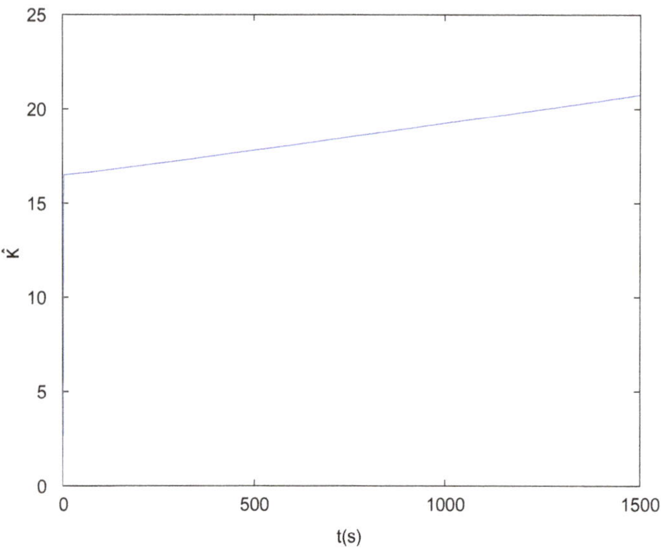


**Figure 7. The magnitude of the controller's switching part with traditional adaptive method.**

In the simulation, the initial condition of Equation 33 is set as $x(0)=[5\;6\;7]^T$, $\hat{K}(0)=0$, and $\gamma=2$. Let $d(t)=3\varepsilon(t-40)\sin(t)+10\varepsilon(t-80)\sin(0.2t)$, where $\varepsilon(t)$ represents the step function. After a stable sliding surface is determined by $C=[-3.2\;-1.64\;1]^T$, the controller formed as Equation 25 with the adaptive law in Theorem 2 can stabilize Rössler system in Equation 33. Simulation results are shown in Figures 2-6, where the control input is turned on at $t=10$ s.

Figures 2 and 3 show that Rössler system can be stabilized by the proposed controller. Figure 4 shows the response of the sliding surface. Figures 5 and 6 show that the magnitude of the controller's switching part can adapt with the external disturbances. From Figure 5, we can see that the information of disturbances can be abstracted, which may be useful for further applications.

If we only use the conventional adaptive law $\dot{\hat{K}}=|S|$, the switching part can only increase as shown in Figure 7, which is not acceptable in real applications.

## 5. Conclusions

This paper was motivated by the demand to reduce the amplitude of the switching part of the sliding mode controller. A new adaptive sliding mode control scheme is proposed. Through the discretization of the sliding mode controller, several corresponding discrete time version equations about the sliding motion are derived. And a new relationship about the sliding motion is obtained. This relationship is consistent with Filippov's equivalent control theory. With this relationship, an adaptive scheme of the magnitude of the controller's switching part is proposed based on the two new quantities $\mu^+$ and $\mu^-$, which are first introduced in this paper. Then, the magnitude of the controller's switching part can be adjusted adaptively with a predefined robust margin according to the disturbances. Simulation results of Rössler system show the effectiveness of the proposed control scheme. In the future, the application in chaos synchronization and secure communication may also be studied. This adaptive method may be more suitable to be applied in digital control systems, where $\mu^+$ and $\mu^-$ can be obtained easily.


#### Acknowledgments

The original version of this paper was finished in 2010. It was first presented at the 2nd International Conference on Electronics, Communications and Control (ICECC2012), which was held in Zhoushan, China, October 2012. The conference was later rejected by IEEE in 2013 and the copyright of the paper was returned to the authors. A revised and extended version of this manuscript was then submitted to this journal. We are grateful for the works that the reviewers and the editors have done.

#### Funding

This work was supported by the Scientific Research Fund of Heilongjiang Provincial Education Department [grant number 11541055].



#### Author details

Liqun Shen[1]
E-mail: liqunshen@gmail.com
Chengwei Li[1]
E-mail: chengweili@hit.edu.cn
Mao Wang[2]
E-mail: wangmao0451@sina.com
[1] School of Electrical Engineering and Automation, Harbin Institute of Technology, Harbin 150001, P.R. China.
[2] Space Control and Inertial Technology Research Center, Harbin Institute of Technology, Harbin 150001, P.R. China.


#### References


Bartolini, G. (1989). Chattering phenomena in discontinuous control systems. *International Journal of Systems Science, 20*, 2471–2481. http://dx.doi.org/10.1080/00207728908910327

Bartolini, G., Ferrara, A., & Usani, E. (1998). Chattering avoidance by second-order sliding mode control. *IEEE Transactions on Automatic Control, 43*, 241–246. http://dx.doi.org/10.1109/9.661074

Bartolini, G., & Pydynowski, P. (1993). Asymptotic linearization of uncertain nonlinear systems by means of continuous control. *International Journal of Robust and Nonlinear Control, 3*, 87–103. http://dx.doi.org/10.1002/(ISSN)1099-1239

Chen, D., Liu, Y., Ma, X., & Zhang, R. (2012). Control of a class of fractional-order chaotic systems via sliding mode. *Nonlinear Dynamics, 67*, 893–901.

Chen, D., Zhang, R., Ma, X., & Liu, S. (2012). Chaotic synchronization and anti-synchronization for a novel class of multiple chaotic systems via a sliding mode control scheme. *Nonlinear Dynamics, 69*, 35–55. http://dx.doi.org/10.1007/s11071-011-0244-7

Chen, D., Zhang, R., Sprott, J. C., Chen, H., & Ma, X. (2012). Synchronization between integer-order chaotic systems and a class of fractional-order chaotic systems via sliding mode control. *Chaos: An Interdisciplinary Journal of Nonlinear Science, 22*(2), 023130.

Chen, D., Zhao, W., Ma, X., & Zhang, R. (2011). No-chattering sliding mode control chaos in Hindmarsh–Rose neurons with uncertain parameters. *Computers & Mathematics with Applications, 61*, 3161–3171.

Dadras, S., & Momeni, H. R. (2009). Control uncertain Genesio–Tesi chaotic system: Adaptive sliding mode approach. *Chaos, Solitons & Fractals, 42*, 3140–3146.

Dudley, R. M. (2002). *Real analysis and probability*. Cambridge: Cambridge University Press. http://dx.doi.org/10.1017/CBO9780511755347

Edwards, C., & Spurgeon, S. (1998). *Sliding mode control: Theory and applications*. London: Taylor & Francis.

Fang, L., Li, T., Li, Z., & Li, R. (2013). Adaptive terminal sliding mode control for anti-synchronization of uncertain chaotic systems. *Nonlinear Dynamics, 74*, 991–1002. http://dx.doi.org/10.1007/s11071-013-1017-2

Filippov, A. F. (1960). Differential equations with discontinuous right-hand side. *Matematicheskii sbornik, 93*, 99–128.

Hung, J. Y., Gao, W., & Hung, J. C. (1993). Variable structure control: A survey. *IEEE Transactions on Industrial Electronics, 40*, 2–22. http://dx.doi.org/10.1109/41.184817

Li, W., & Chang, K. (2009). Robust synchronization of drive-response chaotic systems via adaptive sliding mode control. *Chaos, Solitons & Fractals, 39*, 2086–2092.

Roopaei, M., Sahraei, B. R., & Lin, T. (2010). Adaptive sliding mode control in a novel class of chaotic systems. *Communications in Nonlinear Science and Numerical Simulation, 15*, 4158–4170. http://dx.doi.org/10.1016/j.cnsns.2010.02.017

Royden, H. L. (1988). *Real analysis* (3rd ed.). New York, NY: Macmillan.

Rudin, W. (1987). *Real and complex analysis* (3rd ed.). New York, NY: McGraw-Hill.

Slotine, J. J. E. (1984). Sliding controller design for non-linear systems. *International Journal of Control, 40*, 421–434. http://dx.doi.org/10.1080/00207178408933284

Slotine, J. J. E., & Sastry, S. S. (1983). Tracking control of non-linear systems using sliding surfaces, with application to robot manipulators. *International Journal of Control, 38*, 465–492. http://dx.doi.org/10.1080/00207178308933088

Utkin, V. I. (1977). Variable structure systems with sliding modes. *IEEE Transactions on Automatic Control, 22*,

212–222.
http://dx.doi.org/10.1109/TAC.1977.1101446

Utkin, V. I. (1978). Sliding modes and their application in variable structure systems. Moscow: MIR.

Yau, H., Chen, C., & Chen, C. (2000). Sliding mode control of chaotic systems with uncertainties. *International Journal of Bifurcation and Chaos, 10*, 1139–1147.

Yau, H. (2004). Design of adaptive sliding mode controller for chaos synchronization with uncertainties. *Chaos, Solitons & Fractals, 22*, 341–347.

Yoo, D. S., & Chung, M.-J. (1992). A variable structure control with simple adaptation laws for upper bounds on the norm of the uncertainties. *IEEE Transactions on Automatic Control, 37*, 860–865.

Young, K. D., Utkin, V. I., & Ozguner, U. (1999). A control engineer's guide to sliding mode control. *IEEE Transactions on Control Systems Technology, 7*, 328–342.
http://dx.doi.org/10.1109/87.761053

# Permissions

All chapters in this book were first published in CE, by Cogent OA; hereby published with permission under the Creative Commons Attribution License or equivalent. Every chapter published in this book has been scrutinized by our experts. Their significance has been extensively debated. The topics covered herein carry significant findings which will fuel the growth of the discipline. They may even be implemented as practical applications or may be referred to as a beginning point for another development.

The contributors of this book come from diverse backgrounds, making this book a truly international effort. This book will bring forth new frontiers with its revolutionizing research information and detailed analysis of the nascent developments around the world.

We would like to thank all the contributing authors for lending their expertise to make the book truly unique. They have played a crucial role in the development of this book. Without their invaluable contributions this book wouldn't have been possible. They have made vital efforts to compile up to date information on the varied aspects of this subject to make this book a valuable addition to the collection of many professionals and students.

This book was conceptualized with the vision of imparting up-to-date information and advanced data in this field. To ensure the same, a matchless editorial board was set up. Every individual on the board went through rigorous rounds of assessment to prove their worth. After which they invested a large part of their time researching and compiling the most relevant data for our readers.

The editorial board has been involved in producing this book since its inception. They have spent rigorous hours researching and exploring the diverse topics which have resulted in the successful publishing of this book. They have passed on their knowledge of decades through this book. To expedite this challenging task, the publisher supported the team at every step. A small team of assistant editors was also appointed to further simplify the editing procedure and attain best results for the readers.

Apart from the editorial board, the designing team has also invested a significant amount of their time in understanding the subject and creating the most relevant covers. They scrutinized every image to scout for the most suitable representation of the subject and create an appropriate cover for the book.

The publishing team has been an ardent support to the editorial, designing and production team. Their endless efforts to recruit the best for this project, has resulted in the accomplishment of this book. They are a veteran in the field of academics and their pool of knowledge is as vast as their experience in printing. Their expertise and guidance has proved useful at every step. Their uncompromising quality standards have made this book an exceptional effort. Their encouragement from time to time has been an inspiration for everyone.

The publisher and the editorial board hope that this book will prove to be a valuable piece of knowledge for researchers, students, practitioners and scholars across the globe.

# List of Contributors

**Adikanda Parida and Debashis Chatterjee**
Department of Electrical Engineering, Jadavpur University, Kolkata, India

**Ling Lu, Yuanyuan Zou and Yugang Niu**
Key Laboratory of Advanced Control and Optimization for Chemical Process, East China University of Science & Technology, Ministry of Education, Shanghai 200237, China

**Ahmad Jazlan**
School of Electrical, Electronics and Computer Engineering, University of Western Australia, 35 Stirling Highway, Crawley, Perth, Western Australia 6009, Australia
Faculty of Engineering, Department of Mechatronics Engineering, International Islamic University Malaysia, Jalan Gombak, 53100 Kuala Lumpur, Malaysia

**Victor Sreeram and Roberto Togneri**
School of Electrical, Electronics and Computer Engineering, University of Western Australia, 35 Stirling Highway, Crawley, Perth, Western Australia 6009, Australia

**Hamid Reza Shaker**
Center for Energy Informatics, University of Southern Denmark, Campusvej 55, DK-5230 Odense M, Denmark

**Diem Dang Huan**
Faculty of Basic Sciences, Bacgiang Agriculture and Forestry University, Bacgiang, 21000 Vietnam
Vietnam National University, Hanoi, 144 Xuan Thuy Street, Cau Giay, Hanoi, 10000 Vietnam

**Hongjun Gao**
School of Mathematical Science, Nanjing Normal University, Nanjing, 210023 P.R. China

**Adriana Salinas and Rafael Kelly**
Laboratorio de Robótica, Departamento de Electrónica y Telecomunicaciones, CICESE, Carretera Ensenada-Tijuana No. 3918, Zona Playitas, Ensenada 22860, Baja California, Mexico

**Javier Moreno-Valenzuela**
Instituto Politécnico Nacional-CITEDI, Department of Systems and Control, Ave. Instituto Politécnico Nacional No. 1310, Nueva Tijuana, Tijuana 22435, Baja California, Mexico

**Hamdati Al Shehhi and Igor Boiko**
The Petroleum Institute, Abu Dhabi, UAE

**Suresh Rasappan**
Department of Mathematics, Vel Tech University, No. 42 Avadi-Vel Tech Road, Avadi, Chennai 600062 Tamilnadu, India

**Yilun Shang**
Einstein Institute of Mathematics, Hebrew University of Jerusalem, Jerusalem 91904, Israel; SUTD-MIT International Design Center, Singapore University of Technology and Design, Singapore 138682, Singapore

**Rinchin W. Mosobi, Toko Chichi and Sarsing Gao**
Department of Electrical Engineering, North Eastern Regional Institute of Science and Technology, Nirjuli, Arunachal Pradesh 791 109, India

**Agamirza E. Bashirov**
Department of Mathematics, Eastern Mediterranean University, Mersin 10, Turkey
Institute of Cybernetics, ANAS, Baku, Azerbaijan

**Noushin Ghahramanlou**
Department of Mathematics, Eastern Mediterranean University, Mersin 10, Turkey

**Quan Wang and Yuanyuan Zou**
Key Laboratory of Advanced Control and Optimization for Chemical Processes, Ministry of Education, East China University of Science and Technology, Shanghai, 200237, China

**Shaobo Wang**
Automation Division of Shanghai Electric Group Co., Ltd., Shanghai, China

**Arif Iqbal, G.K. Singh and Vinay Pant**
Department of Electrical Engineering, Indian Institute of Technology Roorkee, Roorkee 247667, Uttarakhand, India

**Jiaye Fu and Yugang Niu**
Key Laboratory of Advanced Control and Optimization for Chemical Process (East China University of Science & Technology), Ministry of Education, Shanghai 200237, China

**Hassan Qudrat-Ullah**
School of Administrative Studies, York University, 4700 Keele Street, Toronto, Ontario, Canada M9V 3K7

**M.K. Bhanarkar and P.M. Korake**
Communication Electronics Research Laboratory, Department of Electronics, Shivaji University, Kolhapur 416004, India

**Liqun Shen and Chengwei Li**
School of Electrical Engineering and Automation, Harbin Institute of Technology, Harbin 150001, P.R. China

**Mao Wang**
Space Control and Inertial Technology Research Center, Harbin Institute of Technology, Harbin 150001, P.R. China

# Index

www.ingramcontent.com/pod-product-compliance
Lightning Source LLC
Chambersburg PA
CBHW082048190326
41458CB00010B/3482

* 9 7 8 1 6 3 2 4 0 6 2 6 2 *